DE MONTFORT
UNIVERSITY

Caythorpe Campus
Library

~~Reference only~~

PC898

Clinical Dissection Guide for Large Animals

Horse, Ox, Sheep, Goat, Pig

Clinical Dissection Guide for Large Animals

Horse, Ox, Sheep, Goat, Pig

GHEORGHE M. CONSTANTINESCU, D.V.M., Ph.D.

Associate Professor of Veterinary Anatomy,
Department of Veterinary Biomedical Sciences,
College of Veterinary Medicine,
University of Missouri—Columbia
Columbia, Missouri

with 504 *illustrations* in 319 plates

Illustrations by
Gheorghe M. Constantinescu, D.V.M., Ph.D.

Mosby
Year Book

St. Louis Baltimore Boston Chicago London Philadelphia Sydney Toronto

Mosby Year Book
Dedicated to Publishing Excellence

Editor: Robert W. Reinhardt
Assistant editor: Melba Steube
Project manager: John A. Rogers
Production editor: Catherine M. Vale
Manuscript editor: Kathy Lumpkin
Production: Jeanne Genz, Catherine M. Vale
Illustration reproduction: Top Graphics
Text designer: Laura Steube
Cover designer: Julia Taugner
Cover illustration: Gheorghe M. Constantinescu

Printed in the United States of America

Mosby–Year Book, Inc.
11830 Westline Industrial Drive
St. Louis, Missouri 63146

Library of Congress Cataloging-in-Publication Data

Constantinescu, Gheorghe M.
    Clinical dissection guide for large animals: horse, ox, sheep, goat, pig / Gheorghe M.
Constantinescu : illustrations by Gheorghe M. Constantinescu.
        p.   cm.
    Includes index.
    ISBN 0-8016-2564-5
    1. Veterinary dissection—Laboratory manuals.   2. Veterinary
anatomy—Laboratory manuals.   3. Veterinary dissection—Atlases.
4. Veterinary anatomy—Atlases. I. Title.
    [DNLM: 1. Anatomy, Veterinary—atlases. 2. Anatomy, Veterinary—
laboratory manuals. 3. Animals, Domestic—anatomy & histology—
atlases. 4. Animals, Domestic—anatomy & histology—laboratory
manuals. 5. Dissection methods—atlases. 6. Dissection—
veterinary—atlases.   SF 762 C758c]
SF762.C66   1991
636.089′1—dc20
DNLM/DLC   90-13774
for Library of Congress   CIP

CL/MV/MV   9   8   7   6   5   4   3   2   1

In memory of

my father,

and to

my mother,

in gratitude for my education.

To my wife,

Ileana Anghelina,

for her full and unlimited support, understanding, devotion, and love;
I am grateful to her for sharing her life with me.

To my children,

Alexandru Rāzvan and **Adina Elizabeth,**

for their support and love.

Foreword

It is a pleasure and a honor for me, as the former professor of Veterinary Anatomy of Dr. Gheorghe M. Constantinescu, to write the foreword for his *Clinical Dissection Guide for Large Animals*.

A short biographic sketch about Dr. Constantinescu seems relevant. Dr. Constantinescu was born in Bucharest, Romania, on January 20, 1932. His father was an M.D., and his mother earned a B.L. (Bachelor of Letters in Romanian, French, and German) and a B.A.Ed. He graduated from the Faculty of Veterinary Medicine in Bucharest in 1955 (with a D.V.M.) and received his Ph.D. in Veterinary Anatomy in 1964. During his first and second years, he was the only student who earned the highest grade in Anatomy: 10. He was a brilliant student, and I asked him to be an honorary preparator—to help me with teaching and in the laboratory for the last 4 years out of the 5 years of the D.V.M. program.

After graduation, Dr. Constantinescu worked for 3 years as a laboratory chief in Veterinary Anatomy in Bucharest, then as a researcher in the Zootechnical Institute and was the first Veterinarian of the Zoological Park in Bucharest. After 3 years as a veterinarian and Vice-President of the Agricultural Council of the Panciu county, he was named Associate Professor of Veterinary Anatomy of the Faculty of Veterinary Medicine in Timisoara, Romania, from 1965 to 1982. During this last period, he was Scientific Secretary of the Agronomic Institute in Timisoara and Associate Dean. He was appointed as an Associate Professor of Veterinary Anatomy at the College of Veterinary Medicine, University of Missouri-Columbia, U.S.A., in 1984, where he is presently employed.

Dr. Constantinescu is the author or coauthor of more than 200 publications, including more than 30 books and chapters. He is a talented anatomical and medical illustrator, as well as a sculptor. He is married and has a son and a daughter.

The *Clinical Dissection Guide for Large Animals* that I am presenting is a remarkable work—a world premiere in Veterinary Anatomy, which accumulates the author's knowledge, experience, and talent and which is the first dissection guide to discuss all the large domestic animals in one volume. It is one of the few examples in which the author of a veterinary anatomical book is also the illustrator. All of the illustrations are original; they cannot be seen in any other book.

The style of the text is vivid and original, filled both with practical directions on using specimens and, especially, with sentences that serve to reinforce the importance of major anatomical structures. This saves students time, because they will not have to return to descriptive anatomy discussion during a dissection. The impressive number of illustrations (504) printed by Mosby–Year Book, Inc. in an excellent manner contributes to the high quality of this guide.

I warmly recommend the *Clinical Dissection Guide for Large Animals* to students, practitioners, and breeders. My most sincere congratulations to Dr. Gheorghe M. Constantinescu and to Mosby–Year Book, Inc. for this superb work.

Vasile Gheție, D.V.M., Ph.D., Dr.h.c.

Member of the Romanian Academy
Professor Emeritus of Veterinary Anatomy
Faculty of Veterinary Medicine
Bucharest, Romania

Preface

"Hic locus est ubi mors gaudet succurrere vitae"*

This book is directed principally to professional students in colleges of veterinary medicine. The purpose of this text is to provide the student with a self-directed and detailed guide for the anatomic dissection of the major species of large domestic animals—the horse, ox, sheep, goat, and pig.

The guide to large animal dissection is organized as a progressive series of anatomical illustrations, accompanied by keyed descriptive text. The impetus to prepare such a profusely illustrated guide came from questions and concerns raised by veterinary students during anatomy teaching laboratories. The book's organizational format reflects the presentation techniques that have proven to be most helpful to the student (and the teacher) during the author's 30 years of experience in teaching professional level veterinary anatomy.

This guide presents regional and topographical dissection approaches; it starts with surface features and progresses to deeper structures, stopping along the way for appropriate comments on principal anatomical entities for each region. All of the anatomical drawings and accompanying descriptions are original to this text and are based on specimens dissected personally by the author.

The primary dissection models are the horse and the ox, but appropriate attention is paid to other species when important interspecific anatomical differences are encountered. Although elements of clinical anatomy are mentioned throughout the entire book, the "clinical" anatomy of the horse is especially well developed, with one chapter devoted to structural landmarks and approaches to major structures that are subject to clinical intervention.

The nomenclature used in this guide generally conforms to the most recent edition of *Nomina Anatomica Veterinaria* (1983). For structures not included in the *Nomina Anatomica Veterinaria,* the author has provided especially detailed descriptions and illustrations.

Although this dissection guide is dedicated principally to the veterinary student, it also provides relevant anatomical information that will prove helpful to other groups. Students and veterinarians in meat inspection and necropsy services will find details about the location, relationships (topography), and shape of lymph nodes, of the structures of the digestive, respiratory, and urogenital apparatuses, and of the cardiovascular and central nervous systems. The chapter on clinical anatomy of the horse is dedicated primarily to students in equine clinical rotations and to veterinarians in equine practice. They will find detailed landmarks and approaches for nerve block or local anesthesia, for collection of blood samples, for rectal exploration, for percussion and auscultation of the lungs and heart, and for other structures subject to clinical intervention. Likewise, undergraduate-school and graduate-school educators in veterinary science, biomedical science, and animal science can gain extensive familiarity with the anatomy of large domestic animals through this guide. Finally, the professional equine breeder and cattle breeder will discover in this book new and useful information about the anatomy of large animals.

---

*"Here is the place where the death is glad to be of use to the life."

ACKNOWLEDGEMENTS

The author wishes to thank his former students Phil G. Buhman, Mark R. Crabill, Ruth M. Halenda, Joe B. Scott, Mary K. Voelker, and Jennifer K. Wanner for reading and offering suggestions on individual chapters. The author particularly appreciates Michelle B. Freeman, Robin E. Wharton, and W. Wayne Fry for their careful reading of the entire text and for their thoughtful suggestions.

Most of all, the author greatly appreciates the suggestions, editorial advice, and support of his colleague, Brian L. Frappier, D.V.M., Ph.D., Instructor of Anatomy in the Department of Veterinary Biomedical Sciences of the College of Veterinary Medicine, University of Missouri-Columbia, who carefully read and made helpful comments on the entire book.

For Chapter 9, "Landmarks and approaches of main anatomic structures of the horse susceptible to clinical intervention," the author is grateful for suggestions and professional advice from Eleanor M. Green, D.V.M., diplomate of the American College of Veterinary Internal Medicine and diplomate of the American Board of Veterinary Practitioners and whose specialty is equine practice and who is Associate Professor of Equine Medicine and Surgery of the College of Veterinary Medicine in Columbia-Missouri.

Sincere thanks also go to Donna L. Heavin for typing the manuscript.

Last, but not least, the author is deeply grateful to Dr. H. Richard Adams, D.V.M., Ph.D., Professor and Chairman of the Department of Veterinary Biomedical Sciences at the College of Veterinary Medicine, University of Missouri—Columbia, for his full support and encouragement.

The author also extends his appreciation and gratitude to the Publisher, to the editorial team of Mosby–Year Book, Inc., and especially to the managing editor, Mr. Robert W. Reinhardt.

The author wishes to encourage readers to make known their suggestions and criticism.

Gheorghe M. Constantinescu
Columbia, Missouri, U.S.A., 1991

Contents

Introduction

How to use this book

This guide was written to be used by people from many different disciplines who are interested in large animal anatomy. For veterinary students in the anatomy laboratory, the book addresses five major species. Anatomical structures are described level by level, from superficial to deep; therefore instructors can tell students how deep to make their dissections. Students should not expect their specimens to look exactly like the illustrations; however, despite natural variations, many similarities will be found. Students should try to put their specimens in the same position as shown in the illustrations to expose the same structures. They should also check against the skeleton as often as needed, compare the same structure with that of the other species, and start the dissection from the suggested landmarks.

Some interpretations given to certain structures in veterinary anatomy are incorrect, because they have been perpetuated from the classic texts without personal verification. To avoid such misinterpretation, every structure described has been dissected and checked against books and scientific papers. If a certain structure is not described or illustrated here, two possibilities exist: a variation has not been accounted for, or a misinterpretation may be present. The author apologizes beforehand and asks that he be notified of the details of any such situation.

Because of the expense involved in printing a book with colored pictures, it was necessary to use black and white illustrations in this guide. The author suggests that readers color the arteries red, the veins blue, and the nerves yellow to make these structures more distinct in comparison with other structures.

Most of all, students should not confuse this dissection guide with a textbook. The guide also cannot replace handouts or lectures. It is useful in the laboratory only for dissecting specimens following its directions, even though certain descriptions are stated under titles like "Remember" or "Notice" for refreshing the student's memory.

In addition, the book is also useful to second- to fourth-year students and to practitioners in large animal clinics (see the details in the preface).

The following abbreviations are used in the text and illustrations:

A.—artery
Aa.—arteries
A.V.—artery and vein
A.V.N.—artery, vein, and nerve
br.—branch
fig.—figure
figs.—figures
ggl.—ganglion
lig.—ligament
ligg.—ligaments

ln.—lymph node
lnn.—lymph nodes
M.—muscle
Mm.—muscles
N.—nerve
Nn.—nerves
p.—part
V.—vein
Vv.—veins

Chapter 1

Neck and thoracic inlet

NECK
Horse

Caution. Because there is only a small amount of loose connective tissue, skin the neck carefully.

If you are dissecting the parotid region as a unit, incise the skin at the edge of the mane from the poll to the cranial extent of the withers, as well as along the caudal border of the mandible and the cranial border of the shoulder. Leaving 2 to 3 cm of skin on the dorsal midline, including the poll, reflect the skin ventrally. Another technique involves incising the skin on the ventral midline and reflecting the skin dorsally, as shown in Fig. 1-1. To dissect the parotid region separately, refer to the penultimate chapter of this book.

The first structures visible ventrally after skinning are the cutaneus colli M., which is adherent to the skin, and the superficial cervical fascia, which surrounds and protects the most superficial structures. The fascia is pierced by cutaneous cervical nerves and vessels supplying the skin.

Notice that the superficial cervical fascia covers all the superficial muscles of the neck, except the cutaneus colli M., which is developed within the ventral portion of the fascia. The symmetrical structures thereof start from a common median raphe. The fascia is very reduced in the area of the sternocephalicus (sternomandibularis), cleidomastoideus, and omotransversarius (cleidotransversarius) Mm. The cutaneus colli M. is well developed in the horse.

Dissect and reflect the cutaneus colli M. ventrally. The omohyoideus, sternocephalicus, cleidomastoideus, and omotransversarius Mm. and the external jugular V. are now exposed. The external jugular V. runs in the jugular groove, which is limited dorsally by the cleidomastoideus M. and ventrally by the sternocephalicus M.

Caution. Find and dissect carefully the cervical branch of the facial N. on the lateral aspect of the cranial half of the external jugular V.

Through the transparency of the superficial cervical fascia, lymphatic vessels are visible, especially in the caudal half of the area; they belong to the superficial cervical lnn. The fascia is perforated at the same time by the dorsal cutaneous branches of C_2 to C_6 Nn.

Dissect and remove the superficial cervical fascia only, in the visible areas of the splenius and rhomboideus cervicis Mm. Leave a portion of this fascia, which connects the trapezius cervicalis M. to the omotransversarius M., 10 cm cranial to the scapula. The apparent area of the splenius and rhomboideus cervicis Mm. and the dorsal branch of the accessory N. are now exposed.

Locate the ventral cutaneous branches of C_2 to C_6 Nn. Their apparent origins indicate the limit between the cleidomastoideus and the omotransversarius Mm. Make only a superficial incision on the line that unites the apparent origins of these nerves.

Caution. Do not try to completely separate the two muscles.

Locate the linguofacial V. and the parotid gland, which is covered by the parotidoauricularis M. Also locate the aponeurosis of the cleidomastoideus M. (the cranial attachment of the muscle), which covers the aponeurosis of the splenius M.

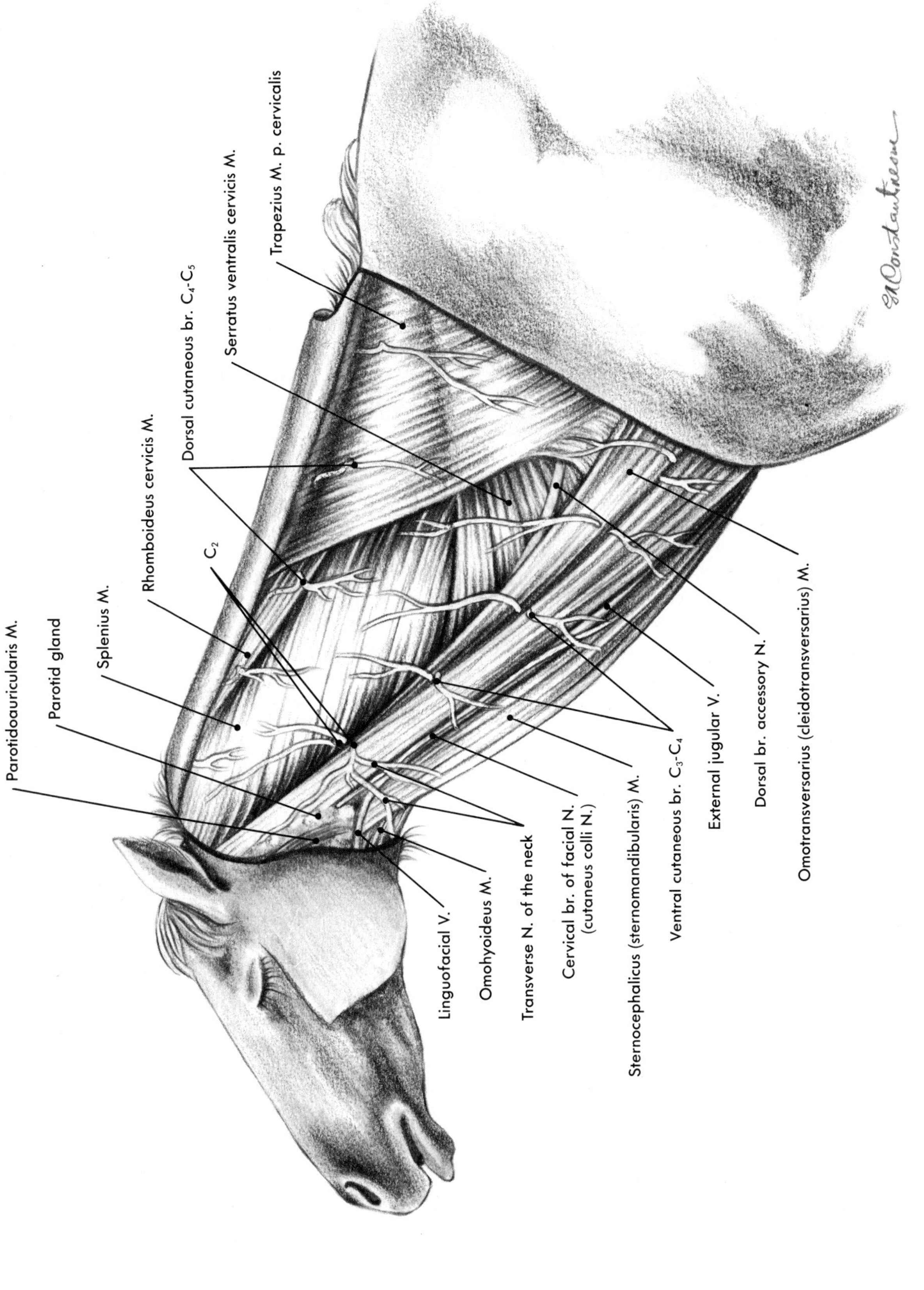

Parotidoauricularis M.

Parotid gland

Splenius M.

Rhomboideus cervicis M.

C₂

Dorsal cutaneous br. C₄-C₅

Serratus ventralis cervicis M.

Trapezius M. p. cervicalis

Linguofacial V.

Omohyoideus M.

Transverse N. of the neck

Cervical br. of facial N.
(cutaneus colli N.)

Sternocephalicus (sternomandibularis) M.

Ventral cutaneous br. C₃-C₄

External jugular V.

Dorsal br. accessory N.

Omotransversarius (cleidotransversarius) M.

Fig. 1-1

Superficial level lateral neck.

On the ventral aspect of the neck, first dissect the superficial cervical muscles. Keeping intact the ventral cutaneous branches of C_2 to C_6 Nn., dissect the common origin of both sternocephalicus Mm. on the manubrium sterni. Toward the head these two muscles diverge, with the sternothyroideus and the sternohyoideus Mm. between them. These four muscles (two symmetrical pairs) also originate from the manubrium sterni, in a manner similar to the sternocephalicus M., that is, by way of a common body (Fig. 1-2). In the middle of the neck, this common body is continued by a short, thin tendon, from which the two sternothyroideus Mm. (laterally) and the two sternohyoideus Mm. (medially) arise. Dorsal to these four muscles, in the cranial half of the neck, the tracheal rings can be distinguished.

Remove the fascia pretrachealis (the middle or visceral fascia) from the cervical part of the trachea to expose all the rings. In the cranial third or half of the neck, the two omohyoideus Mm. cross toward the two sternocephalicus Mm. in a divergent angle from the basihyoid to the shoulder. Dissect the two omohyoideus Mm. carefully, saving the transverse N. of the neck (the transversus colli N.), which travels toward the intermandibular space.

Transect the omotransversarius and cleidomastoideus Mm. 20 cm cranial to the shoulder; separate them from the omohyoideus M. and reflect the stumps.

Caution. Save the external jugular V., and at this time, do not transect the superficial cervical fascia that protects it.

Very close to the shoulder on the deep aspect of the omotransversarius M., dissect the group of the superficial cervical lnn. and the prescapular branch of the superficial cervical A.V.

Transect the superficial cervical fascia alongside the external jugular V.; dissect and expose the vein; and separate it from the carotid sheath (vagina carotica), which protects the common carotid A. and the vagosympathetic trunk dorsally and the recurrent laryngeal N. ventrally. The carotid sheath and the structures within it are situated alongside the esophagus and trachea. The deep cervical lnn. (except the caudal group) are now exposed, along with the tracheal trunk (Fig. 1-3).

Incise the splenius M. parallel and 2 cm ventral to the ventral border of the rhomboideus cervicis M. Then make a perpendicular incision from the caudal end of the first incision to 20 cm cranial to the cranial border of the scapula. Reflect the two stumps ventrally. On the deep aspect of the stumps, lateral branches of the dorsal branches of the cervical nerves are visible, accompanied by vessels arising from the deep cervical A.V. The nervous branches intermingle in the superficial dorsal cervical plexus. The semispinalis capitis M. with both its biventer cervicis (dorsal) and complexus (ventral) Mm., as well as the longissimus capitis and the longissimus atlantis Mm. (the second located ventrally from the first), are now exposed.

Incise the semispinalis capitis M. parallel and 4 cm ventral to the splenius M., and separate it from the ligamentum nuchae. Reflect the muscle ventrally to identify the medial branches of the dorsal branches of the cervical nerves, which intermingle in the deep dorsal cervical plexus. Arteries and veins from the deep cervical A.V. accompany the nerves. Both the funiculus nuchae and the lamina nuchae of the ligamentum nuchae are now exposed (Fig 1-4).

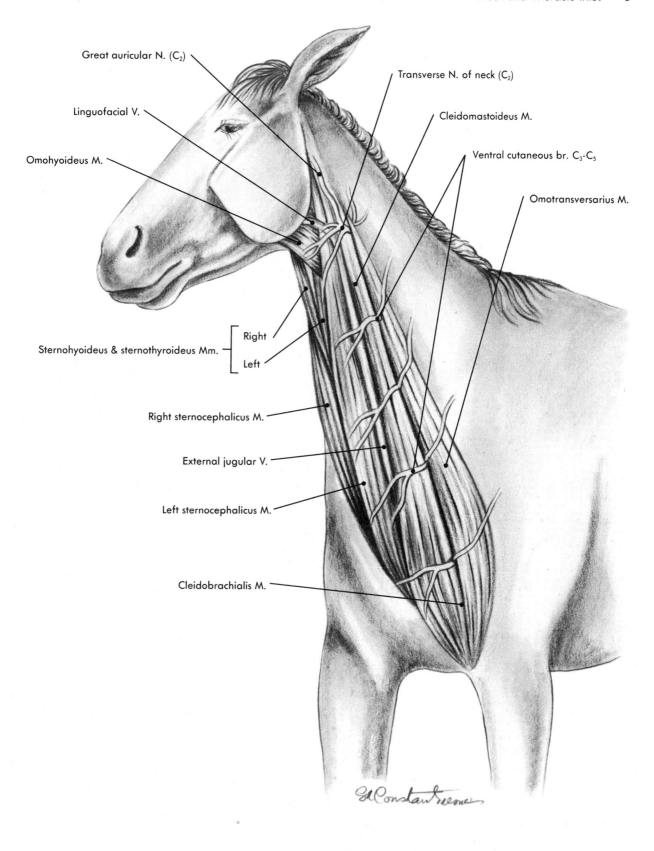

Great auricular N. (C₂)

Linguofacial V.

Omohyoideus M.

Sternohyoideus & sternothyroideus Mm.

Right sternocephalicus M.

External jugular V.

Left sternocephalicus M.

Cleidobrachialis M.

Transverse N. of neck (C₂)

Cleidomastoideus M.

Ventral cutaneous br. C₃-C₅

Omotransversarius M.

[Right
  Left]

Superficial level ventral neck.   **Fig. 1-2**

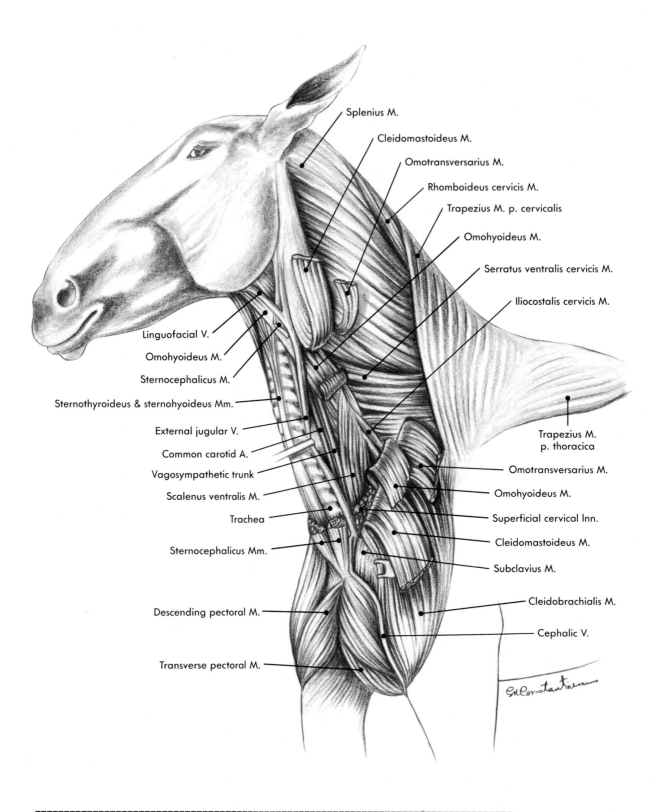

Splenius M.

Cleidomastoideus M.

Omotransversarius M.

Rhomboideus cervicis M.

Trapezius M. p. cervicalis

Omohyoideus M.

Serratus ventralis cervicis M.

Iliocostalis cervicis M.

Linguofacial V.

Omohyoideus M.

Sternocephalicus M.

Sternothyroideus & sternohyoideus Mm.

External jugular V.

Common carotid A.

Vagosympathetic trunk

Scalenus ventralis M.

Trachea

Sternocephalicus Mm.

Descending pectoral M.

Transverse pectoral M.

Trapezius M. p. thoracica

Omotransversarius M.

Omohyoideus M.

Superficial cervical lnn.

Cleidomastoideus M.

Subclavius M.

Cleidobrachialis M.

Cephalic V.

Fig. 1-3   Deep level ventral neck.

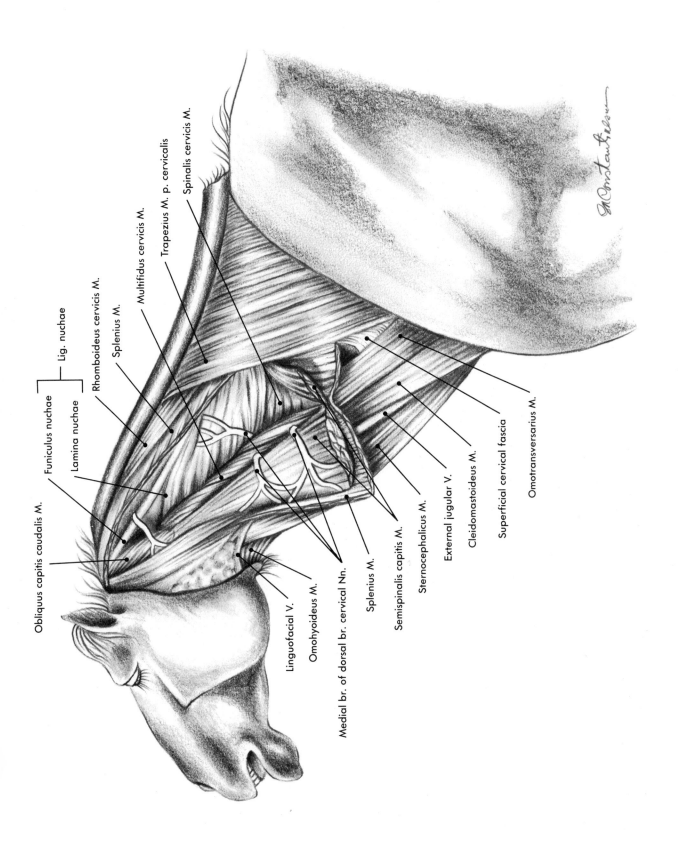

Lig. nuchae

Funiculus nuchae

Lamina nuchae

Obliquus capitis caudalis M.

Rhomboideus cervicis M.

Splenius M.

Multifidus cervicis M.

Trapezius M. p. cervicalis

Spinalis cervicis M.

Linguofacial V.

Omohyoideus M.

Medial br. of dorsal br. cervical Nn.

Splenius M.

Semispinalis capitis M.

Sternocephalicus M.

External jugular V.

Cleidomastoideus M.

Superficial cervical fascia

Omotransversarius M.

Fig. 1-4

Deep level lateral neck.

Transect the trapezius cervicalis M. parallel and 10 cm ventral to the dorsal border of the neck just above the incision line of the splenius and semispinalis capitis Mm., and continue the incision up to the skin of the shoulder. Reflect by dissection the ventral stumps separating these three muscles. Next, make a perpendicular incision through the splenius, semispinalis capitis, and serratus ventralis cervicis Mm. crossing the first incision parallel to the shoulder. Two muscles will be exposed: the longissimus cervicis M., inserted on the transverse processes of the last four cervical vertebrae, and the spinalis cervicis M., inserted on the spinous processes of the same last four cervical vertebrae (Fig 1-5).

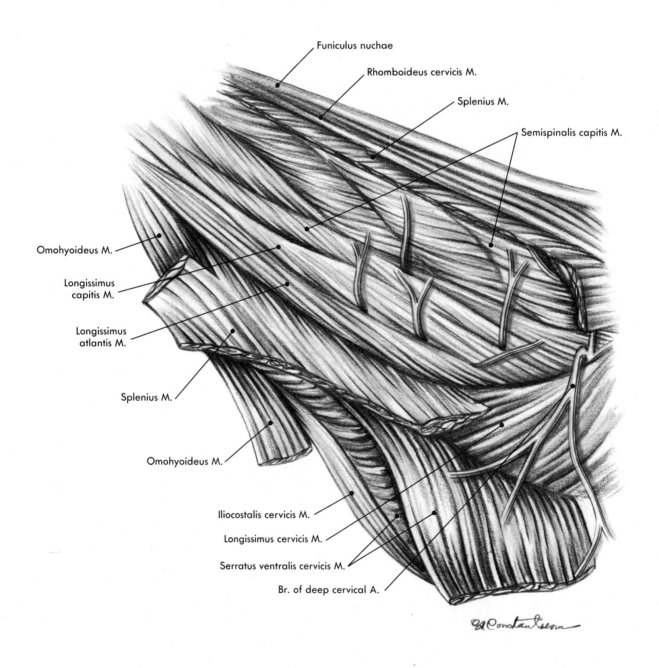

Fig. 1-5   Deep level laterodorsal neck.

Make a vertical incision through the cranial aponeurosis of the semispinalis M., parallel to the previous incision of the splenius. Caudal to the axis, identify the multifidus Mm. on the dorsal aspect of the vertebral column. On the lateral aspect of the axis, identify the obliquus capitis caudalis M. (Fig. 1-6).

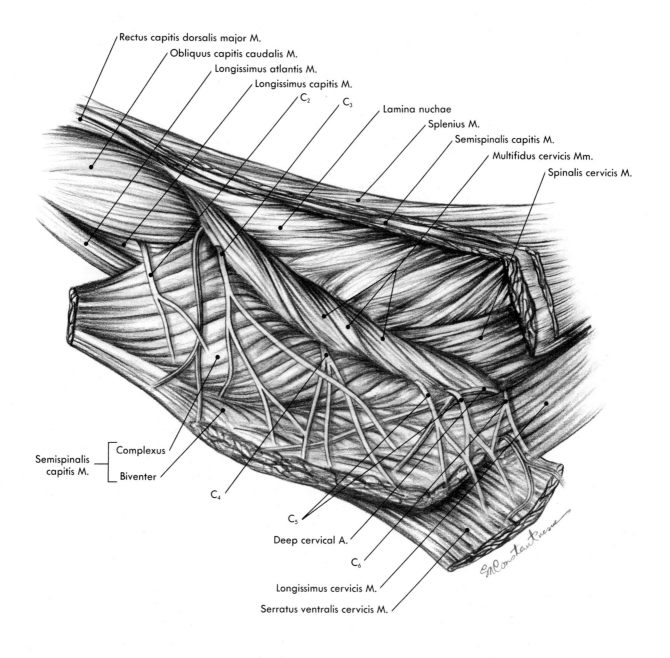

Rectus capitis dorsalis major M.
Obliquus capitis caudalis M.
Longissimus atlantis M.
Longissimus capitis M.
C₂
C₃
Lamina nuchae
Splenius M.
Semispinalis capitis M.
Multifidus cervicis Mm.
Spinalis cervicis M.

Semispinalis capitis M.
Complexus
Biventer

C₄
C₅
Deep cervical A.
C₆
Longissimus cervicis M.
Serratus ventralis cervicis M.

Deepest level laterodorsal neck.

Fig. 1-6

Continue the cranial incision of the splenius and semispinalis capitis Mm. (through their aponeuroses) as close as possible to the head. This exposes the rectus capitis dorsalis major and the rectus capitis dorsalis minor Mm.

Notice: The rectus capitis dorsalis major M. has a superficial and a deep portion.

Find the atlas and locate the lateral vertebral and the alar foramina by palpation, and make a longitudinal incision through the obliquus capitis caudalis M. The origin of the dorsal branch of the C_1 N. and the occipital branch of the occipital A. are now exposed. Locate the transverse foramen of the atlas and make a longitudinal incision through the obliquus capitis caudalis M. to expose the vertebral A.

The most cranial structure in the area is the obliquus capitis cranialis M., located between the wing of the atlas and the nuchal crest (Fig. 1-7, *A*).

Locate the spinous processes of both the atlas and the axis. A nuchal bursa lies between each process and the funiculus nuchae (named the cranial and the caudal nuchal bursae). These bursae protect the funiculus nuchae during movements of the neck (Fig. 1-7, *B*).

Transect the omohyoideus M. and reflect the stumps to expose the following structures: caudally, the iliocostalis cervicis M. (dorsally) and the scalenus system (ventrally); cranially, the longus capitis, rectus capitis ventralis, and rectus capitis lateralis Mm. At the cranial end of the trachea on the lateral aspect of the first two to three tracheal rings, the corresponding lobe of the thyroid gland can be exposed, as can the parathyroid glands. The scalenus system is represented by the scalenus ventralis (the most developed) and the scalenus medius Mm. From between these two muscles the roots of the brachial plexus emerge. Leave intact the origin of the phrenic N. on the lateral aspect of the scalenus ventralis. *Notice* that the iliocostalis cervicis is located between the serratus ventralis cervicis (dorsally) and the scalenus Mm. (ventrally). These muscles will be better exposed after removal of the thoracic limb.

Transect the longus capitis M. and remove its last two attachments on the cervical vertebrae. On the ventral aspect of the cervical vertebrae, identify the longus colli m. with its cervical part made up of successive V-shaped fascicles, oriented with the two arms facing caudally (Fig. 1-8). *Notice* that this muscle replaced the ventral longitudinal ligament up to T6, inclusively.

Notice that the longus colli M. is protected by the deep cervical fascia (lamina prevertebralis).

Dissect and reflect the attachments on the transverse processes of C_2 to C_5 of all the muscles in the area, and expose the deepest lateral aspect of the cervical region; a specific area between the cranial and caudal articular processes and the transverse processes (with both dorsal and ventral tubercles) of C_2 to C_5 is now exposed. The strong, deep cervical fascia protects the intertransversarii cervicis Mm., which in horse are made up of three distinct muscles: intertransversarius dorsalis, intertransversarius medius, and intertransversarius ventralis Mm. Between the intertransvarsarius dorsalis and the intertransversarius medius Mm., dissect the dorsal branch of the corresponding cervical nerve, which is accompanied by an external branch of the vertebral A. Ventral to the intertransversarius ventralis M. and close to the ventral tubercle of the transverse process of the corresponding cervical vertebra, dissect the ventral branch of the vertebral A. and the ventral branch of the corresponding cervical nerve.

Dorsal to the intertransversarius Mm. and on the lateral aspect of the lamina nuchae and the spinalis cervicis M., the multifidus cervicis Mm. are now exposed, covered by the same deep cervical fascia as the intertransversarii Mm. The multifidus cervicis Mm. are made up of a medial part, a lateral part, and a deep part (Fig. 1-9).

Palpate the transverse foramina of the vertebrae and expose the vertebral A.

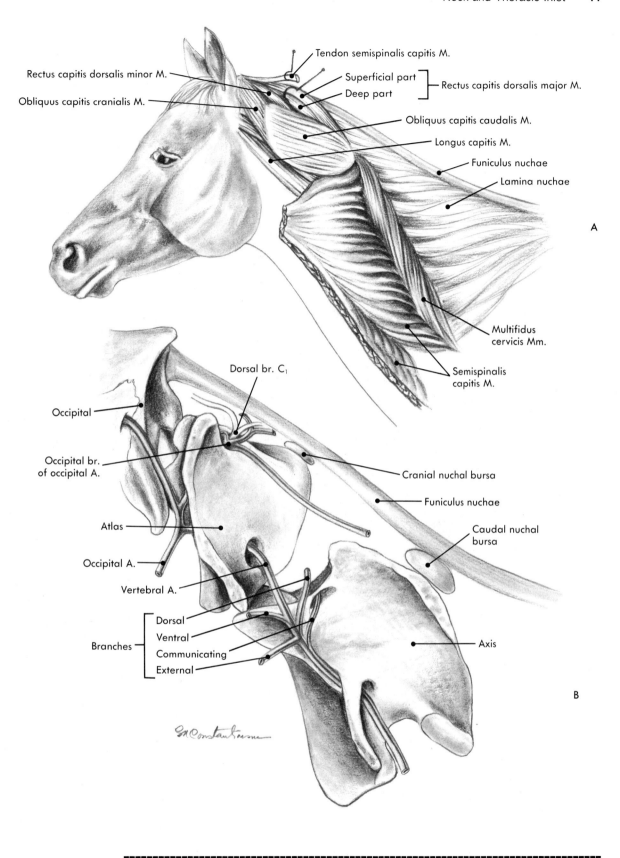

Tendon semispinalis capitis M.

Rectus capitis dorsalis minor M.

Superficial part

Deep part } — Rectus capitis dorsalis major M.

Obliquus capitis cranialis M.

Obliquus capitis caudalis M.

Longus capitis M.

Funiculus nuchae

Lamina nuchae

A

Multifidus cervicis Mm.

Semispinalis capitis M.

Dorsal br. C₁

Occipital

Cranial nuchal bursa

Occipital br. of occipital A.

Funiculus nuchae

Atlas

Caudal nuchal bursa

Occipital A.

Vertebral A.

Dorsal

Ventral

Branches { Communicating

External

Axis

B

A, Deepest level laterodorsal neck. **B,** Atlantoaxial deepest structures.

Fig. 1-7

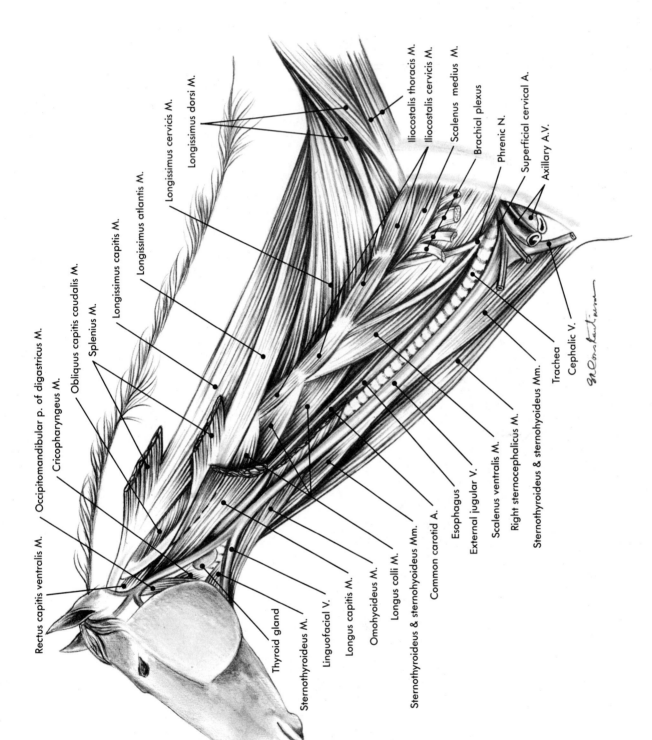

Rectus capitis ventralis M.

Occipitomandibular p. of digastricus M.

Cricopharyngeus M.

Obliquus capitis caudalis M.

Splenius M.

Longissimus capitis M.

Longissimus atlantis M.

Longissimus capitis M.

Longissimus cervicis M.

Longissimus dorsi M.

Iliocostalis thoracis M.

Iliocostalis cervicis M.

Scalenus medius M.

Brachial plexus

Phrenic N.

Superficial cervical A.

Axillary A.V.

Cephalic V.

Trachea

Sternothyroideus & sternohyoideus Mm.

Right sternocephalicus M.

Scalenus ventralis M.

External jugular V.

Esophagus

Common carotid A.

Sternothyroideus & sternohyoideus Mm.

Longus colli M.

Omohyoideus M.

Longus capitis M.

Linguofacial V.

Sternothyroideus M.

Thyroid gland

Deep level lateral neck.

Fig. 1-8

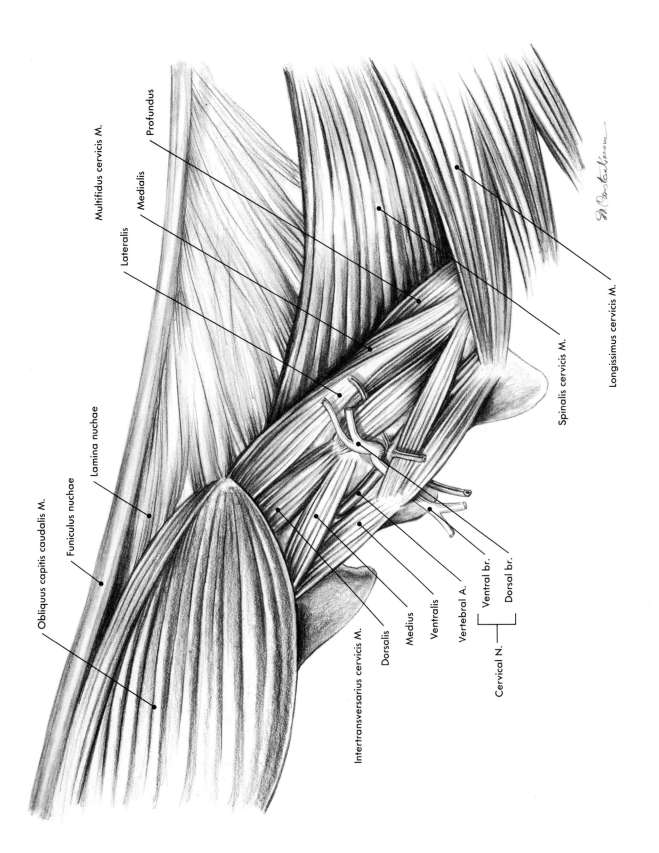

Multifidus cervicis M.

Profundus

Medialis

Lateralis

Lamina nuchae

Funiculus nuchae

Obliquus capitis caudalis M.

Spinalis cervicis M.

Longissimus cervicis M.

Intertransversarius cervicis M.

Dorsalis

Medius

Ventralis

Vertebral A.

Cervical N. { Ventral br. / Dorsal br.

Fig. 1-9

Intertransversarii and multifidus cervicis Mm.

Large ruminants

Incise the skin on the middorsal line of the neck from the intercornual protuberance to the cranial angle of the scapula. From the cranial end of the incision make another one, caudal to the base of the ear and perpendicular to the ventral midline. From the caudal end of the first incision, make a third incision to the manubrium sterni surrounding the shoulder joint.

When skinning the cervical region in the large ruminants, the very thin and adherent cutaneus colli M. may remain attached to the skin, together with the superficial cervical fascia. The dorsal and the ventral cutaneous branches of C_2 to C_6 Nn. perforate the fascia and the cutaneus colli M.

After removing the superficial cervical fascia, the following structures are now exposed in the dorsal cervical area. The trapezius cervicalis M. is located dorsocaudally; from the shoulder toward the head in an oblique position, two muscles are exposed: the cleidooccipitalis M. (dorsally and superficially at the origin) and the cleidomastoideus M. (ventrally). Between the trapezius M. and the cleidooccipitalis M., in a caudocranial order, four muscles along with adjacent nerves will be exposed: the omotransversarius M. accompanied by ventral branches of C_5 N.; the serratus ventralis cervicis M. with the accessory N.; the splenius M.; and the most cranial extent of the rhomboideus cervicis M. In some specimens, the omotransversarius M. and the accessory N. can be seen between the cleidooccipitalis and the cleidomastoideus Mm. Ventral to the terminal tendon of the cleidomastoideus M., the tendon of the sternomastoideus M. is also exposed.

The sternocephalicus M., consisting of the superficial sternomandibularis M. and the deep sternomastoideus M., the omohyoideus M., and both the sternothyroideus and sternohyoideus Mm. are also exposed in the ventral cervical area.

Locate the jugular groove at the border between the dorsal and the ventral cervical area and expose the external jugular V. by incising and reflecting the superficial cervical fascia. The cleidomastoideus m. forms the dorsal border, the sternomandibularis M. forms the ventral border (Fig 1-10) and the sternomastoideus M. forms the deep wall of the jugular groove. Unlike in the horse, there is not a cervical branch of the facial N. parallel to this vein.

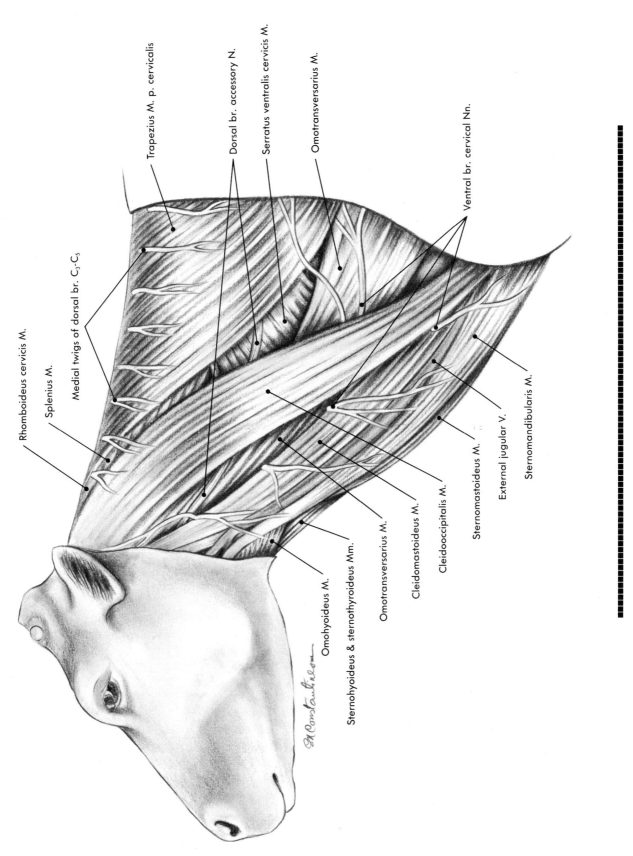

Trapezius M. p. cervicalis

Dorsal br. accessory N.

Serratus ventralis cervicis M.

Omotransversarius M.

Ventral br. cervical Nn.

Rhomboideus cervicis M.

Splenius M.

Medial twigs of dorsal br. C_3-C_5

Sternomandibularis M.

External jugular V.

Sternomastoideus M.

Cleidooccipitalis M.

Cleidomastoideus M.

Omotransversarius M.

Sternohyoideus & sternothyroideus Mm.

Omohyoideus M.

Fig. 1-10

First level lateral neck.

Transect the cleidooccipitalis and the cleidomastoideus Mm., and reflect the stumps by dissecting them away from the adjacent structures. The splenius, omotransversarius and scalenus ventralis Mm. and the dorsal branch of the accessory N. are now widely exposed.

Transect the sternomastoideus M. and dissect it carefully to separate the muscle from the external jugular V. (superficially) and the carotid sheath (deeply). The omohyoideus M. is now widely exposed.

Reflect the external jugular V. dorsally and the sternomandibularis M. ventrally, revealing its medial aspect with the ventral branch of the accessory N. Dissect the middle cervical fascia (lamina pretrachealis) to expose the trachea and the esophagus. Also dissect the carotid sheath to expose the common carotid A. accompanied by the internal jugular V., the vagosympathetic trunk, and the recurrent laryngeal Nn (Fig 1-11). The deep cervical lnn. are also exposed.

Transect the trapezius cervicalis M. 2 cm ventral and parallel to the dorsal midline and reflect it caudally.

Transect the omotransversarius M. and reflect the stumps. Carefully dissect the medial aspect of its caudal stump, which is adherent to the superficial cervical ln. and to the prescapular branches of the superficial cervical A.V.

Identify the following structures: the rhomboideus cervicis M.; the serratus ventralis cervicis, splenius, and omohyoideus Mm.; the longissimus capitis, the longissimus atlantis, the longus atlantis, the longus capitis, the scalenus ventralis, the scalenus dorsalis, and the iliocostalis cervicis Mm. Branches of the cervical nerves may also be seen (Fig. 1-12).

Notice that the longus atlantis is part of the intertransversarii Mm.

Transect the splenius M. parallel to the ventral border of rhomboideus cervicis M. Continue the incision cranially, following the middorsal line, the wing of the atlas, and the ventral border of the splenius, which is in contact with the dorsal border of the longissimus capitis M. Reflect the splenius M. ventrally.

Make similar incisions through the semispinalis capitis M. and reflect it ventrally. Branches of the deep cervical and vertebral Aa. are found between the splenius and semispinalis capitis Mm., as well as on the medial aspect of semispinalis capitis M.

Transect the cranial insertion of the longissimus capitis M. and reflect the muscle caudally.

The following structures are now exposed: the funiculus and lamina nuchae; the rectus capitis dorsalis major and minor Mm.; the spinalis cervicis, multifidus, and intertransversarii cervicis Mm.; the obliquus capitis caudalis, longissimus atlantis, and longus atlantis Mm.; the caudal articular process of the axis; and two strong branches of C_2 and C_3 Nn.

After reflecting the forelimb as far caudally as possible or after removing the limb, dissect the serratus ventralis cervicis, the iliocostalis cervicis, and the scaleni Mm., ventral to the structures already described. Ventral branches of nerves C_4 and C_5 and the origin of the phrenic N. lie obliquely on the scalenus ventralis M. The brachial plexus emerges between the scalenus ventralis and medius Mm. The scalenus medius M. can be observed by reflecting the scalenus dorsalis M. dorsally (Fig 1-13)

Remember! Among the large domestic animals only the large ruminants and the pig constantly show a scalenus dorsalis M., a scalenus medius M., and a scalenus ventralis M. In the goat the scalenus dorsalis M. is not always present. The horse and the sheep have only the scalenus medius M. and the scalenus ventralis M.

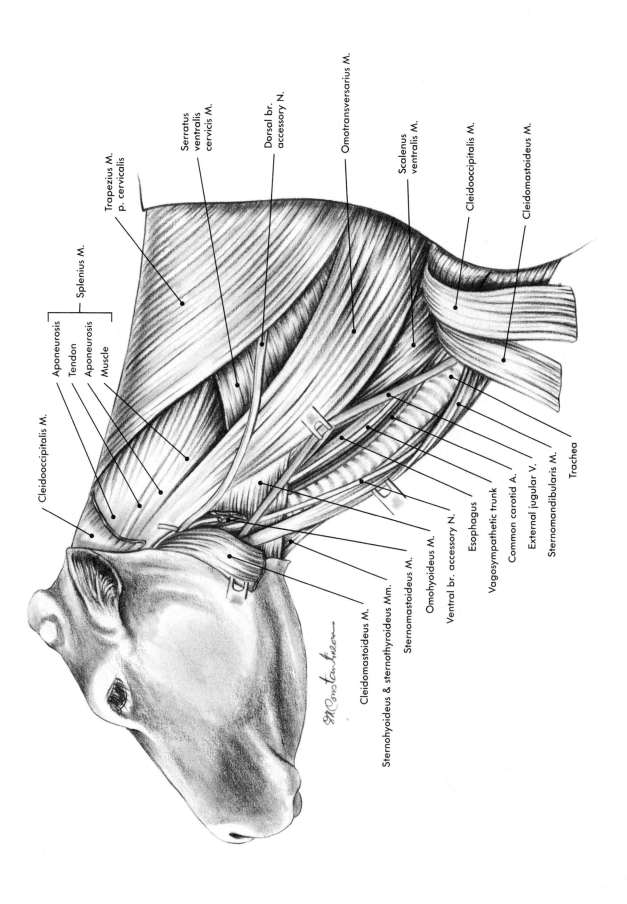

Trapezius M.
p. cervicalis

Splenius M.

Aponeurosis
Tendon
Aponeurosis
Muscle

Cleidooccipitalis M.

Cleidomastoideus M.

Sternohyoideus & sternothyroideus Mm.

Sternomastoideus M.

Omohyoideus M.

Ventral br. accessory N.

Esophagus

Vagosympathetic trunk

Common carotid A.

External jugular V.

Sternomandibularis M.

Trachea

Serratus
ventralis
cervicis M.

Dorsal br.
accessory N.

Omotransversarius M.

Scalenus
ventralis M.

Cleidooccipitalis M.

Cleidomastoideus M.

Fig. 1-11

Second level lateral neck.

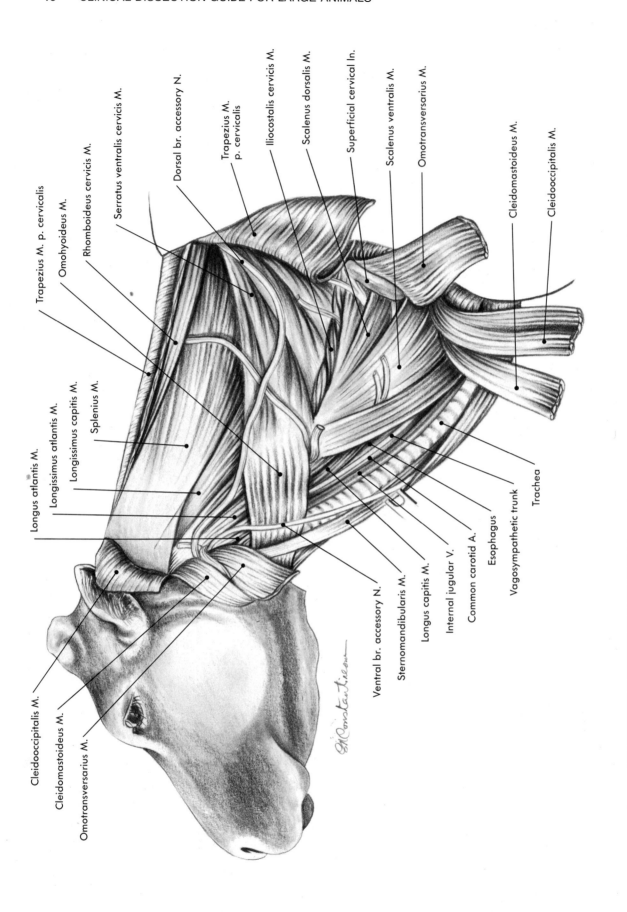

Cleidooccipitalis M.

Cleidomastoideus M.

Omotransversarius M.

Longus atlantis M.

Longissimus atlantis M.

Longissimus capitis M.

Splenius M.

Trapezius M. p. cervicalis

Omohyoideus M.

Rhomboideus cervicis M.

Serratus ventralis cervicis M.

Dorsal br. accessory N.

Trapezius M. p. cervicalis

Iliocostalis cervicis M.

Scalenus dorsalis M.

Superficial cervical ln.

Scalenus ventralis M.

Omotransversarius M.

Cleidomastoideus M.

Cleidooccipitalis M.

Ventral br. accessory N.

Sternomandibularis M.

Longus capitis M.

Internal jugular V.

Common carotid A.

Esophagus

Vagosympathetic trunk

Trachea

Third level lateral neck.

Fig. 1-12

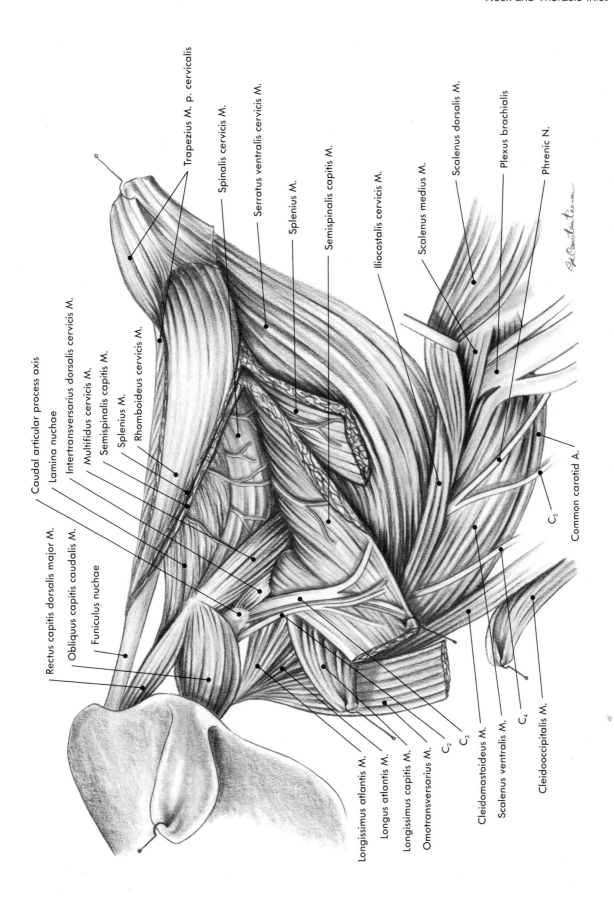

Trapezius M. p. cervicalis

Spinalis cervicis M.

Serratus ventralis cervicis M.

Splenius M.

Semispinalis capitis M.

Iliocostalis cervicis M.

Scalenus medius M.

Scalenus dorsalis M.

Plexus brachialis

Phrenic N.

Caudal articular process axis

Lamina nuchae

Intertransversarius dorsalis cervicis M.

Multifidus cervicis M.

Semispinalis capitis M.

Splenius M.

Rhomboideus cervicis M.

Rectus capitis dorsalis major M.

Obliquus capitis caudalis M.

Funiculus nuchae

Longissimus atlantis M.

Longus atlantis M.

Longissimus capitis M.

Omotransversarius M.

C₂

C₃

Cleidomastoideus M.

Scalenus ventralis M.

C₄

Cleidooccipitalis M.

C₅

Common carotid A.

Deepest level laterodorsal neck.

Fig. 1-13

Sheep
■■■■■■■■■
(Figs. 1-14
to 1-16)

Skin an area similar to that in the large ruminants. Dissect the superficial and deep structures of the neck.

Caution. Leave intact the branches of C_2, C_3, and C_4 nerves intersecting the external jugular vein and the cervical br. (VII), the occipital and linguofacial Vv., and the omo-hyoideus, sternothyroideus, and sternohyoideus Mm. (Fig. 1-14).

Caution. Leave the origin of the phrenic N. lying on the scalenus ventralis M. intact.

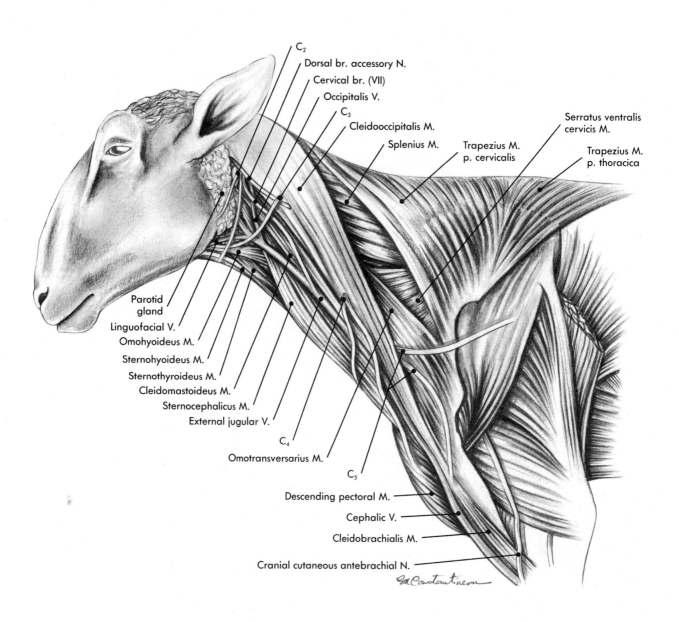

Fig. 1-14

Superficial level lateral neck.

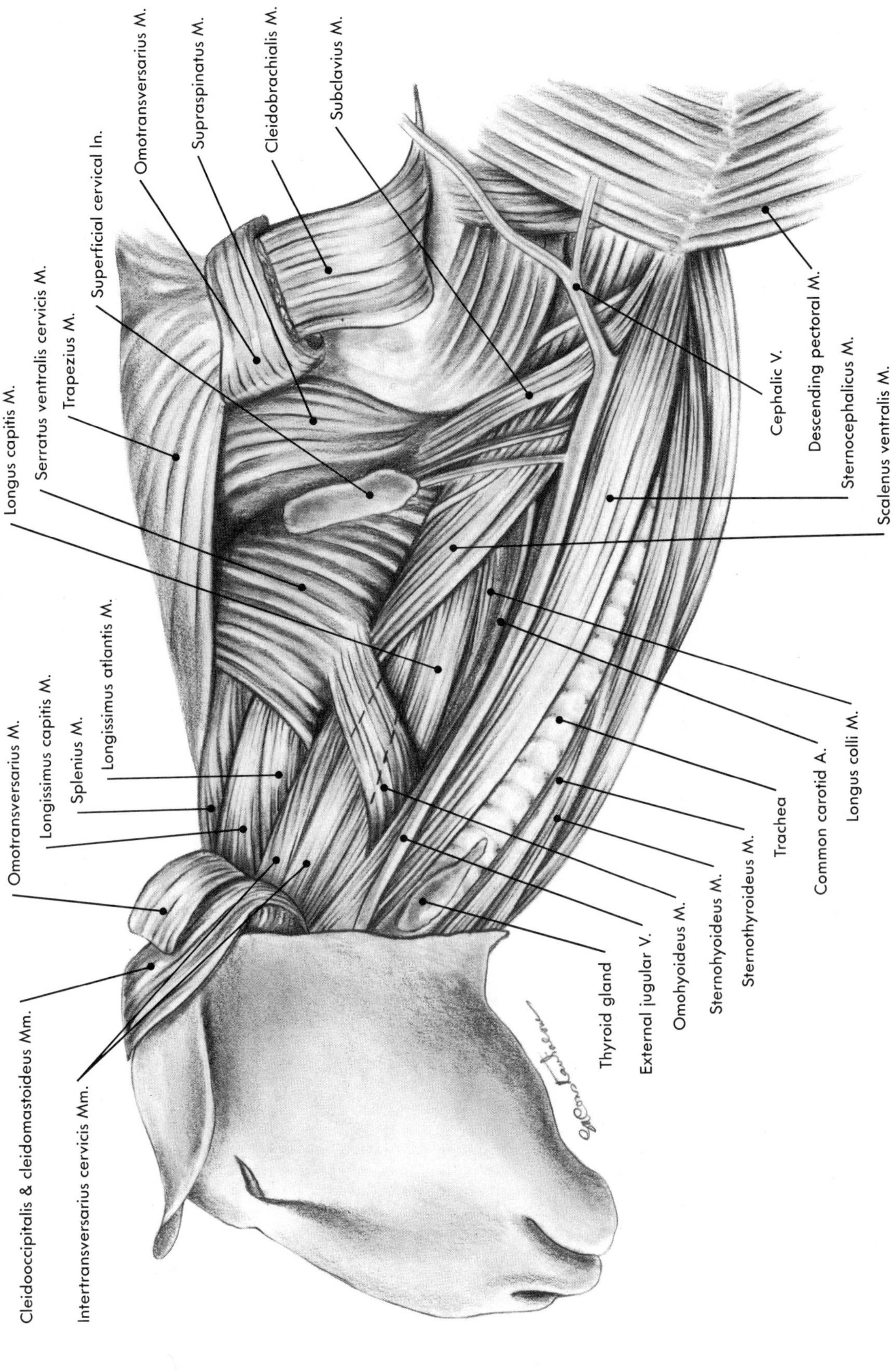

Cleidooccipitalis & cleidomastoideus Mm.

Intertransversarius cervicis Mm.

Omotransversarius M.

Longissimus capitis M.

Splenius M.

Longissimus atlantis M.

Longus capitis M.

Serratus ventralis cervicis M.

Trapezius M.

Superficial cervical ln.

Omotransversarius M.

Supraspinatus M.

Cleidobrachialis M.

Subclavius M.

Cephalic V.

Descending pectoral M.

Sternocephalicus M.

Scalenus ventralis M.

Longus colli M.

Common carotid A.

Trachea

Sternothyroideus M.

Sternohyoideus M.

Omohyoideus M.

External jugular V.

Thyroid gland

Fig. 1-15

Deep level lateral neck.

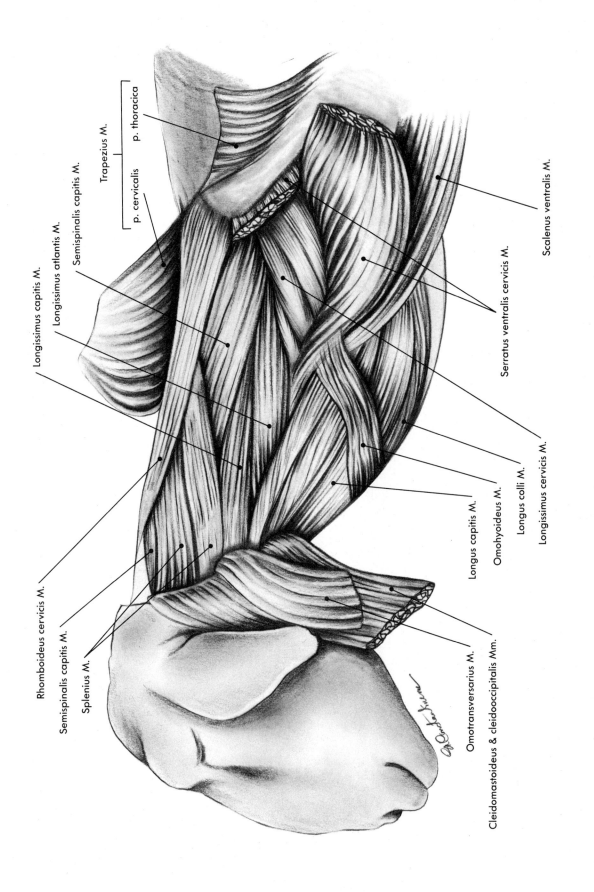

Trapezius M. { p. thoracica / p. cervicalis

Semispinalis capitis M.

Longissimus atlantis M.

Longissimus capitis M.

Rhomboideus cervicis M.

Semispinalis capitis M.

Splenius M.

Omotransversarius M.

Cleidomastoideus & cleidooccipitalis Mm.

Longus capitis M.

Omohyoideus M.

Longus colli M.

Longissimus cervicis M.

Serratus ventralis cervicis M.

Scalenus ventralis M.

Fig. 1-16   Deep level laterodorsal neck.

Skin the cervical area, reflecting the skin dorsally within the following boundaries: from the manubrium sterni along the ventral midline cranially, as far as the cervical appendix (tassel) and caudal to the base of the ear and up to the middorsal line of the neck; from the manubrium sterni along a vertical line crossing the humerus and along the spina scapularis to the highest point of the thorax.

Dissect the area and a topography similar to that of the sheep will be exposed, displaying the superficial structures (Fig. 1-17).

Goat
■■■■■■■■
(Figs. 1-17 to 1-21)

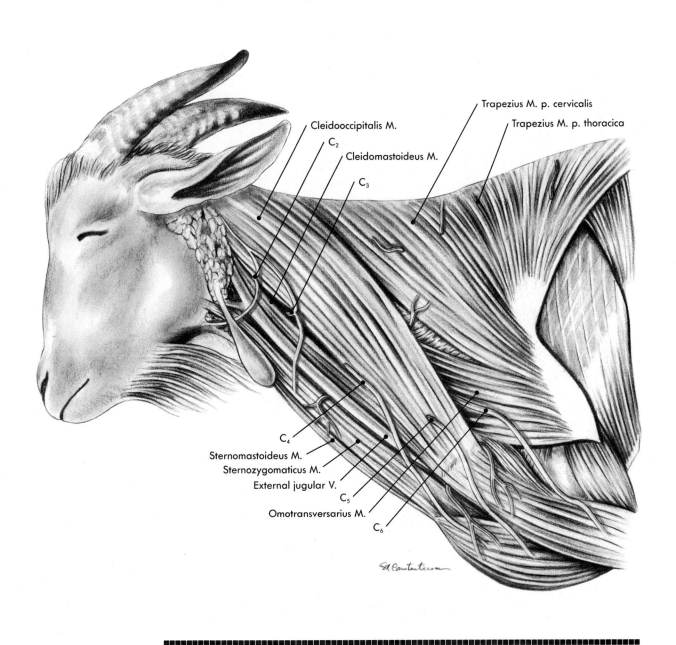

Cleidooccipitalis M.

C$_2$

Cleidomastoideus M.

C$_3$

Trapezius M. p. cervicalis

Trapezius M. p. thoracica

C$_4$

Sternomastoideus M.

Sternozygomaticus M.

External jugular V.

C$_5$

Omotransversarius M.

C$_6$

Superficial level lateral neck.

Fig. 1-17

It is necessary first to dissect the thoracic inlet and, subsequently, the ventral structures of the neck before starting the last step in the dissection of the goat neck (the third level of the dorsal part) (Figs. 1-18 to 1-21).

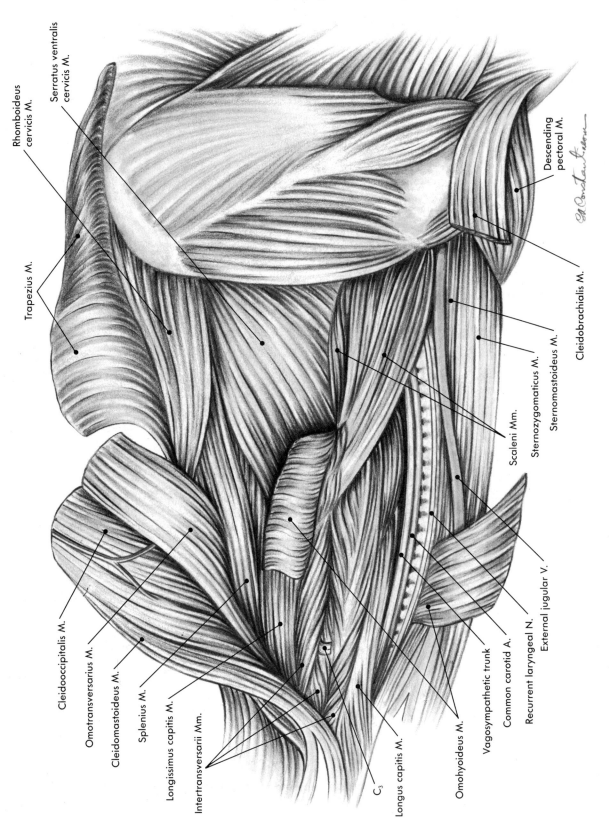

Serratus ventralis cervicis M.

Rhomboideus cervicis M.

Trapezius M.

Descending pectoral M.

Cleidobrachialis M.

Sternomastoideus M.

Sternozygomaticus M.

Scaleni Mm.

Cleidooccipitalis M.

Omotransversarius M.

Cleidomastoideus M.

Splenius M.

Longissimus capitis M.

Intertransversarii Mm.

C₃

Longus capitis M.

Omohyoideus M.

Vagosympathetic trunk

Common carotid A.

Recurrent laryngeal N.

External jugular V.

Fig. 1-18

Deep level lateral neck.

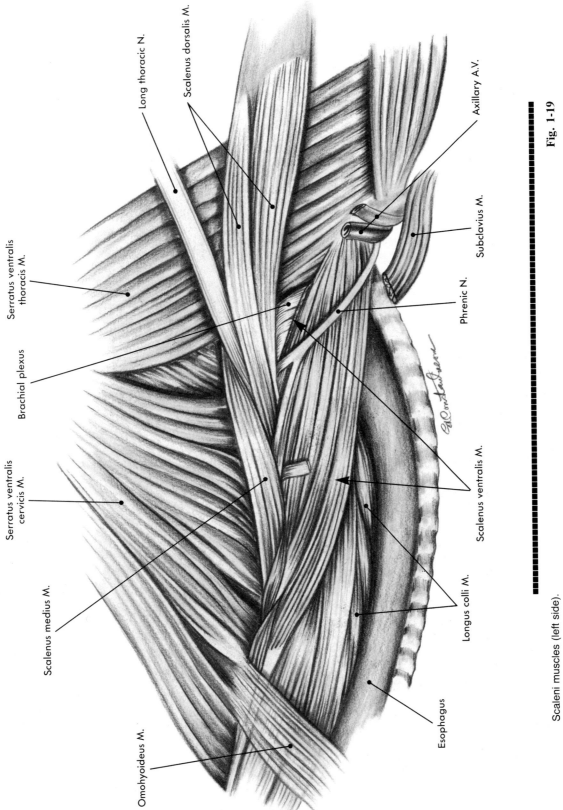

Long thoracic N.

Scalenus dorsalis M.

Axillary A.V.

Serratus ventralis thoracis M.

Subclavius M.

Brachial plexus

Phrenic N.

Serratus ventralis cervicis M.

Scalenus ventralis M.

Scalenus medius M.

Longus colli M.

Omohyoideus M.

Esophagus

Fig. 1-19

Scaleni muscles (left side).

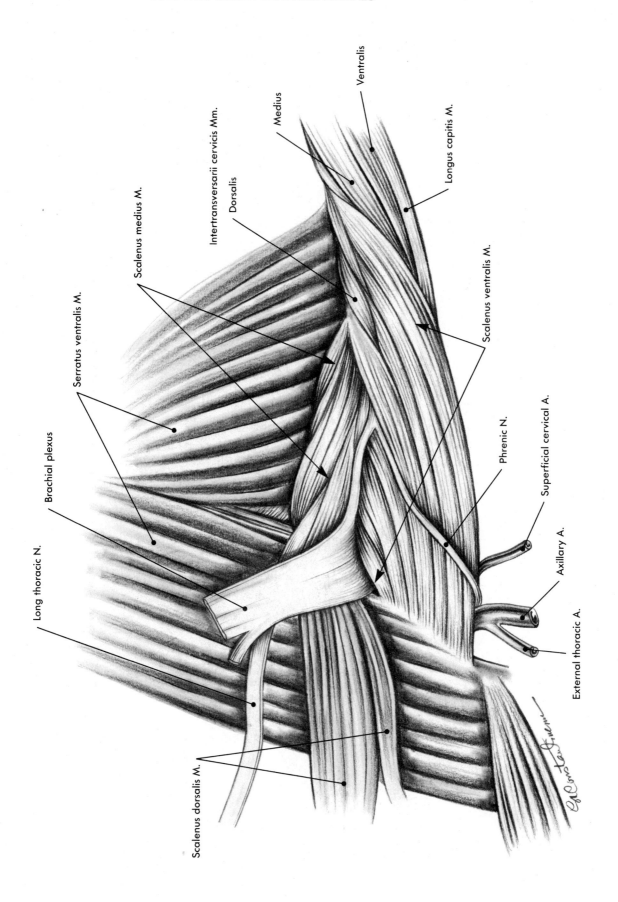

Long thoracic N.

Brachial plexus

Serratus ventralis M.

Scalenus medius M.

Intertransversarii cervicis Mm.

Dorsalis

Medius

Ventralis

Longus capitis M.

Scalenus ventralis M.

Phrenic N.

Superficial cervical A.

Axillary A.

External thoracic A.

Scalenus dorsalis M.

Scaleni muscles (right side).

Fig. 1-20

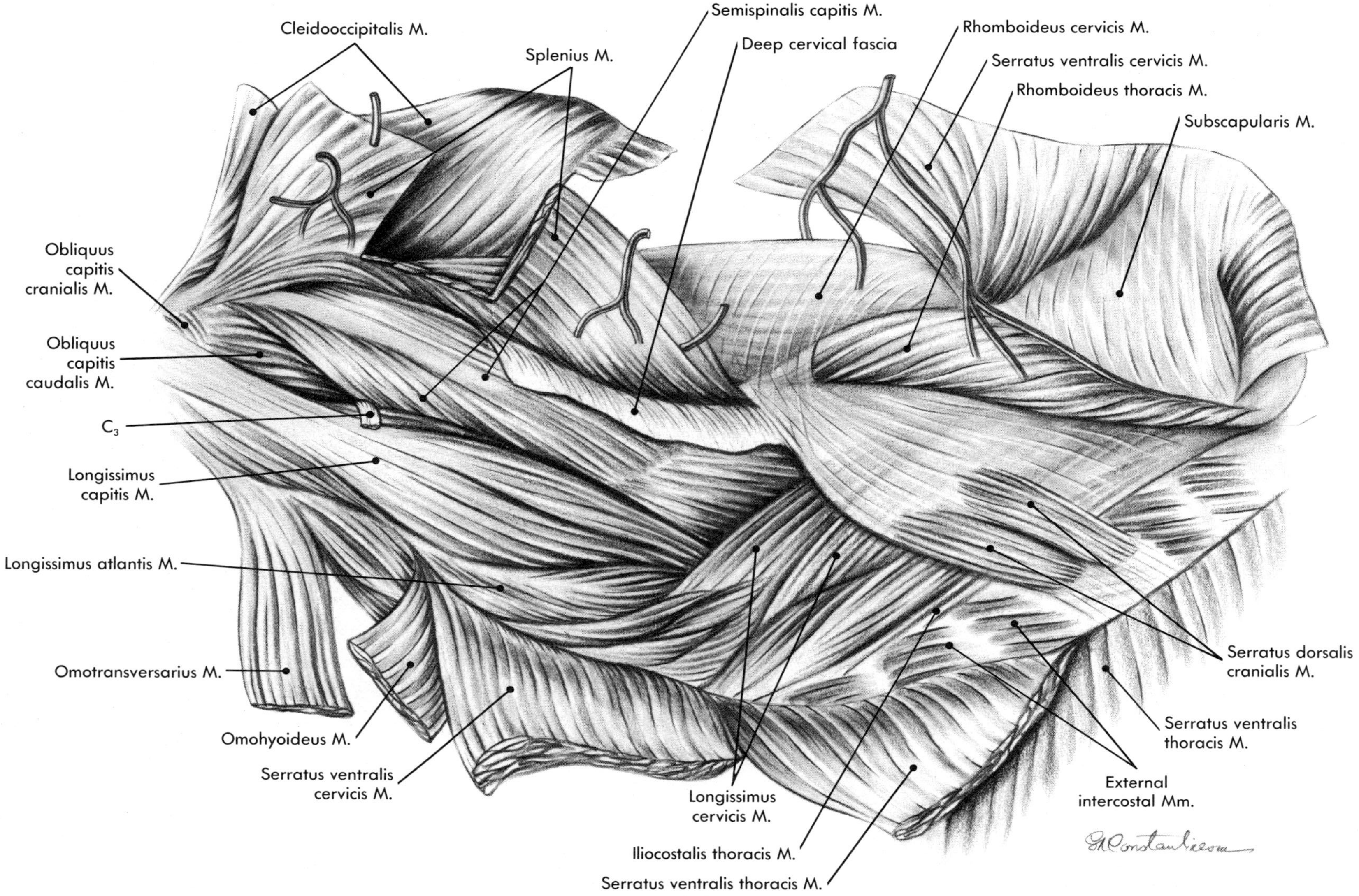

Rhomboideus cervicis M.

Serratus ventralis cervicis M.

Rhomboideus thoracis M.

Subscapularis M.

Serratus dorsalis cranialis M.

Serratus ventralis thoracis M.

External intercostal Mm.

Semispinalis capitis M.

Deep cervical fascia

Longissimus cervicis M.

Iliocostalis thoracis M.

Serratus ventralis thoracis M.

Splenius M.

Cleidooccipitalis M.

Obliquus capitis cranialis M.

Obliquus capitis caudalis M.

C₃

Longissimus capitis M.

Longissimus atlantis M.

Omotransversarius M.

Omohyoideus M.

Serratus ventralis cervicis M.

Fig. 1-21

Deep level laterodorsal neck.

Pig
■■■■
(Figs. 1-22
to 1-25)

Make a ventral incision in the skin on the midline from the manubrium sterni to the intermandibular space, then continue the incision over the caudal border of the mandible, caudal to the base of the ear, joining the middorsal line. Extend the incision from the manubrium sterni, surrounding the cranial aspect of the shoulder, over the spina scapularis and its tuberosity, and to the highest point of the thorax.

Dissect the superficial structures of the neck (Fig. 1-22).

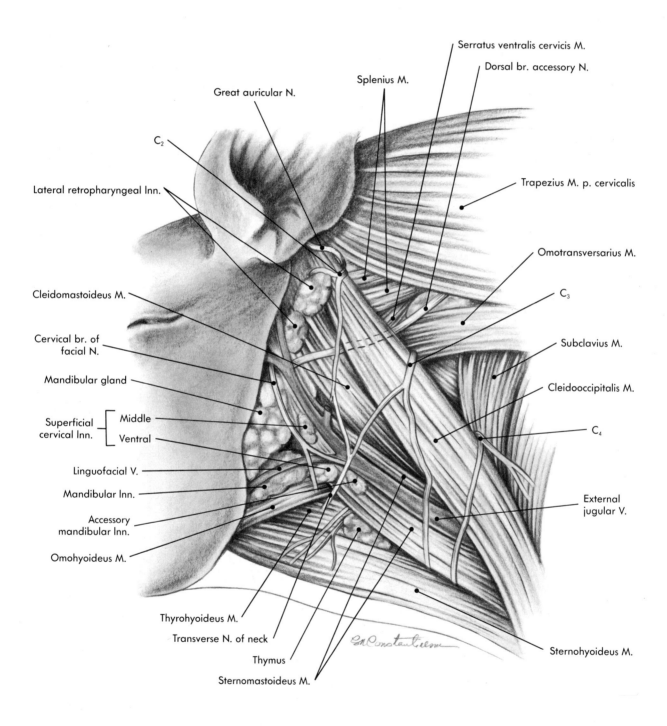

Great auricular N.
Serratus ventralis cervicis M.
Dorsal br. accessory N.
Splenius M.
C₂
Lateral retropharyngeal lnn.
Trapezius M. p. cervicalis
Omotransversarius M.
Cleidomastoideus M.
C₃
Cervical br. of facial N.
Subclavius M.
Mandibular gland
Cleidooccipitalis M.
Superficial cervical lnn.   Middle
                             Ventral
C₄
Linguofacial V.
Mandibular lnn.
Accessory mandibular lnn.
External jugular V.
Omohyoideus M.
Thyrohyoideus M.
Transverse N. of neck
Thymus
Sternohyoideus M.
Sternomastoideus M.

Fig. 1-22   Superficial level lateral neck.

Carefully dissect and cranially reflect the parotid salivary gland to expose the deep structures (Figs. 1-23 to 1-25).

Caution. Leave intact the dorsal branch of the accessory N., which crosses the lateral side of cleidomastoideus M.

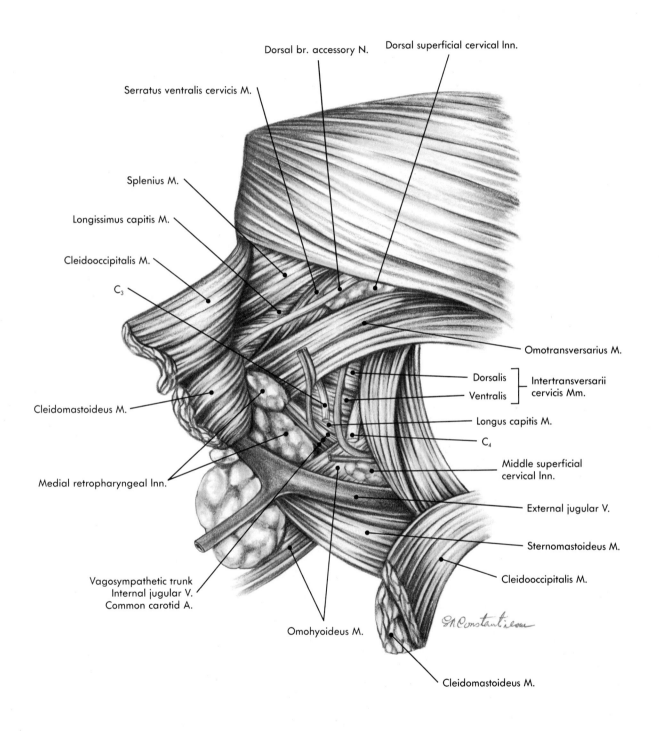

Dorsal br. accessory N.

Dorsal superficial cervical Inn.

Serratus ventralis cervicis M.

Splenius M.

Longissimus capitis M.

Cleidooccipitalis M.

C₃

Omotransversarius M.

Dorsalis } Intertransversarii
Ventralis } cervicis Mm.

Longus capitis M.

C₄

Cleidomastoideus M.

Middle superficial cervical Inn.

Medial retropharyngeal Inn.

External jugular V.

Sternomastoideus M.

Cleidooccipitalis M.

Vagosympathetic trunk
Internal jugular V.
Common carotid A.

Omohyoideus M.

Cleidomastoideus M.

Deep level lateral neck.

Fig. 1-23

Remove the thoracic limb in a manner similar to that for previous species. Some of the muscles dissected in Fig. 1-22 and 1-23 will be transected, and others exposed (Fig. 1-24).

To remove the thoracic limb, it is necessary first to study Figs. 1-34 and 1-35.

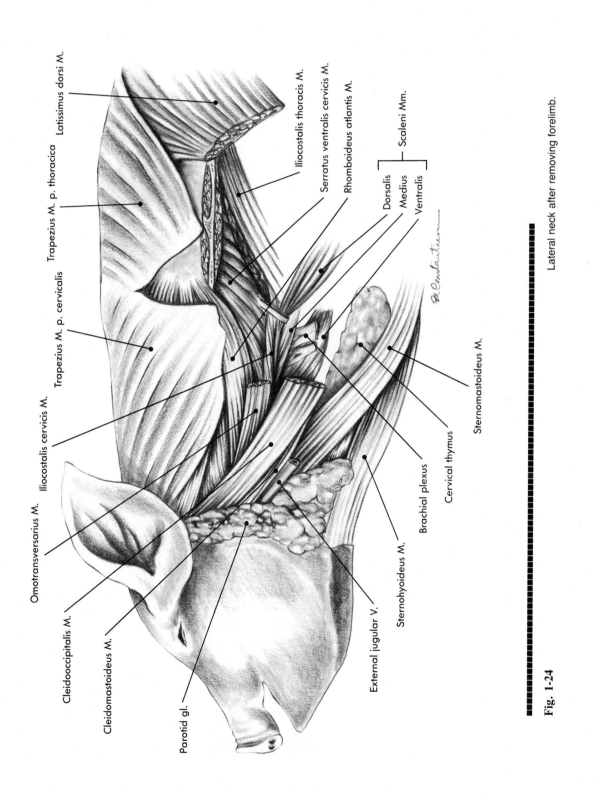

Fig. 1-24

Lateral neck after removing forelimb.

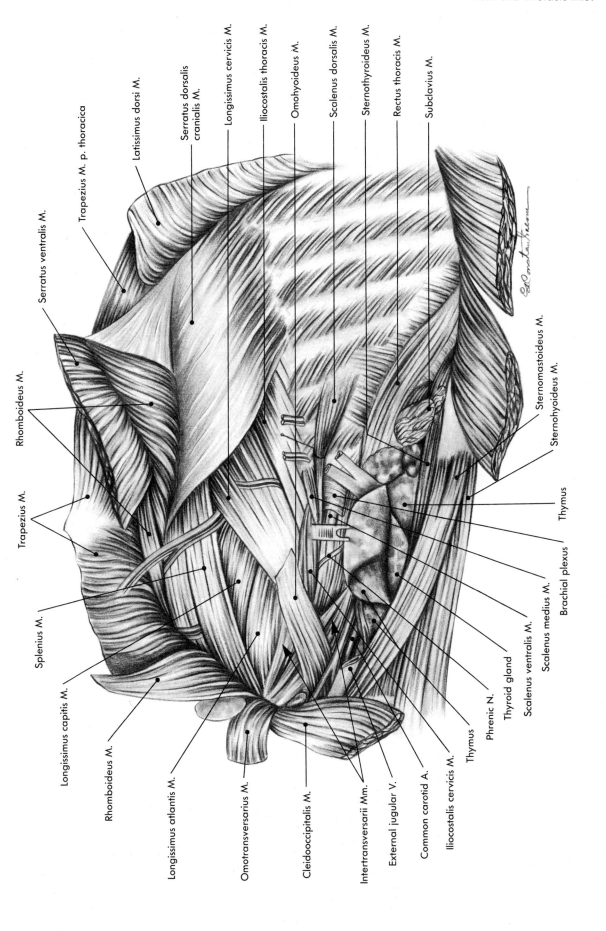

Latissimus dorsi M.

Serratus dorsalis
cranialis M.

Longissimus cervicis M.

Iliocostalis thoracis M.

Omohyoideus M.

Scalenus dorsalis M.

Sternothyroideus M.

Rectus thoracis M.

Subclavius M.

Trapezius M. p. thoracica

Serratus ventralis M.

Rhomboideus M.

Trapezius M.

Splenius M.

Longissimus capitis M.

Rhomboideus M.

Longissimus atlantis M.

Omotransversarius M.

Cleidooccipitalis M.

Intertransversarii Mm.

External jugular V.

Common carotid A.

Iliocostalis cervicis M.

Thymus

Phrenic N.

Thyroid gland

Scalenus ventralis M.

Scalenus medius M.

Brachial plexus

Thymus

Sternohyoideus M.

Sternomastoideus M.

Fig. 1-25

Deepest structures lateral neck (base of scapula is reflected).

THORACIC
INLET
Horse
■■■■■■■■■

By lateral approach continue the incision of the carotid sheath and the middle cervical fascia (visceral, or pretracheal, lamina) toward the thoracic inlet to dissect and isolate the common carotid A. and the external jugular V. The cephalic and axillary Vv. join the corresponding external jugular V. before entering the thorax in the bijugular trunk, or the jugular confluent, which is the origin of the cranial vena cava (Fig. 1-26).

Dissect and isolate the descending pectoral M. from the cleidobrachialis and transverse pectoral Mm. Transect the descending pectoral and the cleidobrachialis Mm. transversely, then isolate the subclavius M. Join the dissection of the transverse pectoral M. to the blunt dissection of the loose connective tissue ventral and caudal to the brachial plexus. Transect the subclavius and the deep pectoral Mm.—the latter together with the superficial thoracic V. and the lateral thoracic N. Retain intact the axillary vessels coming from the thorax through the thoracic inlet, surrounding the first rib.

Dissect the common carotid A. and *notice* the common origin of both right and left arteries from the bicarotid trunk. Carefully dissect the vagosympathetic trunk and *notice* the separation of the cervical sympathetic trunk from the vagus N. The sympathetic trunk leads to the middle and/or the caudal cervical (cervicothoracic) ganglion. *Notice* the exchange of fibers between the middle cervical ggl. and the vagus N. From the thoracic cavity *notice* the axillary A. exiting with the superficial cervical and the external thoracic Aa.

Dissect the roots of the brachial plexus, the scalenus ventralis, scalenus medius, and iliocostalis cervicis Mm. Identify the origins of the phrenic N. running on the lateral aspect of the scalenus ventralis M. and dissect them. Transect both scaleni Mm. from the first rib. Leaving intact the origins of the phrenic N., transect and remove half of the scalenus ventralis M. to allow dissection deep to the first rib, to better expose the middle cervical ggl.

Identify the origin of the long thoracic and lateral thoracic Nn. from the brachial plexus. Dissect the recurrent laryngeal N., the trachea, the esophagus, and the caudal deep cervical lnn. that surround the vessels and viscera at the thoracic inlet (Fig. 1-27).

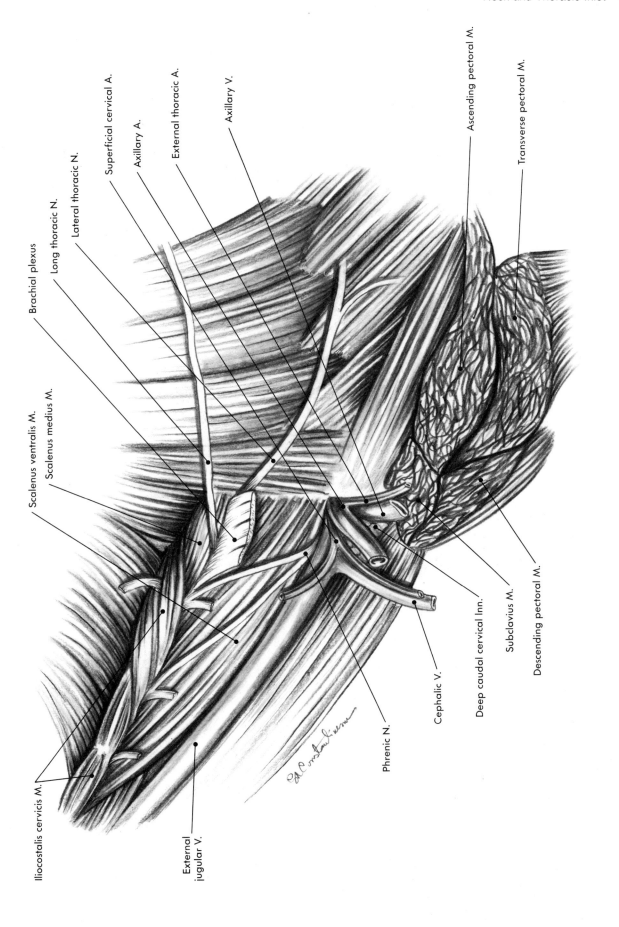

Ascending pectoral M.

Transverse pectoral M.

Axillary V.

External thoracic A.

Axillary A.

Superficial cervical A.

Lateral thoracic N.

Long thoracic N.

Brachial plexus

Scalenus ventralis M.

Scalenus medius M.

Iliocostalis cervicis M.

External jugular V.

Phrenic N.

Cephalic V.

Deep caudal cervical lnn.

Subclavius M.

Descending pectoral M.

Fig. 1-26

Thoracic inlet (left side).

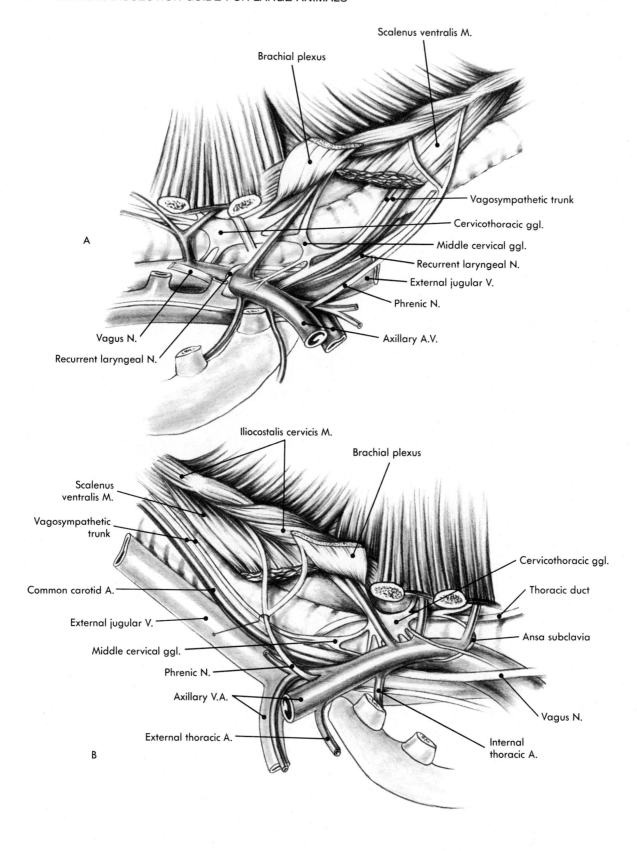

Fig. 1-27   **A,** Thoracic inlet (right side, deep aspect). **B,** Thoracic inlet (left side, deep aspect).

Take the horse as a model and perform the dissection of the thoracic inlet in ruminants and pigs.

Large ruminants
(Figs. 1-28 to 1-30)

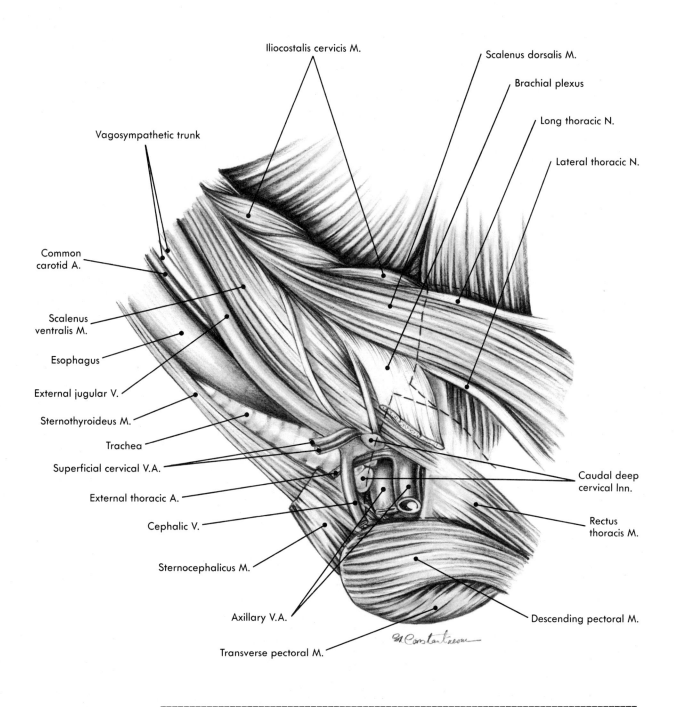

Iliocostalis cervicis M.

Scalenus dorsalis M.

Brachial plexus

Long thoracic N.

Lateral thoracic N.

Vagosympathetic trunk

Common carotid A.

Scalenus ventralis M.

Esophagus

External jugular V.

Sternothyroideus M.

Trachea

Superficial cervical V.A.

External thoracic A.

Cephalic V.

Sternocephalicus M.

Axillary V.A.

Transverse pectoral M.

Caudal deep cervical lnn.

Rectus thoracis M.

Descending pectoral M.

Thoracic inlet (left side). *Dotted lines,* outline of omotransversarius (dorsal) and the brachiocephalic (ventral) Mm.

Fig. 1-28

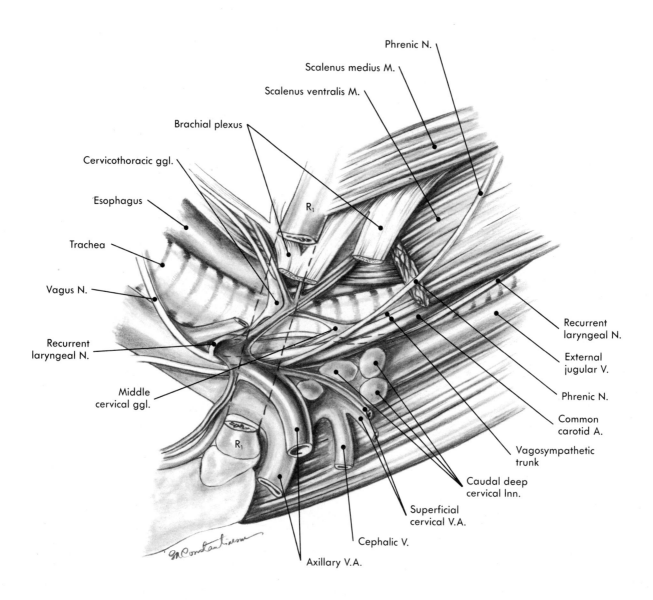

Fig. 1-29   Thoracic inlet (right side, deep aspect). *Dotted lines,* outline of Rib I.

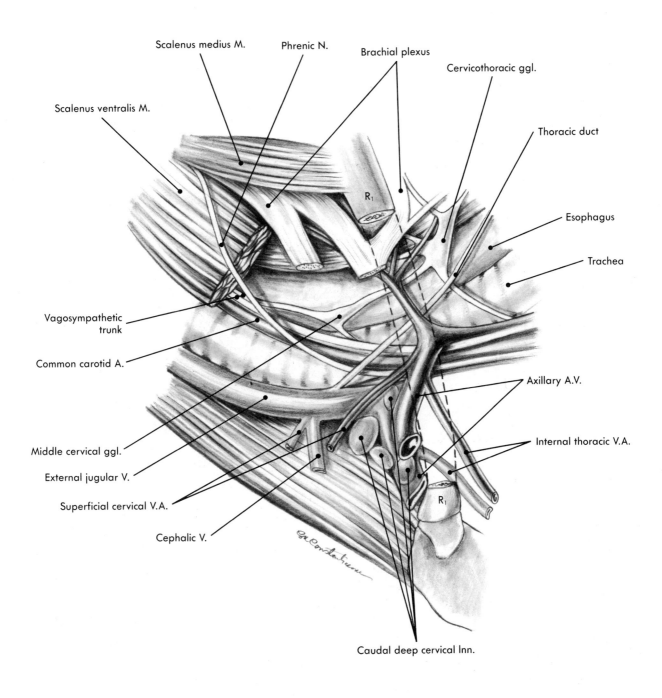

Scalenus medius M.

Phrenic N.

Brachial plexus

Cervicothoracic ggl.

Scalenus ventralis M.

Thoracic duct

R₁

Esophagus

Trachea

Vagosympathetic trunk

Common carotid A.

Axillary A.V.

Middle cervical ggl.

Internal thoracic V.A.

External jugular V.

R₁

Superficial cervical V.A.

Cephalic V.

Caudal deep cervical lnn.

Thoracic inlet (left side, deep aspect). *Dotted lines*, outline of Rib I.

Fig. 1-30

Sheep
■■■■■■■■
(Fig. 1-31)

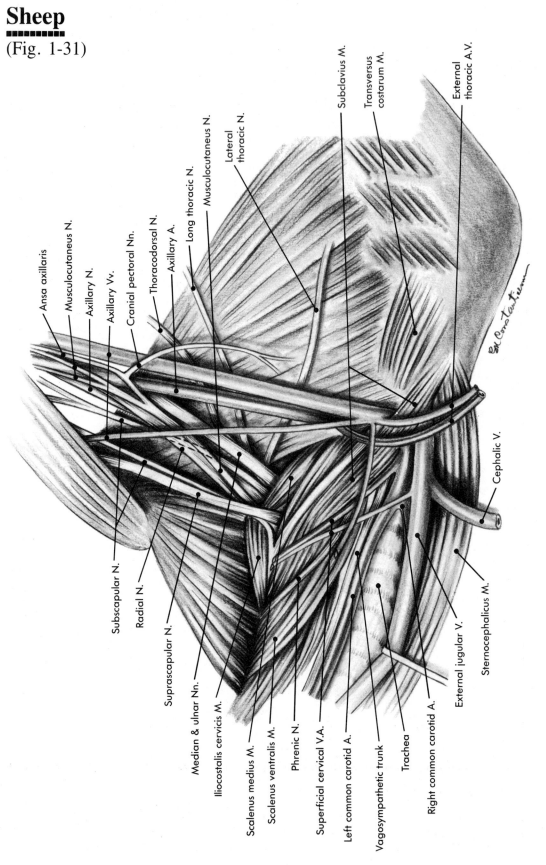

Fig. 1-31

Thoracic inlet, brachial plexus, and axillary vessels (left side).

Goat
■■■■■■■■
(Figs. 1-32
and 1-33)

Fig. 1-32

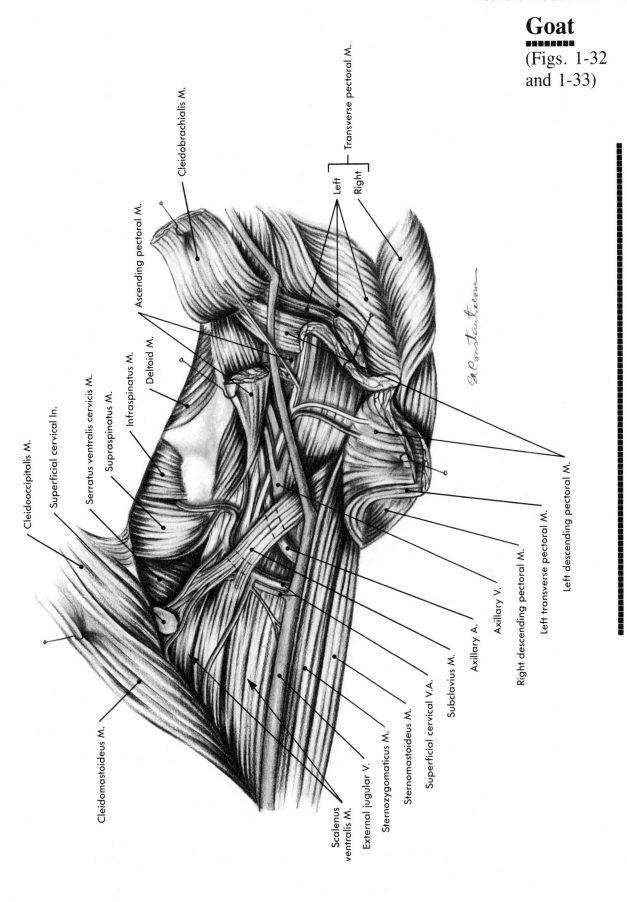

Cleidobrachialis M.

Transverse pectoral M.

Left

Right

Ascending pectoral M.

Deltoid M.

Infraspinatus M.

Supraspinatus M.

Serratus ventralis cervicis M.

Superficial cervical ln.

Cleidooccipitalis M.

Left descending pectoral M.

Left transverse pectoral M.

Right descending pectoral M.

Axillary V.

Axillary A.

Subclavius M.

Superficial cervical V.A.

Sternomastoideus M.

Sternozygomaticus M.

External jugular V.

Scalenus ventralis M.

Cleidomastoideus M.

Thoracic inlet (left lateral aspect).

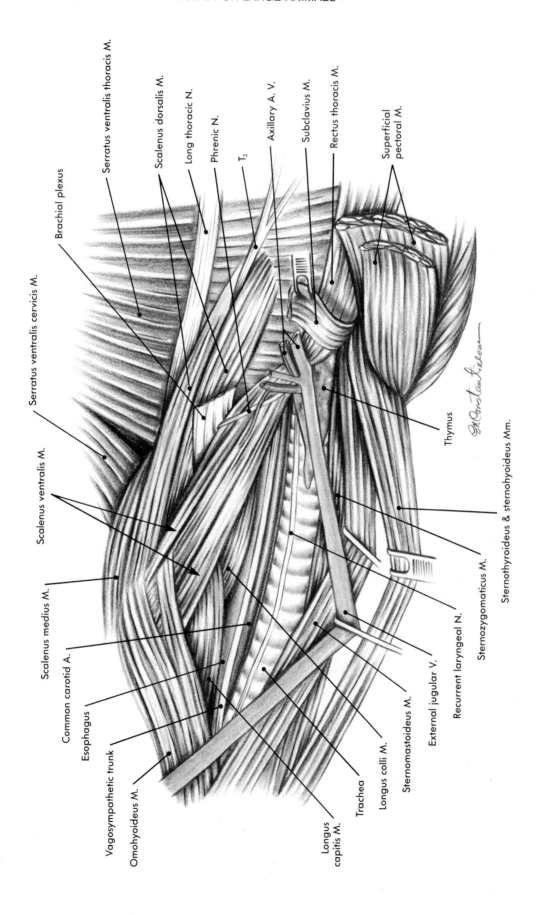

Serratus ventralis thoracis M.

Scalenus dorsalis M.

Long thoracic N.

Phrenic N.

T₂

Axillary A. V.

Subclavius M.

Rectus thoracis M.

Superficial pectoral M.

Brachial plexus

Serratus ventralis cervicis M.

Scalenus ventralis M.

Thymus

Scalenus medius M.

Common carotid A.

Esophagus

Vagosympathetic trunk

Omohyoideus M.

Longus capitis M.

Trachea

Longus colli M.

Sternomastoideus M.

External jugular V.

Recurrent laryngeal N.

Sternozygomaticus M.

Sternothyroideus & sternohyoideus Mm.

Fig. 1-33

Thoracic inlet, deep structures (left lateral aspect).

Pig
(Figs. 1-34 and 1-35)

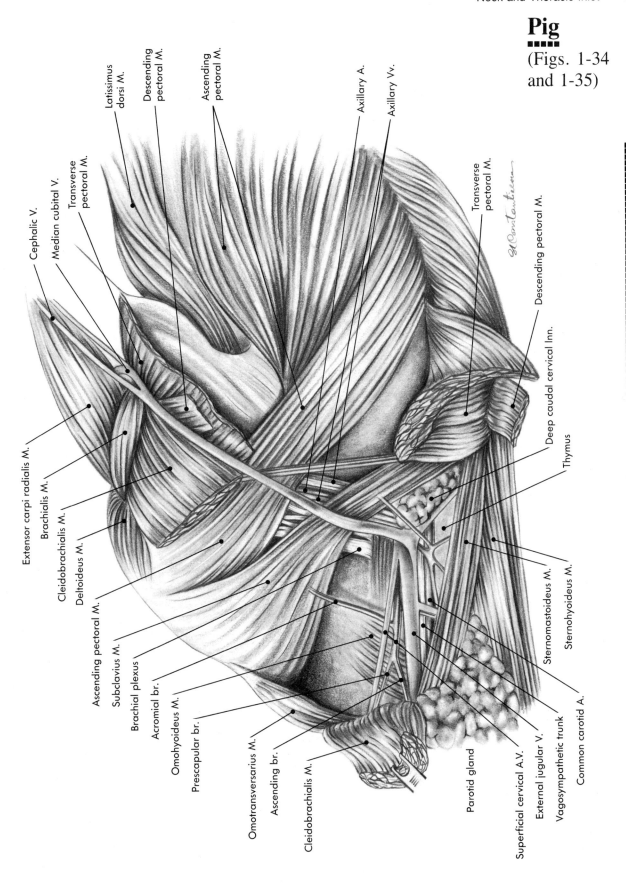

Fig. 1-34

Thoracic inlet, deep structures (left side).

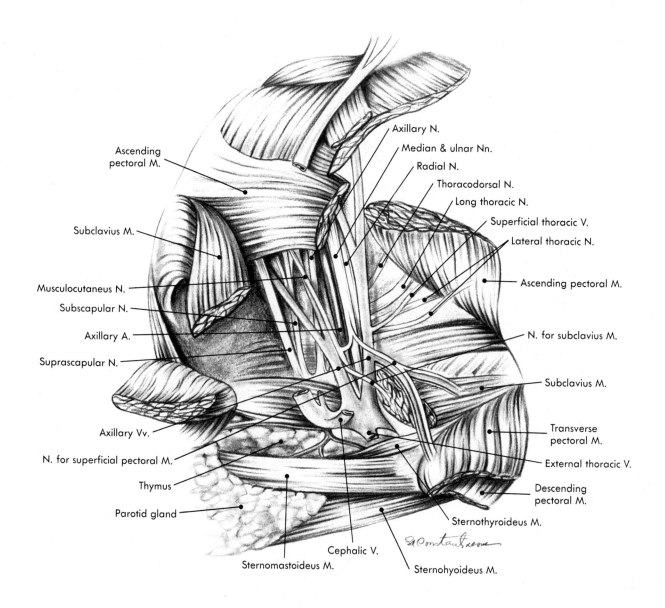

Ascending pectoral M.

Subclavius M.

Musculocutaneus N.

Subscapular N.

Axillary A.

Suprascapular N.

Axillary Vv.

N. for superficial pectoral M.

Thymus

Parotid gland

Sternomastoideus M.

Cephalic V.

Sternohyoideus M.

Axillary N.

Median & ulnar Nn.

Radial N.

Thoracodorsal N.

Long thoracic N.

Superficial thoracic V.

Lateral thoracic N.

Ascending pectoral M.

N. for subclavius M.

Subclavius M.

Transverse pectoral M.

External thoracic V.

Descending pectoral M.

Sternothyroideus M.

Fig. 1-35

Thoracic inlet, brachial plexus, and axillary vessels (left side).

Chapter 2
Thorax and thoracic viscera

THORAX
Horse

Continue the incision of the skin from the cranial extent of the withers to the spinous process of the last thoracic vertebra, following the dorsal midline of the specimen (use the last rib as a landmark). Make a transverse incision in the skin along the last rib to the corresponding costochondral junction, and continue the incision ventrally and perpendicularly up to the midventral line. Extend the incision of the skin from the cranial border of the withers to the manubrium sterni and then on the midline to the ventral border of the pectoral region. Reflect the skin ventrally off the pectoral region, shoulder, arm, thorax, and abdomen, ventral to the hypochondrium (the costal arch). Skin the area carefully to separate the skin from the cutaneus Mm. and the superficial fasciae.

After examining the cutaneus omobrachialis and the cutaneus trunci Mm., reflect them ventrally. The thoracolumbar fascia and the latissimus dorsi M. are pierced by the dorsolateral cutaneous branches of the thoracic nerves, similar to the dorsal cutaneous branches in the cervical area (see Fig. 1-1). The same fascia and the intercostal and external abdominal oblique Mm. are pierced by the ventrolateral cutaneous branches of the intercostal nerves (in two rows). Identify the intercostobrachial N. at the level of the ventral border of the latissimus dorsi M. and the caudal border of the triceps brachii or tensor fasciae antebrachii Mm. (Fig. 2-1).

Palpate the spine of the scapula and its tuberosity, and make an incision separating the two parts of the trapezius M. (cervical and thoracic parts), from the tuberosity of the spine of the scapula to the withers. Reflect the two parts dorsally. A second technique is also suggested: Transect the vertebral attachment of the two parts of the trapezius M. and reflect them ventrally. The scapular cartilage is now exposed. The caudal extent of it is overlapped by the aponeurosis of the latissimus dorsi M., which protects the trapezius M. against potential injuries by the cartilage during walking.

Notice that both parts of the trapezius M. are protected by the corresponding superficial fascia on each side.

Both rhomboideus cervicis and thoracis Mm. are now exposed, as well as part of the serratus ventralis cervicis, splenius, supraspinatus, and infraspinatus Mm. (Fig. 2-2).

At about 15 cm ventral to the junction of the ventral border of the latissimus dorsi and the triceps brachii Mm., find the apparent origin of the superficial thoracic V. at the dorsal border of the pectoralis profundus (ascendens) M. Dissect the vein and the muscle, and find the lateral thoracic N., which parallels the vein. Between the latissimus dorsi and the pectoralis profundus Mm., the serratus ventralis thoracis M. (covered by its fascia) and the origin of the external abdominal oblique M. (covered by the tunica flava abdominis [the abdominal tunic]) are now exposed, along with the external intercostal Mm. and the serratus dorsalis caudalis (expiratorius) M.

Transect the latissimus dorsi M. at the caudal border of the triceps brachii M. and reflect it dorsally. The serratus ventralis thoracis and the serratus dorsalis cranialis (inspiratorius) Mm. are now exposed.

Notice that the serratus dorsalis M. has a muscular portion (ventrally) and an aponeurosis (dorsally).

Make two vertical parallel incisions, 10-15 cm apart, through the serratus dorsalis M. to expose the insertion of its aponeurosis on the ribs. The aponeurosis separates the iliocostalis thoracis M. from the group of longissimus, semispinalis, and spinalis thoracis Mm. Make the incisions deeper through the last three muscles to expose the multifidus thoracis, the levatores costarum and the intertransversarii thoracis Mm. *Notice* that the latter are usually difficult to identify because they are fused with the multifidus thoracis Mm. (Fig. 2-3).

In the shoulder area, identify the subclavius, cleidobrachialis, pectoralis descendens, deltoideus, and triceps brachii Mm. Identify the clavicular intersection of the brachiocephalicus M. and the cephalic V., which lies parallel to the deltoid branch of the superficial cervical A. in the lateral pectoral groove between the pectoralis descendens and the cleidobrachialis Mm.

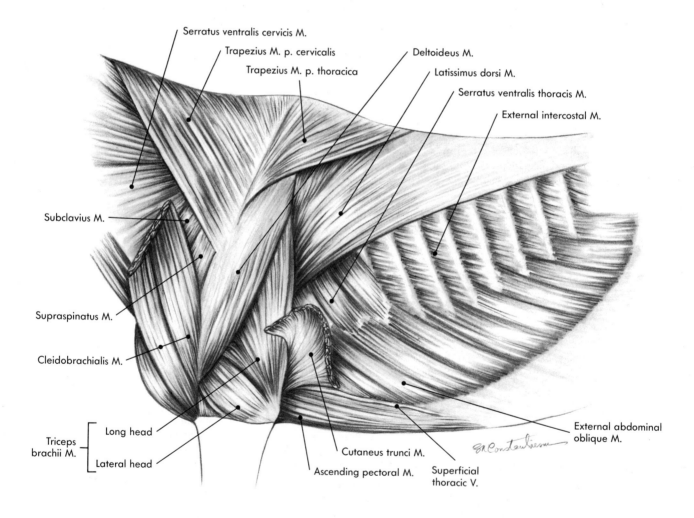

Fig. 2-1   Thorax (superficial structures, left side).

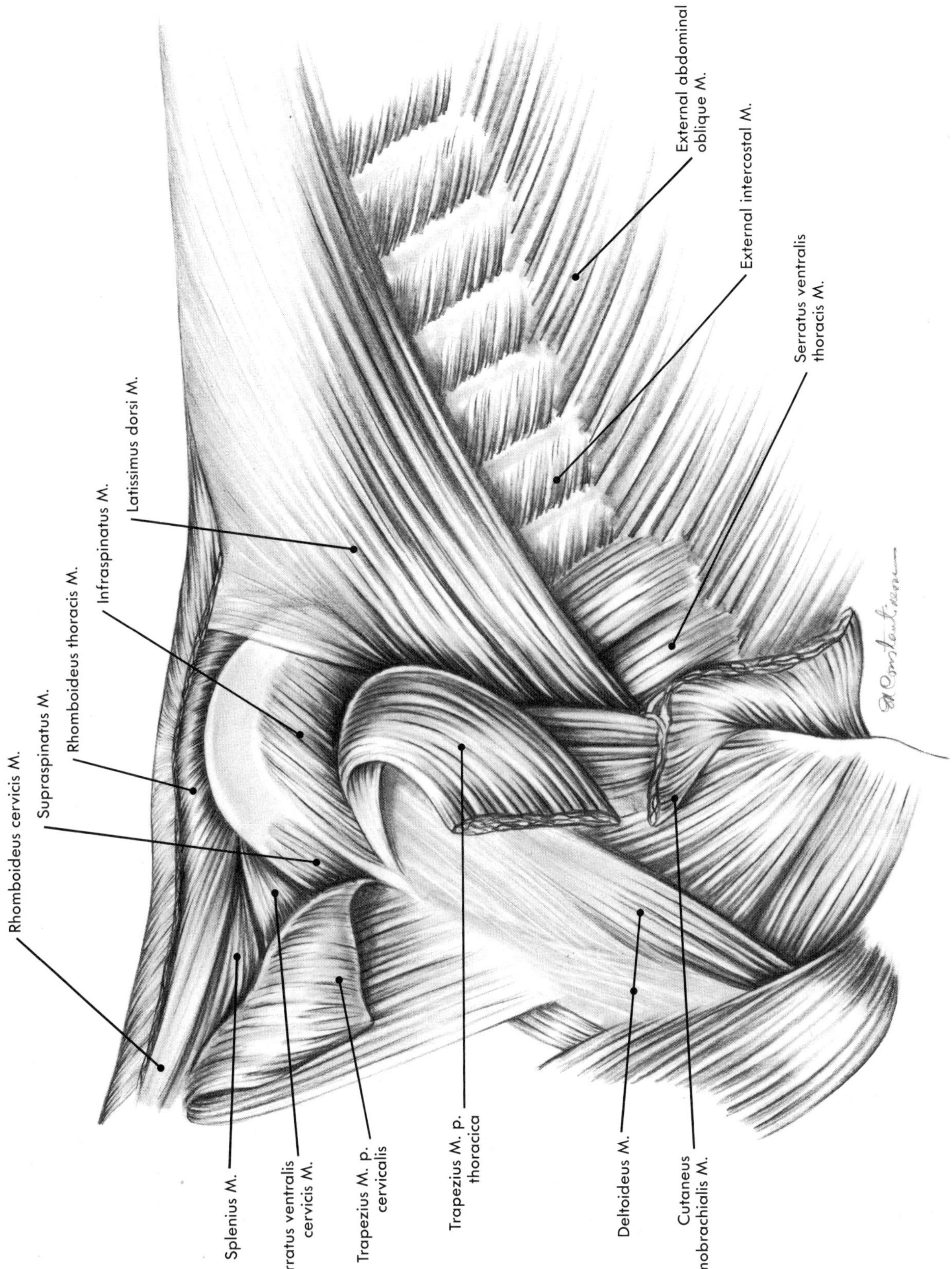

Rhomboideus cervicis M.

Supraspinatus M.

Rhomboideus thoracis M.

Infraspinatus M.

Latissimus dorsi M.

External abdominal oblique M.

External intercostal M.

Serratus ventralis thoracis M.

Splenius M.

Serratus ventralis cervicis M.

Trapezius M. p. cervicalis

Trapezius M. p. thoracica

Deltoideus M.

Cutaneus omobrachialis M.

Fig. 2-2

Deep dorsoscapular structures (left side).

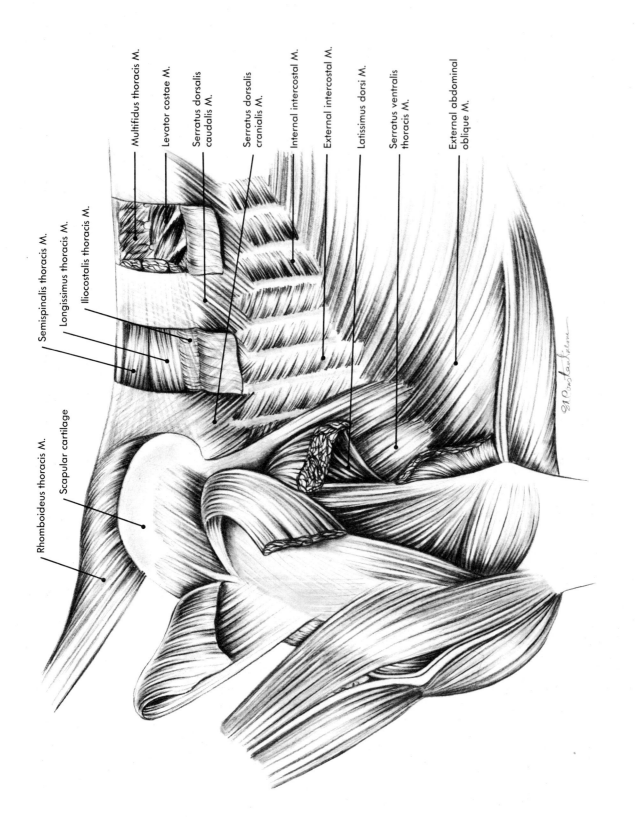

Multifidus thoracis M.

Levator costae M.

Serratus dorsalis caudalis M.

Serratus dorsalis cranialis M.

Internal intercostal M.

External intercostal M.

Latissimus dorsi M.

Serratus ventralis thoracis M.

External abdominal oblique M.

Semispinalis thoracis M.

Longissimus thoracis M.

Iliocostalis thoracis M.

Rhomboideus thoracis M.

Scapular cartilage

Deep dorsal structures (left side).

Fig. 2-3

Transect the scapula and the corresponding muscles transversely with a saw, dorsal to the tuberosity of the spine. Cut through the subclavius, supraspinatus, infraspinatus, deltoideus, triceps brachii, tensor fasciae antebrachii (on the lateral aspect of the shoulder) and subscapularis and serratus ventralis cervicis and thoracis Mm. (both on the medial aspect of the shoulder).

Separate the forelimb from the origin of the serratus ventralis Mm. (cervicis and thoracis) and from the costal insertions of the serratus ventralis thoracis M. There is an abundance of loose connective tissue between the limb and the serratus ventralis thoracis M., and a blunt dissection is suggested. For an easier procedure, push the free extremity of the corresponding limb medially.

The area of the pectoral and arm muscles has already been exposed and dissected and the structures identified. Dissect the vessels located in the lateral pectoral groove (the cephalic V. and the deltoid branch of the superficial cervical A.), and transect them in the middle of the groove.

Expose the brachial plexus exiting from between the middle scalenus M. (dorsally) and the ventral scalenus M. (ventrally). The only liaisons between the body and the limb are now the brachial plexus and the axillary vessels. Transect them as close as possible to the body, leaving them on the limb. Take off the thoracic limb, and wrap and store it for a later dissection.

The scalenus Mm., the iliocostalis cervicis M., and the origins of the sternocephalicus (sternomandibularis), sternothyroideus, and sternohyoideus Mm. are now entirely exposed, along with the trachea, the esophagus, and the nerves and vessels entering and exiting the thoracic cavity (Fig. 2-4).

You can remove the opposite limb by transecting the rhomboideus cervicis and thoracis Mm. (around the free edge of the scapular cartilage), the dorsoscapular ligament, and the serratus ventralis cervicis and thoracis Mm. and by following the same procedure already conducted on the other limb for the ventral structures.

Reflect the remnant of the scapula with the scapular cartilage dorsolaterally to expose the dorsoscapular lig. This structure lines the rhomboideus thoracis M. and is thicker than an ordinary fascia. The yellow color is due to the elastic tissue that comes from the supraspinous lig. and the ligamentum nuchae (Fig. 2-5, *A*).

Notice that the dorsoscapular lig. is also a result of the intermingling aponeuroses of three muscles: splenius, semispinalis capitis, and serratus dorsalis cranialis Mm.

Put the scapula back in its normal position. Transect the rhomboideus thoracis M. carefully, and expose the dorsoscapular lig., then remove the muscle. Identify the connections of the ligament. Transect and reflect the supraspinous lig. dorsally, at the level of the spinous processes of T_2 and T_3 (the highest point of the withers) to expose the supraspinous bursa, surrounded by a large amount of connective tissue (Fig. 2-5, *B*).

Notice that at this level the supraspinous lig. is continuous with the funiculus nuchae.

Identify the dorsolateral cutaneous branches of the first thoracic nerves and the branches of the dorsoscapular A.V., which course between the scapula and the serratus ventralis M. Dissect the fascia of the serratus ventralis thoracis M. and the abdominal tunic, and notice their interdigitations. In an oblique position and ventral to the serratus ventralis thoracis M., identify the rectus thoracis M. The external intercostal Mm. are present in the rest of the area.

Remove the serratus ventralis thoracis M. from the ribs and the serratus ventralis cervicis M. from the transverse processes of the last four to five cervical vertebrae by leaving 1 to 2 cm of the muscle attached to the vertebrae.

The semispinalis and spinalis thoracis and cervicis, the longissimus lumborum, thoracis, cervicis, capitis and atlantis, and iliocostalis lumborum, thoracis and cervicis Mm. are now entirely exposed and partially overlapped by the serratus dorsalis M. (part of it was removed to expose the multifidus and intertransversarii dorsi and the levatores costarum Mm.).

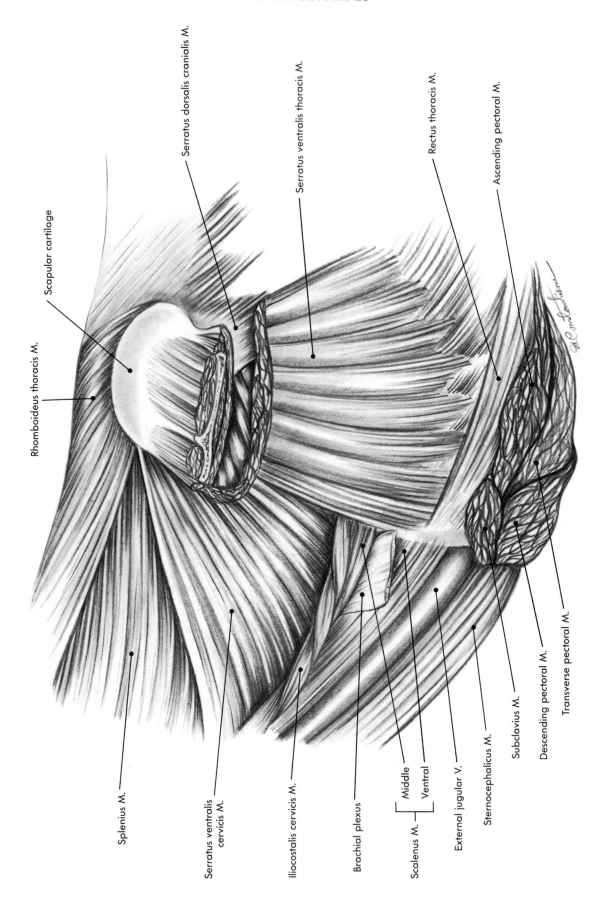

Fig. 2-4   Cervicothoracic area after removing forelimb.

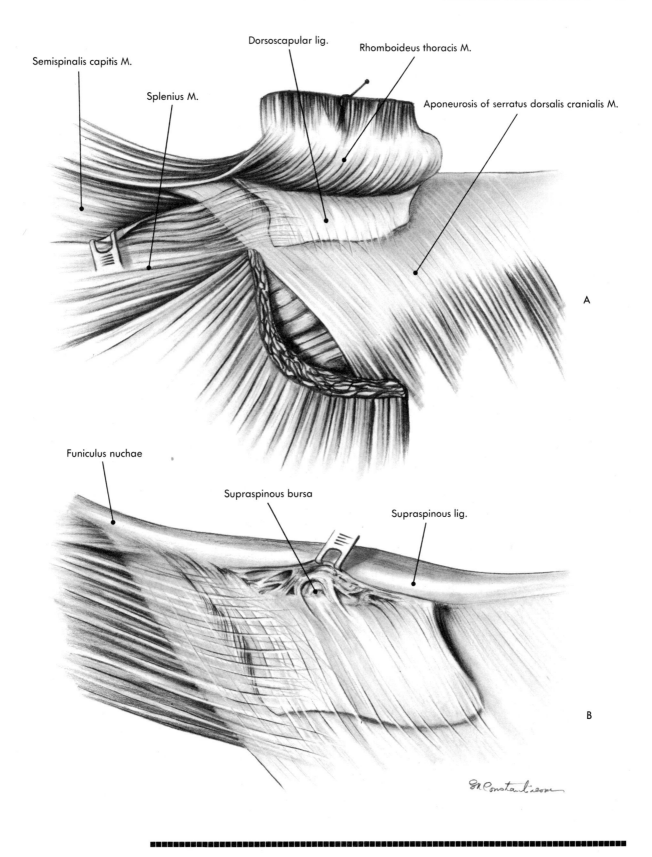

A, Dorsoscapular ligament. B, The supraspinous bursa.

Fig. 2-5

Dissect the external and internal intercostal Mm. and the intercostal A.V.N. in the costal sulcus (at the caudal border of the ribs). Remove the intercostal and other muscles of the lateral aspect of the thoracic wall, starting from the next to the last intercostal space and continuing cranially. Penetrate the thorax, removing also the endothoracic fascia lined by the costal (parietal) pleura. Palpate the area where the costal pleura becomes continuous with the diaphragmatic pleura. This is known as the line of pleural reflection (Fig 2-6, *dotted lines*) and is clinically important because it is the limit of the pleural cavity. The costodiaphragmatic recess is the potential space between the costal and diaphragmatic pleurae.

Caution. Do not cut through the line of pleural reflection when removing the intercostal muscles.

In the following steps, the lungs and the heart, which lies within the pericardium, are exposed. Identify the lobes of the lungs and the wide cardiac notches.

Carefully remove all the ribs except the fifth, ninth, tenth, seventeenth, and eighteenth ribs. First remove the serratus dorsalis and iliocostalis Mm., exposing the dorsal (vertebral) extremities of the ribs as much as possible. Direct the rib cutter with your finger on the medial aspect of each rib that is going to be removed, and transect it. The use of a large, flat rib cutter for this operation is recommended. It is also suggested to disarticulate the costal cartilage from the chondrosternal joints of all the sternal ribs. Remove the other ribs by transecting them dorsal to the pleural reflection.

Pull on the nerves entering and exiting the thorax through the thoracic inlet; this will allow easier dissection.

Caution. The ansa subclavia and the cardiac Nn. are very fine structures and are embedded in fat or loose connective tissue.

Next, dissect the arteries and the veins, saving all the nervous structures, including the cervicothoracic ggl. with its corresponding communicating branches and the thoracic sympathetic trunk.

Remove the left lung by separating it from the left primary bronchus and the left pulmonary vessels for later examination. Look at the left tracheobronchial ln., which is close to the pulmonary hilus, and at the middle tracheobronchial ln. Dissect and identify the left pulmonary ln.

In the cranial mediastinum, on the left side, in a craniocaudal direction and oriented dorsally, the most superficial vessels are the vertebral, deep cervical, and costocervical Vv., which overlap and parallel the corresponding arteries; the cranial extent of the left subclavian A. with the origin of the superficial cervical, axillary, and external and internal thoracic Aa are oriented ventrally. The homologous veins of the superficial cervical, axillary, and external and internal thoracic Aa. are deep to the arteries. Identify the cranial mediastinal lnn., which are especially concentrated toward the thoracic inlet, in continuation with the caudal deep cervical lnn.

The cranial vena cava, bicarotid and brachiocephalic trunks, trachea, and esophagus are large and visible structures (given here in ventrodorsal order). The thoracic duct is observed crossing the esophagus and the trachea caudocranially and dorsoventrally toward the left venous angle.

Remember. The venous angles (right and left) are located at the confluence of the external (in some species, also the internal) jugular veins and the corresponding axillary veins, just cranial to the origin of the cranial vena cava.

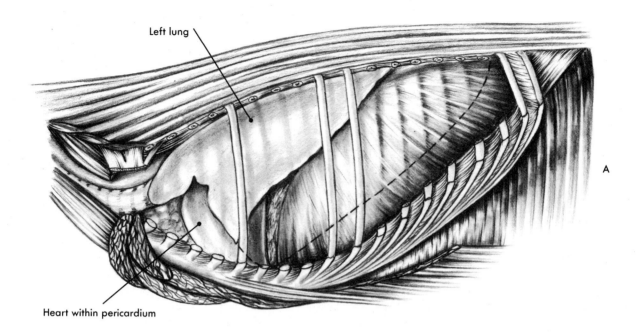

Left lung

Heart within pericardium

A

Dotted line, line of pleural reflection.

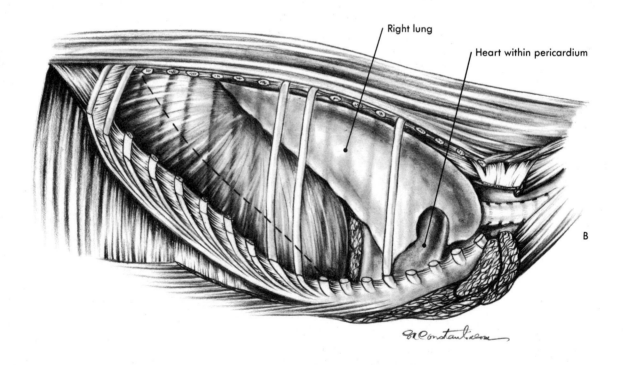

Right lung

Heart within pericardium

B

A, Left thoracic cavity. **B,** Right thoracic cavity.

Fig. 2-6

Continue the dissection of the vagus and phrenic Nn. located alongside the dorsal and ventral borders of the cranial vena cava, respectively, then across the pulmonary trunk. The vagus N. travels toward the aortic arch, giving off the left recurrent laryngeal N., which surrounds the arch and the ligamentum arteriosum and turns back to the cervical area; from this point the vagus N. continues its course toward the esophagus. The phrenic N. travels along the lateral aspect of the left auricula toward the diaphragm, supplying its left half.

Return to the cervicothoracic ggl., and continue the dissection of the thoracic sympathetic trunk, which is close to the costovertebral joints; dissect the thoracic ganglia and their communicating branches.

Caution. Dissect carefully because the caudal cardiac Nn. arise from the cervicothoracic ggl. From the middle cervical ggl., the middle cardiac Nn. are observed.

Continue the dissection of all of these structures in the middle mediastinum. Transect the pericardium and expose the heart, saving the structures previously dissected. Now dissect the cardiac Nn. coming off the vagus N. Identify the pulmonary arteries and veins. Expose the supreme intercostal A. and the corresponding dorsal intercostal Aa. (3 to 5), the dorsal intercostal Aa. (starting with the sixth dorsal intercostal A. from the thoracic aorta), and the bronchoesophageal A. (the only visceral branch of the thoracic aorta).

In the caudal mediastinum, dissect the dorsal and ventral vagal trunks, as a result of the communication between the dorsal and ventral branches (or divisions) of both left and right vagus Nn., respectively. Identify and dissect the greater and the lesser splanchnic Nn.

Remember. In horse, the greater splanchnic N. is a result of the communication between T_7 through T_{15} sympathetic fibers, whereas the lesser splanchnic N. arises from T_{16} through T_{18}. They leave the thoracic cavity to reach the celiac and the cranial mesenteric or renal ganglia, respectively.

Examine the diaphragm with its four crura, the costal and sternal muscular parts, and the centrum tendineum.

Notice that there are three foramina of the diaphragm: aortic, esophageal, and that of the caudal vena cava, named the aortic hiatus (1), the esophageal hiatus (2), and the foramen venae cavae (3), respectively (Fig. 2-7).

Remove the right lung by transecting the right primary bronchus and the right pulmonary vessels for later examination. Examine the right tracheobronchial lnn., and dissect the right pulmonary ln.

On the right side some differences are observed; only the differences will be described.

The deep cervical and the costocervical Vv. may originate from a common trunk (the costocervical V.), whereas the homologous arteries are regularly given off by the costocervical trunk.

The azygos V., present only on the right side, courses along the roof of the thoracic cavity, dorsal to the thoracic duct; at the level of the eighth and seventh rib it bends ventrally, opening into the dorsal wall of the cranial vena cava. At the same level, the thoracic duct crosses the medial aspect of the azygos V. and then the esophagus and passes into the left half of the thoracic cavity.

Identify and examine the cranial mediastinal, sternal, and intercostal lnn.

The right recurrent laryngeal N. originates from the vagus N. and turns around the right subclavian A. and costocervical trunk. Locate the caudal vena cava and the plica venae cavae, a fold of pleura enclosing the right phrenic N. Identify the accessory lobe of the right lung (Fig. 2-8).

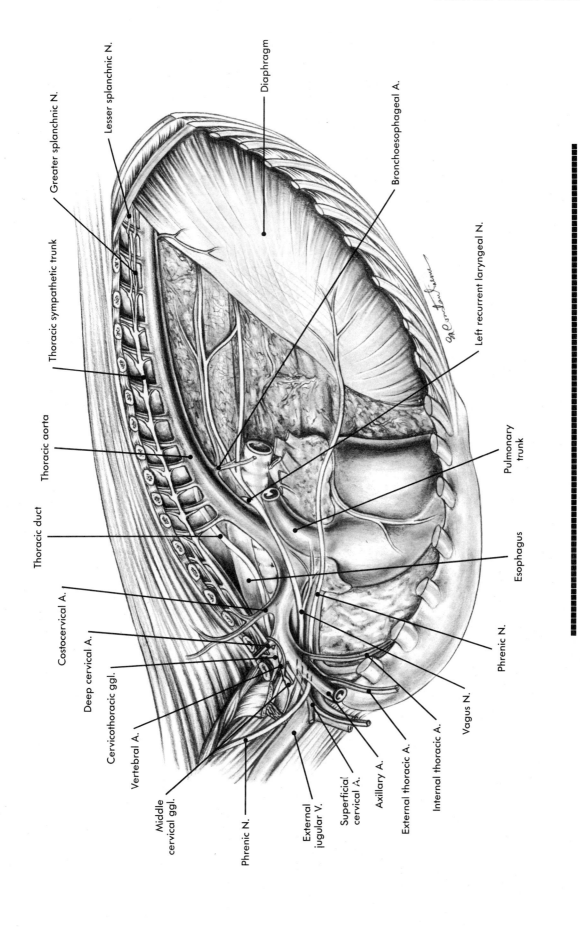

Greater splanchnic N.

Lesser splanchnic N.

Diaphragm

Thoracic sympathetic trunk

Bronchoesophageal A.

Thoracic aorta

Left recurrent laryngeal N.

Thoracic duct

Pulmonary trunk

Costocervical A.

Deep cervical A.

Vertebral A.

Esophagus

Cervicothoracic ggl.

Middle cervical ggl.

Phrenic N.

Phrenic N.

External jugular V.

Superficial cervical A.

Axillary A.

External thoracic A.

Internal thoracic A.

Vagus N.

Heart, vessels, and nerves (left thoracic cavity).

Fig. 2-7

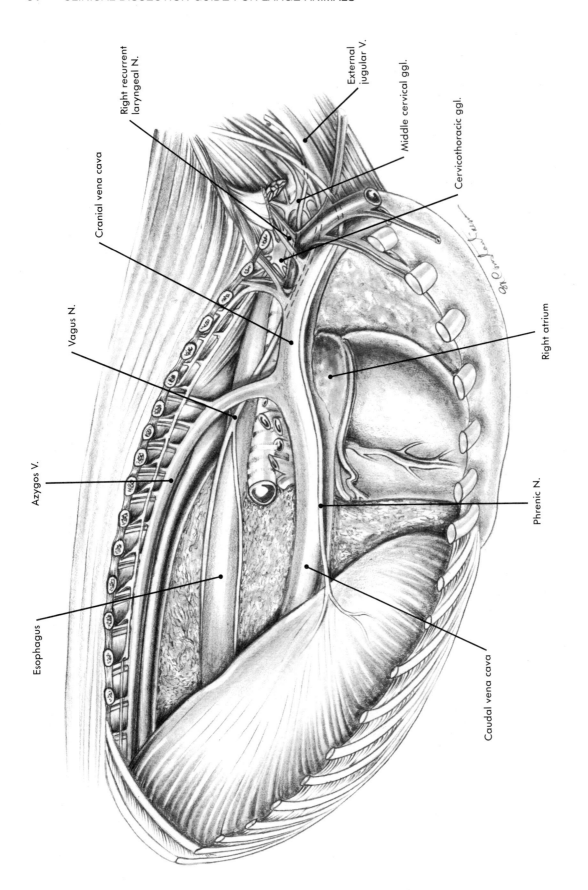

Right recurrent laryngeal N.

External jugular V.

Middle cervical ggl.

Cervicothoracic ggl.

Cranial vena cava

Right atrium

Vagus N.

Azygos V.

Phrenic N.

Esophagus

Caudal vena cava

Fig. 2-8

Heart, vessels, and nerves (right thoracic cavity).

Caution. To dissect all the structures that are located in the median plane and that are included between the two leaves of the mediastinal pleura, it is necessary to remove that pleura.

After removing the right lung *notice* the fenestration of the caudal mediastinal pleura and the possibility of communication between the right and the left pleural cavities.

Make certain you understand the difference between the thoracic and the pleural cavities.

Examine the removed lungs and observe the wide cardiac notches, the absence of the cranial and caudal parts of the cranial lobes, and the absence of the middle lobe of the right lung; neither a polyhedral design of the surface of the lungs nor a tracheal bronchus are observed. Identify the grooves, the so-called impressions of other structures on the medial and costal surfaces of the lungs (Figs. 2-9 and 2-10).

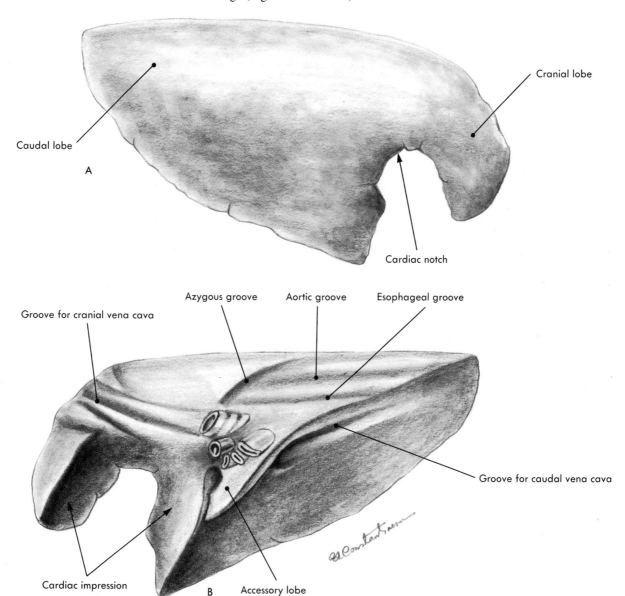

A, Right lung (lateral aspect). **B,** Right lung (medial aspect).

Fig. 2-9

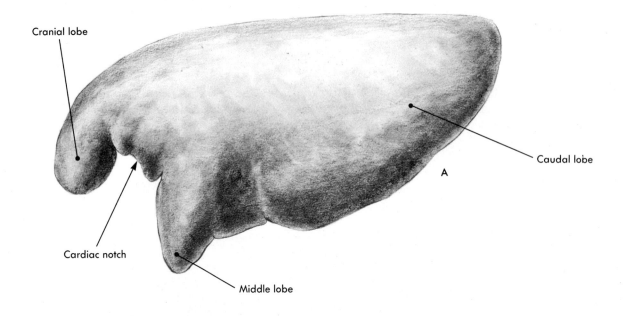

Cranial lobe

Caudal lobe

Cardiac notch

Middle lobe

A

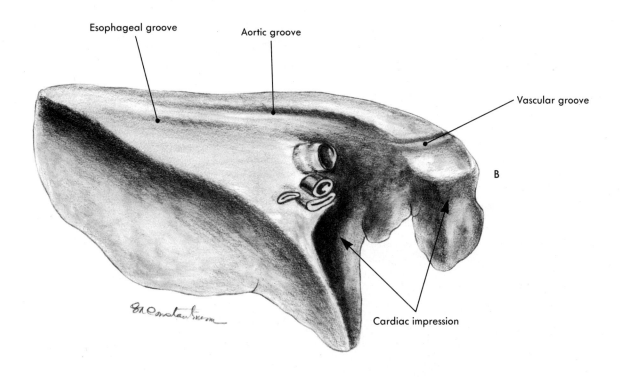

Esophageal groove

Aortic groove

Vascular groove

B

Cardiac impression

Fig. 2-10   **A,** Left lung (lateral aspect). **B,** Left lung (medial aspect).

Transect the cranial vena cava with the right azygous V., the brachiocephalic trunk, and the subclavian A. Transect the caudal vena cava before it enters the right atrium, and remove the heart from the pericardium.

Examine the left (auricular) and the right (atrial) aspects of the heart with the corresponding grooves: the coronary (separating the atrial mass—dorsally—from the ventricular mass—ventrally), the paraconal interventricular (left), and the subsinuosal interventricular (right) grooves. Dissect the coronary vessels and *notice* the specific supply fields of the coronary arteries. Identify the apparent origin of the great vessels: aorta, pulmonary trunk, pulmonary veins, cranial and caudal venae cavae, and azygos V. (Fig. 2-11).

Notice that the horse has a bilateral coronary type of supply to the heart by both right and left coronary Aa.

Take another heart, separate the atria from the ventriculi through the coronary groove and examine the atrioventricular (AV), aortic, and pulmonary valves (Fig. 2-12, *A*), then remove the right ventricular wall, sectioning through the two interventricular grooves, saving the branches of the right and left coronary Aa., examining the interventricular septum (Fig. 2-12, *B*).

Incise the dorsolateral aspect of the cranial vena cava, which opens into the right atrium, and extend the incision throughout the caudal vena cava. Expose and identify the right auricula, the interatrial septum with the fossa ovalis (in the middle) and the opening of the coronary sinus (ventrally), the valves of the caudal vena cava and possibly of the coronary sinus, the intervenous tubercle, the terminal crest, the right atrioventricular valve with its three cusps (angular, parietal, and septal), and the pectinate Mm. (Fig. 2-13).

Identify the conus arteriosus as a basal segment of the right ventricle at the origin of the pulmonary trunk, and then make a longitudinal incision through the conus arteriosus and the pulmonary trunk. Reflect the two parts of the right ventricle and the origin of the pulmonary trunk. Examine the interventricular septum, the three cusps of the right atrioventricular valve connected to the papillary muscles by the chordae tendineae, the trabecula septomarginalis, the trabeculae carneae, and the pulmonary valve with its three valvules (right, left, and intermediate) (Fig. 2-14).

Identify one of the pulmonary Vv. and make a longitudinal incision through the vein and the lateral wall of the left atrium; examine the left auricula and the left atrioventricular valve with its two cusps (parietal and septal). Continue the incision through the left ventricle, and examine the two aforementioned cusps anchored on the papillary muscles by the corresponding chordae tendineae. Continue to expose the interior of the left ventricle; look for the origin of the aorta with the three valvules of the aortic valve (right, left, and septal); and observe the origin of the coronary arteries from the ascending aorta. In addition, examine the trabeculae carneae (Fig. 2-15).

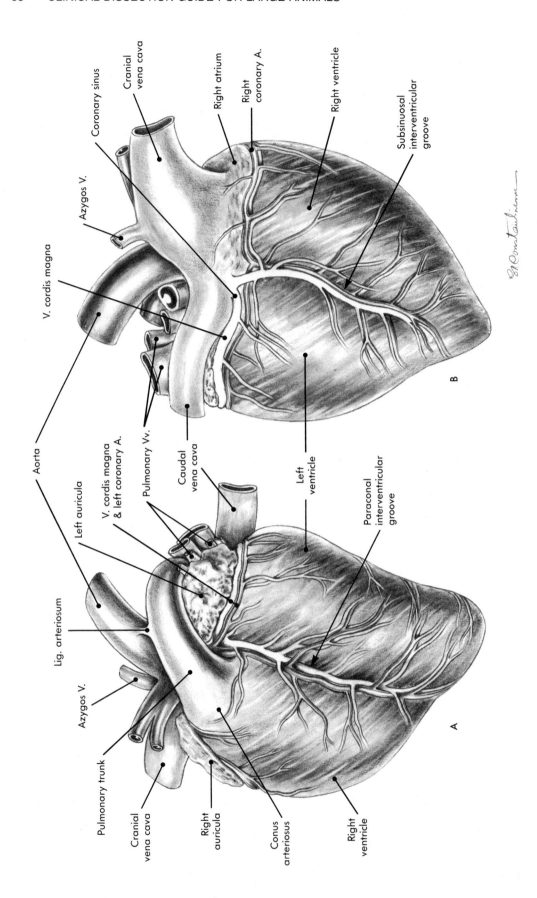

A, Heart (left [auricular] aspect). **B,** Heart (right [atrial] aspect).

Fig. 2-11

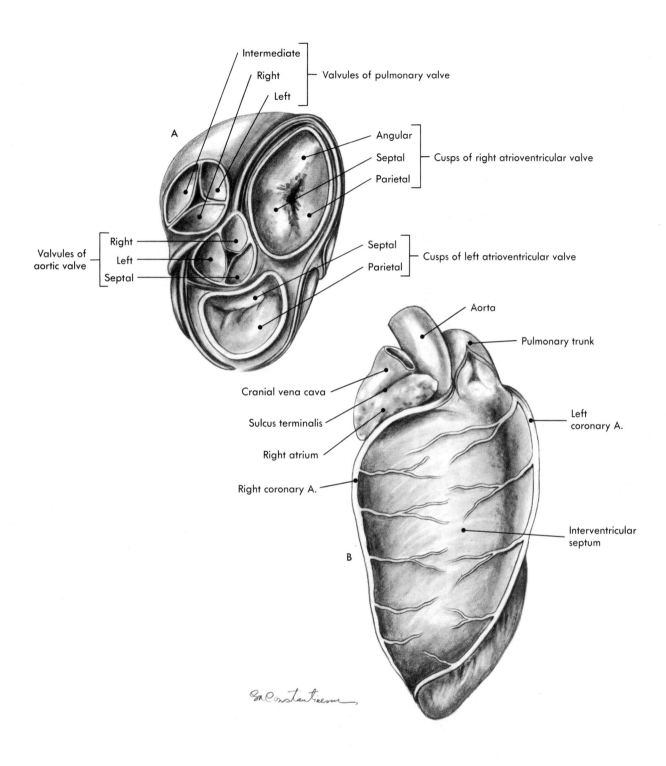

A, Atrioventricular and arterial valves of heart. **B,** Interventricular septum. Right ventricle is removed.

Fig. 2-12

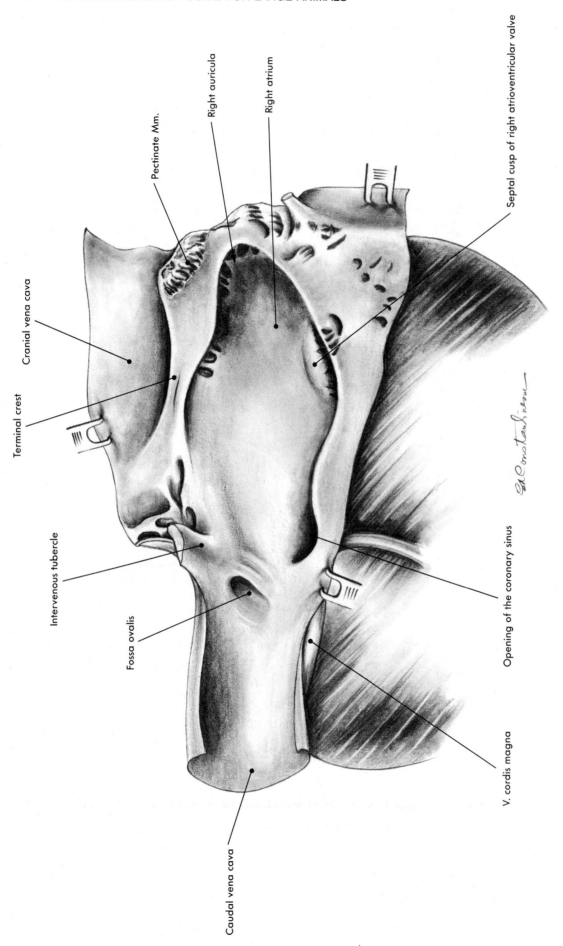

Pectinate Mm.

Right auricula

Right atrium

Septal cusp of right atrioventricular valve

Cranial vena cava

Terminal crest

Intervenous tubercle

Fossa ovalis

Opening of the coronary sinus

V. cordis magna

Caudal vena cava

Right atrium opened.

Fig. 2-13

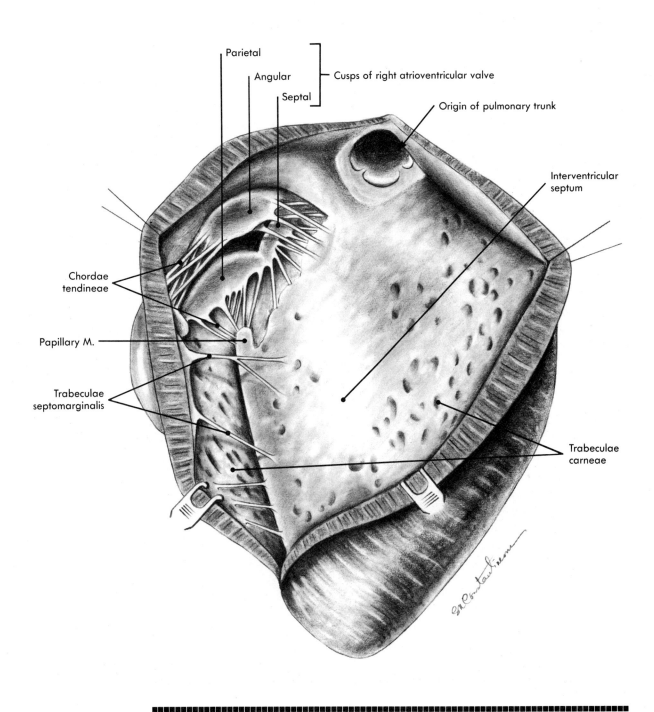

Parietal

Angular

Septal

Cusps of right atrioventricular valve

Origin of pulmonary trunk

Interventricular septum

Chordae tendineae

Papillary M.

Trabeculae septomarginalis

Trabeculae carneae

Right ventricle opened.

Fig. 2-14

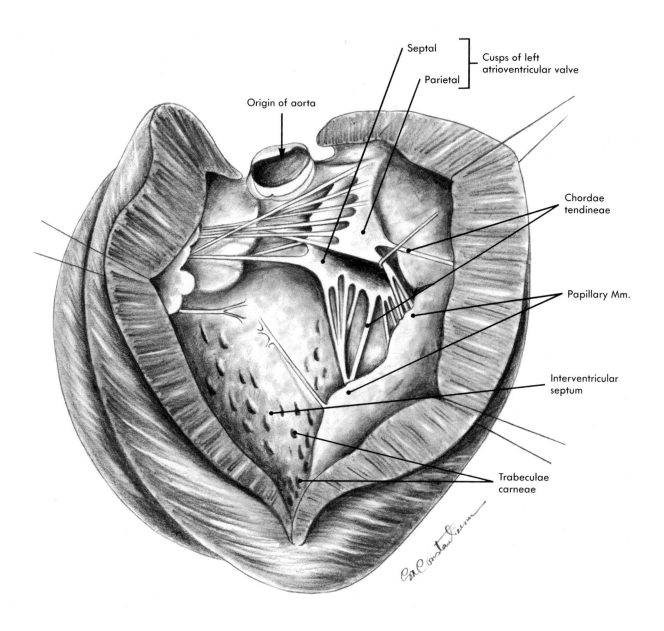

Septal

Parietal

Cusps of left
atrioventricular valve

Origin of aorta

Chordae
tendineae

Papillary Mm.

Interventricular
septum

Trabeculae
carneae

Fig. 2-15

Left ventricle opened.

Continue the skin incision from the dorsal extent of the cranial border of the scapula to the spinous process of the last thoracic vertebra on the dorsal midline. Continue the incision ventrally alongside the last rib to the corresponding costochondral junction and then ventrally up to the midventral line. From the manubrium sterni, continue the incision of the skin on the midventral line to meet the previous incision. Reflect the skin ventrally. Similar to the horse, skin the area carefully to separate the skin from the cutaneus Mm. and the superficial fasciae. Dissect and reflect the cutaneus Mm. and the superficial fasciae ventrally, after examining the cutaneus Nn.

Dissect the thoracic part of the trapezius M., carefully tracing its ventral border to preserve the aponeurosis of the latissimus dorsi M. Dissect the latissimus dorsi M. and separate it from the long head of the triceps brachii and tensor fasciae antebrachii Mm. Trace the cranial border of the latissimus dorsi M., which overlaps the caudal extent of the scapular cartilage. *Notice* the sudden transition between the muscular fibers and the aponeurosis of this muscle. Dissect the pectoralis profundus M. and the visible parts of the serratus ventralis thoracis and external abdominal oblique Mm. protected by fascia and tunica flava abdominis, respectively (Fig. 2-16, *A*).

Transect the scapular insertion of the thoracic part of the trapezius M., and reflect the muscle dorsally, separating it from the aponeurosis of the latissimus dorsi M. Transect the muscular part of the latter, leaving 10 to 15 cm of it attached to the forelimb. Either reflect the remnant of the latissimus dorsi M. dorsally or transect it from the insertions on the supraspinous lig.

The serratus ventralis thoracis, teres major, serratus dorsalis cranialis and caudalis, and external intercostal Mm. are now exposed. Dissect the iliocostalis, longissimus, and semispinalis and spinalis thoracis Mm. (Fig. 2-16, *B*).

Transect the base of the scapula with the corresponding muscles and the superficial pectoral Mm. to remove the forelimb, following a similar procedure to that of the horse. The whole serratus ventralis M., the three portions of the scalenus M., and the rectus thoracis and iliocostalis cervicis Mm. are still attached to the thorax. Transect the serratus ventralis M. as close to the remnant of the scapula as possible, and reflect the base of the scapula dorsally to dissect the deep structures. *Notice* that the rhomboideus cervicis and thoracis Mm. are still attached to scapula. The dorsoscapular lig., the entire serratus dorsalis cranialis M., and the transition of the epaxial muscles from the thoracic to the cervical area are now exposed.

Remove the serratus dorsalis cranialis and serratus ventralis Mm., and carefully dissect the dorsal branches of the thoracic Nn. with their medial and lateral secondary branches. The dorsomedial branches course in a dorsal direction over the longissimus, spinalis, and semispinalis thoracis Mm., whereas the dorsolateral branches travel ventrally over the iliocostalis thoracis M.

Transect the epaxial muscles (which are similar to the horse) to expose the deep muscles of the area, namely the multifidus thoracis, the levatores costarum, and the intertransversarii thoracis Mm. Remove the external intercostal M. from an intercostal space to expose the internal intercostal M. and the intercostal A.V.N.

Large ruminants

■■■■■■■■■■■■■■■■■

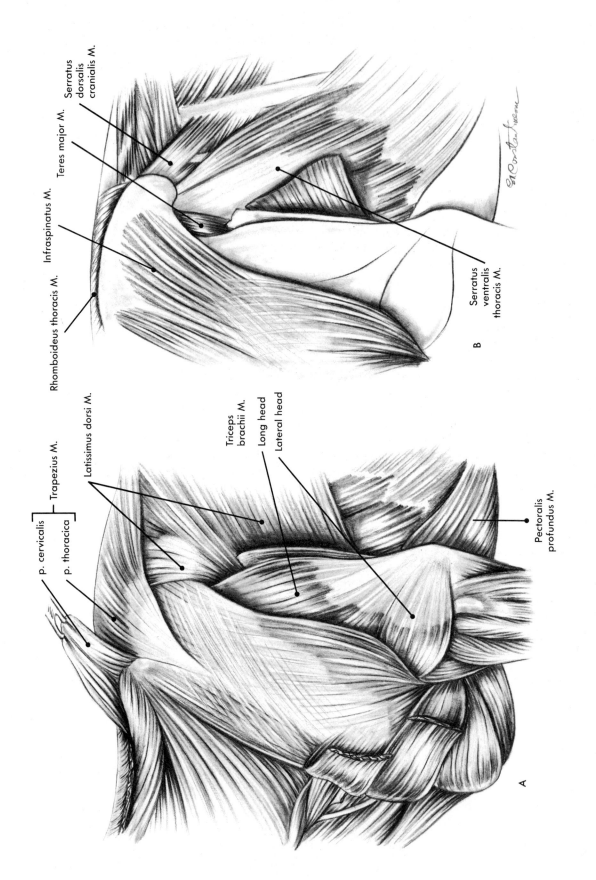

Serratus dorsalis cranialis M.

Teres major M.

Infraspinatus M.

Rhomboideus thoracis M.

Serratus ventralis thoracis M.

B

p. cervicalis
p. thoracica ⎱ Trapezius M.

Latissimus dorsi M.

Triceps brachii M.
Long head
Lateral head

Pectoralis profundus M.

A

Fig. 2-16   **A**, Superficial level shoulder and thorax (left side). **B**, Deep structures shoulder and thorax (left side).

Leaving the first, ninth, and last two ribs in place (RI, RIX, RXII, and RXIII) re-move the other ribs and the intercostal muscles in a similar manner to the horse.

Examine the lungs; identify their lobes, notches, and fissures and the pericardium and diaphragm (Fig. 2-17). Remove the lungs from the thoracic cavity, transecting the bron-chi and vessels as in the horse.

Caution. The right lung has a tracheal bronchus within the cranial mediastinum. Pay special attention to the accessory lobe of the right lung, the plica venae cavae, and the caudal vena cava, which are structures located on the right side.

Caution. The cranial part of the right cranial lobe of the lung passes underneath the trachea from the right into the left half of the thoracic cavity.

Dissect the vessels and nerves on both side, and *notice* the differences from the horse:

(1) On both sides, the three dorsal branches of the subclavian A. and the correspond-ing dorsal veins of the cranial vena cava start in common trunks (the vertebral, deep cervical, and costocervical Aa.Vv.).

(2) In addition to the external jugular Vv. and the axillary Vv., the internal jugular Vv. also contribute to the cranial vena cava.

(3) There are two azygos Vv. in the large ruminants: the left azygous V., which is consistently observed, and the right azygous V., which is usually represented by a branch of the left azygous V. In most specimens the right vein, which is shorter than the left one, has the main trunk in a vertical position (Figs. 2-18 and 2-19).

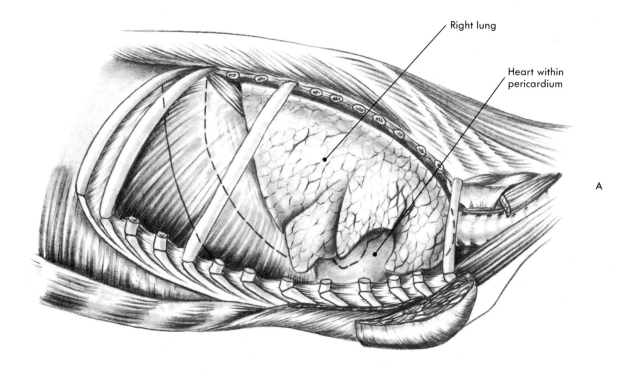

Right lung

Heart within pericardium

A

Dotted line, normal extension of the lung in living specimen;
heavy line, line of pleural reflection.

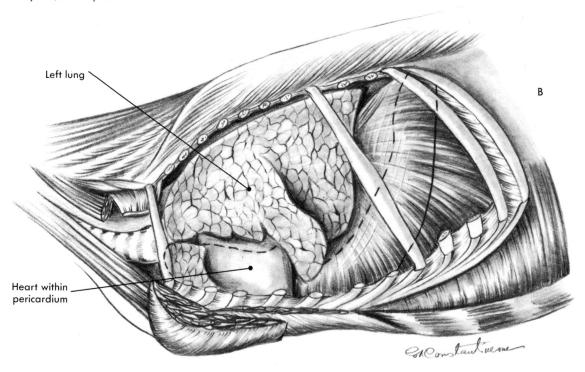

Left lung

B

Heart within pericardium

Fig. 2-17   **A,** Right thoracic cavity. **B,** Left thoracic cavity.

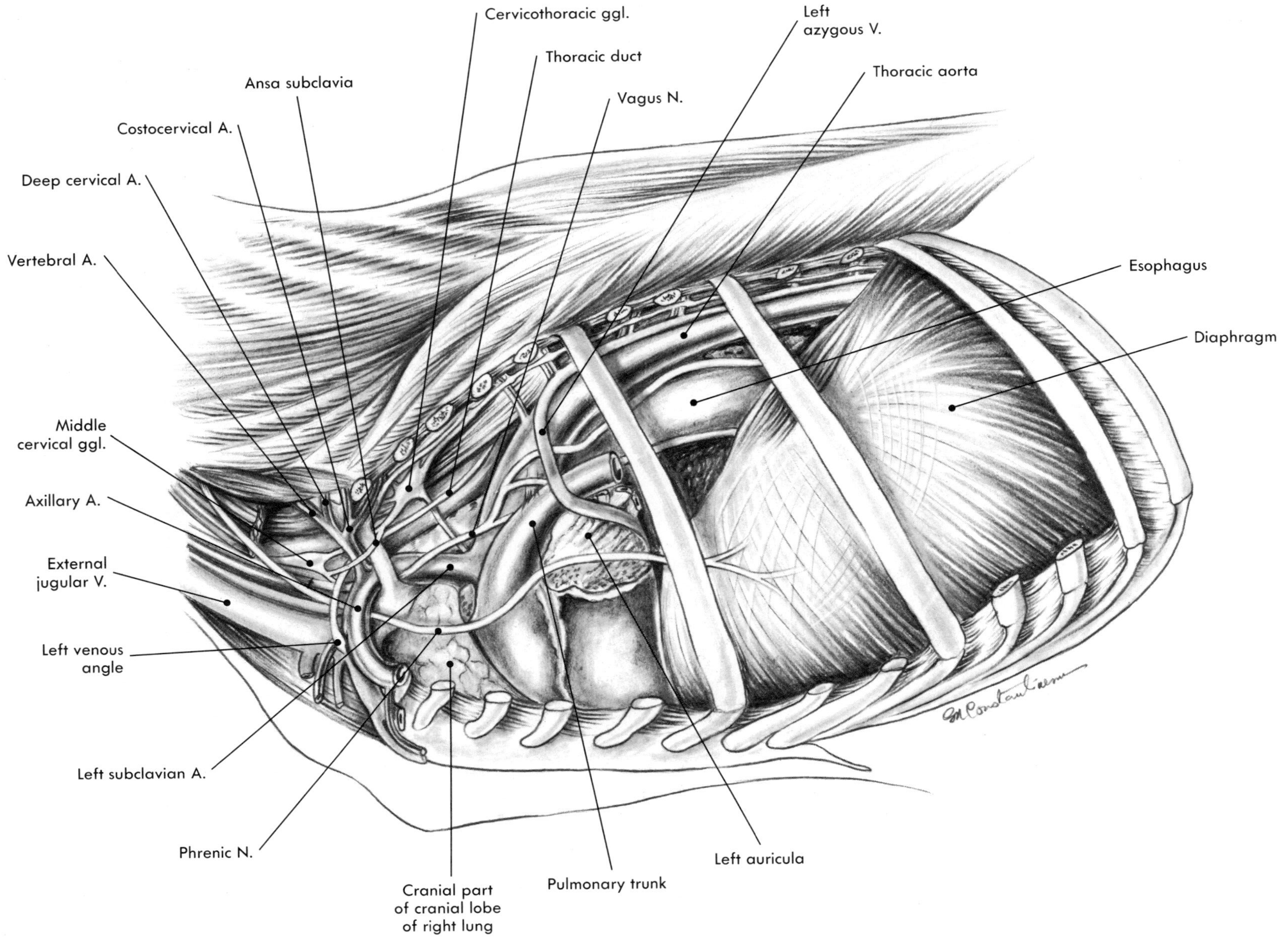

Esophagus

Diaphragm

Thoracic aorta

Left
azygous V.

Vagus N.

Thoracic duct

Cervicothoracic ggl.

Ansa subclavia

Costocervical A.

Deep cervical A.

Vertebral A.

Middle
cervical ggl.

Axillary A.

External
jugular V.

Left venous
angle

Left subclavian A.

Phrenic N.

Cranial part
of cranial lobe
of right lung

Pulmonary trunk

Left auricula

Fig. 2-18

Heart, vessels, and nerves (left thoracic cavity).

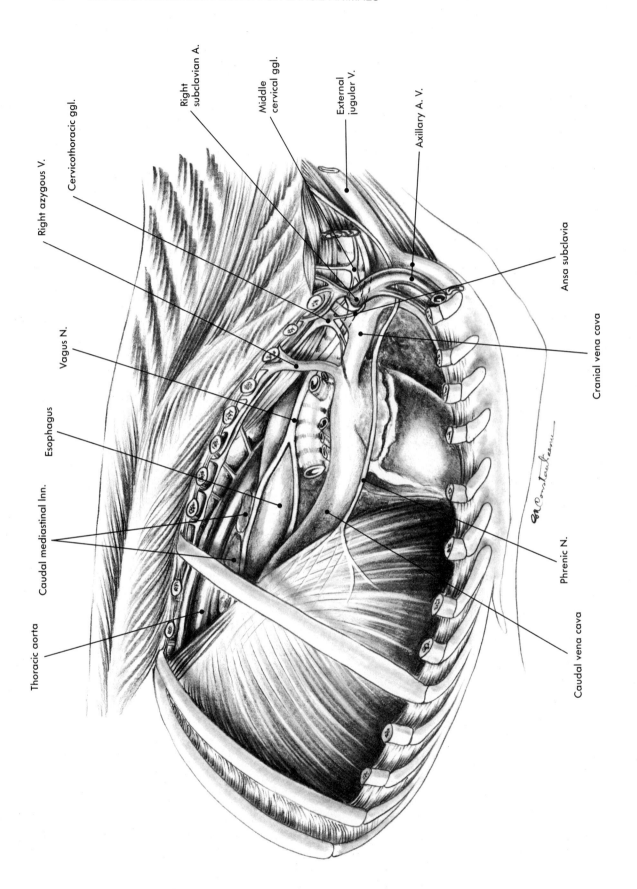

Right azygous V.

Cervicothoracic ggl.

Right subclavian A.

Middle cervical ggl.

External jugular V.

Axillary A. V.

Vagus N.

Esophagus

Ansa subclavia

Caudal mediastinal lnn.

Cranial vena cava

Thoracic aorta

Phrenic N.

Caudal vena cava

Fig. 2-19   Heart, vessels, and nerves (right thoracic cavity).

Remove the heart from the thoracic cavity in a manner similar to that described for the horse. Examine the two lungs and *notice* the different number of lobes and their fissures: the cranial and caudal parts of both cranial lobes (right and left) and the middle (cardiac) lobe of the right lung. Then examine the tracheal bronchus and *notice* the most distinct feature, which differentiates the lungs of the large ruminants from those of the small ruminants—namely the polyhedral design on the surface of *all* the lobes (similar to the pig). Identify the so-called impressions of other structures on both the costal and medial sides of the lungs (Figs. 2-20 and 2-21).

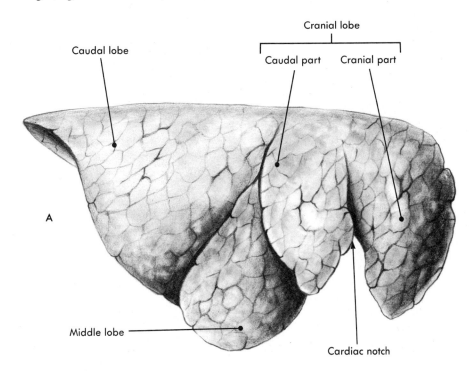

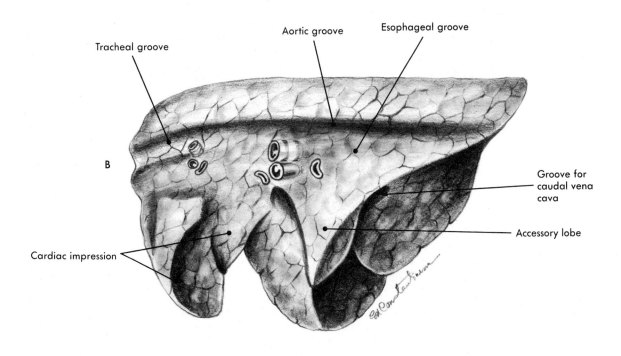

A, Right lung (lateral aspect). B, Right lung (medial aspect).

Fig. 2-20

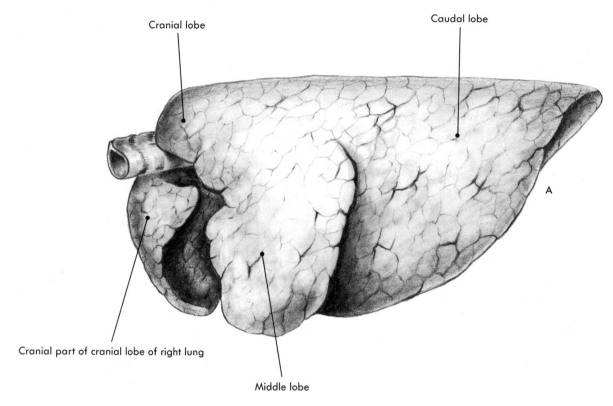

Cranial lobe

Caudal lobe

Cranial part of cranial lobe of right lung

Middle lobe

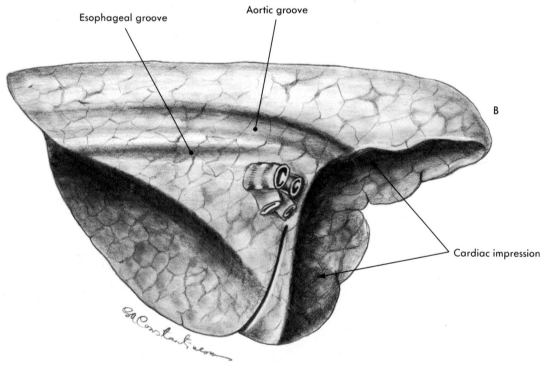

Esophageal groove

Aortic groove

Cardiac impression

Fig. 2-21   **A,** Left lung (lateral aspect). **B,** Left lung (medial aspect).

Examine and identify the external and internal structures of the heart using the horse as a model (Figs. 2-22 to 2-25).

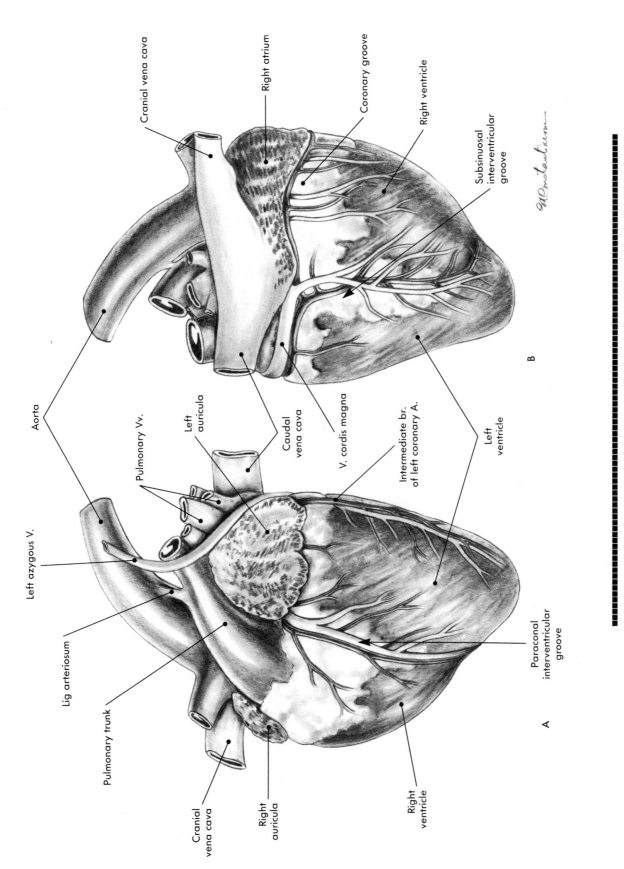

A, Heart (left [auricular] aspect). **B,** Heart (right [atrial] aspect).

Fig. 2-22

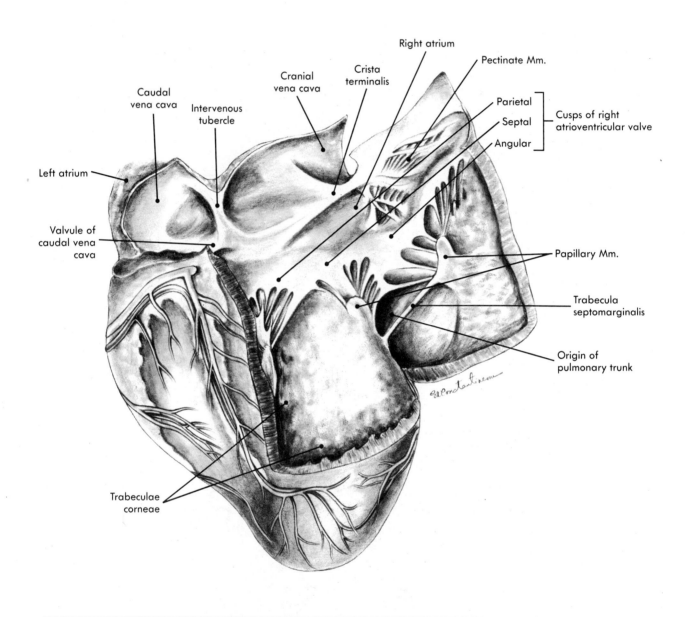

Fig. 2-23   Right heart opened.

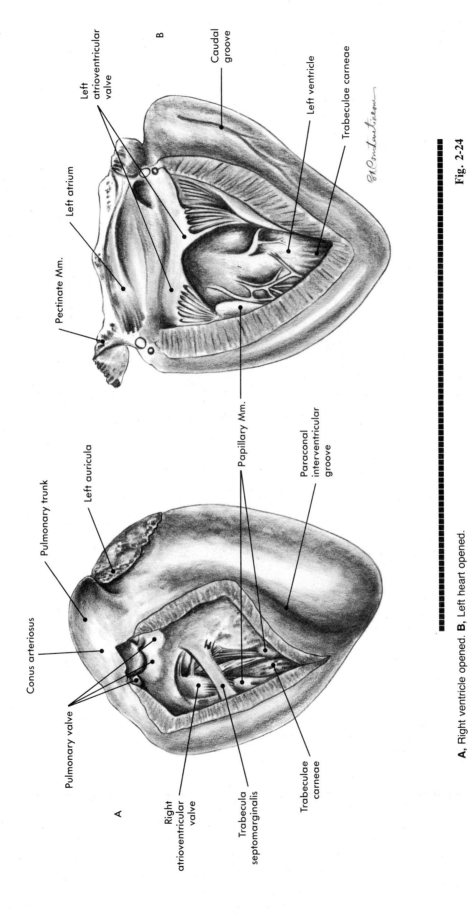

Fig. 2-24

A, Right ventricle opened. **B**, Left heart opened.

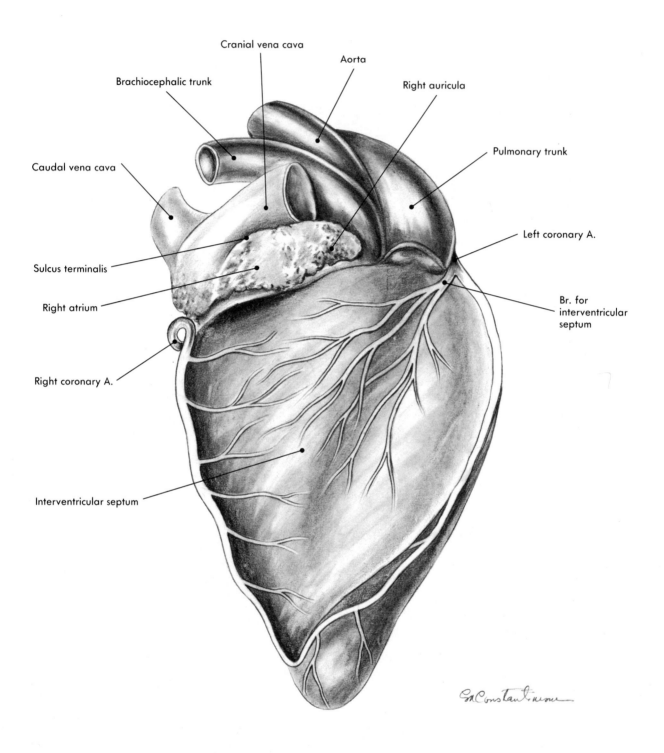

Cranial vena cava

Aorta

Brachiocephalic trunk

Right auricula

Caudal vena cava

Pulmonary trunk

Sulcus terminalis

Left coronary A.

Right atrium

Br. for interventricular septum

Right coronary A.

Interventricular septum

Fig. 2-25   Interventricular septum (right ventricle is removed).

If the thoracic limb has not been removed, skin the thoracic and abdominal (optional) areas. If the limb was previously removed, skin the area caudal to it, up to the pelvic limb and including the thorax and abdomen.

If the thoracic limb is still present, identify and dissect the structures shown in Fig. 2-26. In addition, identify and transect the lateral thoracic N., and then remove the forelimb. Use the procedure described in horse and ox as a guide.

If the thoracic limb was already removed, identify and dissect the muscles still attached to the thorax. Dissect all the other muscles of the thorax, similar to the ox.

Notice that the sheep (and the horse) do not have a scalenus dorsalis M. The other scaleni muscles are very similar to those of the large ruminants.

Sheep
▪▪▪▪▪▪▪▪▪
(Figs. 2-26 and 2-27)

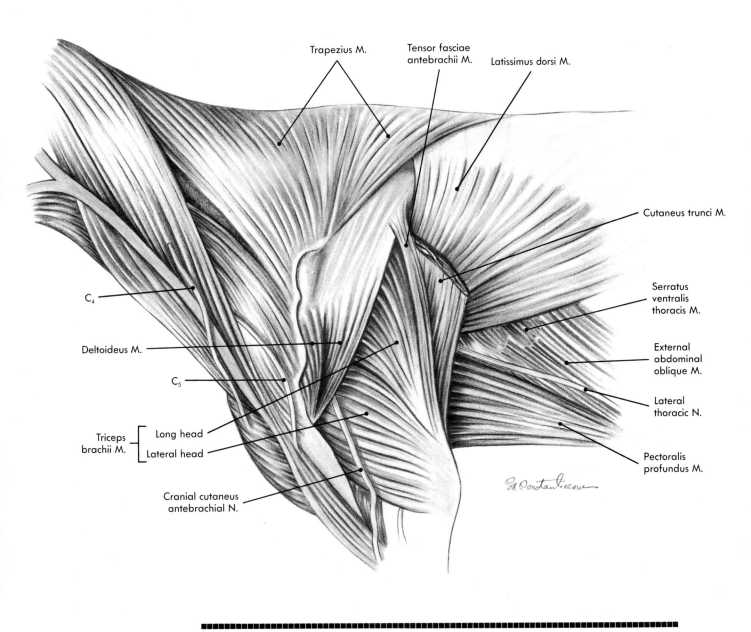

Superficial area of shoulder and thorax (left side).

Fig. 2-26

Examine the lungs and *notice* that, in comparison with those of the large ruminants and the goat, ovine lungs have no polyhedral design on their surface. Examine the interlobar fissures and the left cardiac notch. Reflect the left lung to expose the heart within the pericardium, the vessels related to the heart, the lymph nodes, and the nerves.

Notice that the thoracic duct travels along the esophagus before reaching the left venous angle and that an azygos V. is located on the left side (the left azygous V.). The sheep has also a right azygous V., but it is less developed than the left (Fig. 2-27).

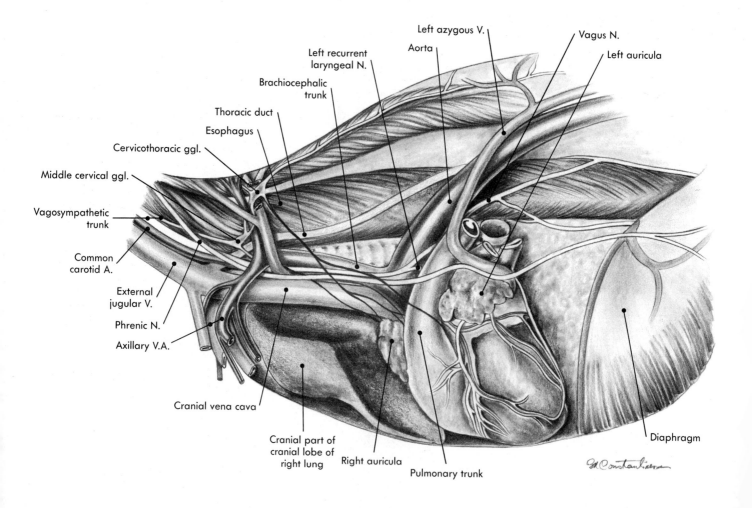

Fig. 2-27   Heart, vessels, and nerves (left thoracic cavity).

The thoracic limb can be removed at this time. If the limb has not been removed, look at the procedure in the previous chapter (Chapter 1, "The Neck and Thoracic Inlet").

After removing the thoracic limb, dissect the thorax exposing the structures shown in Figs. 2-28 to 2-30 and Fig. 2-32. Remove the lungs and the heart and examine them (Figs. 2-31 and 2-33).

Notice that the middle lobe of the right lung and both the right and left caudal pulmonary lobes do not show a polyhedral design.

Goat
■■■■■■■■
(Figs. 2-28
to 2-33)

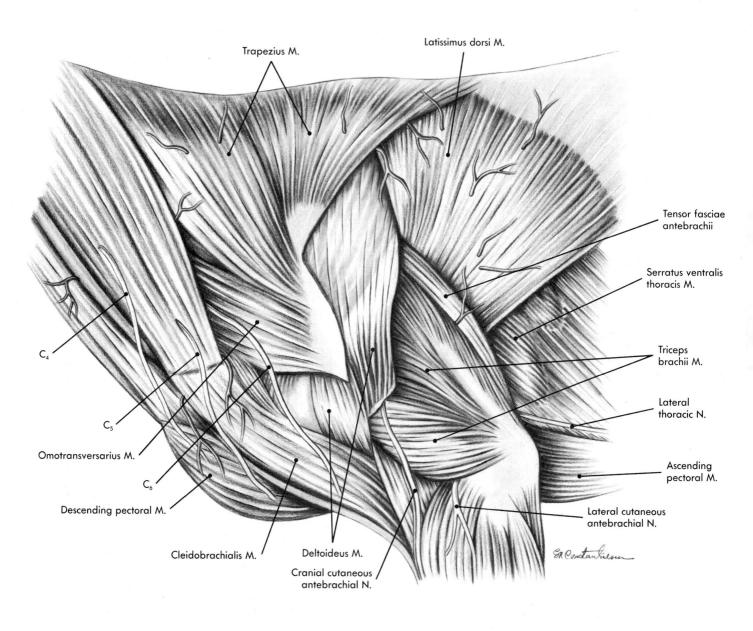

Superficial area of shoulder and thorax (left side).

Fig. 2-28

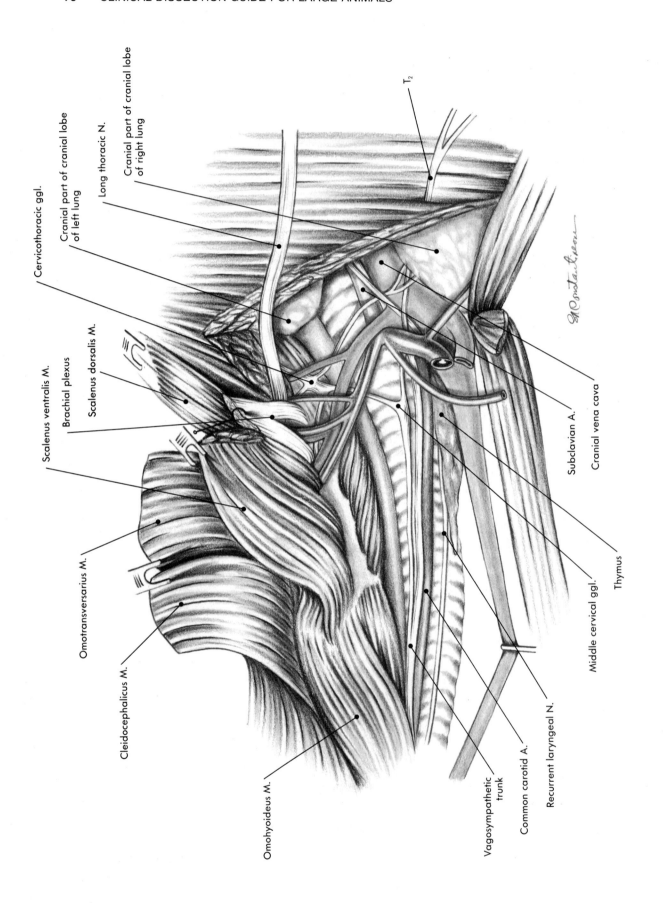

Fig. 2-29 Cranial mediastinum (left side).

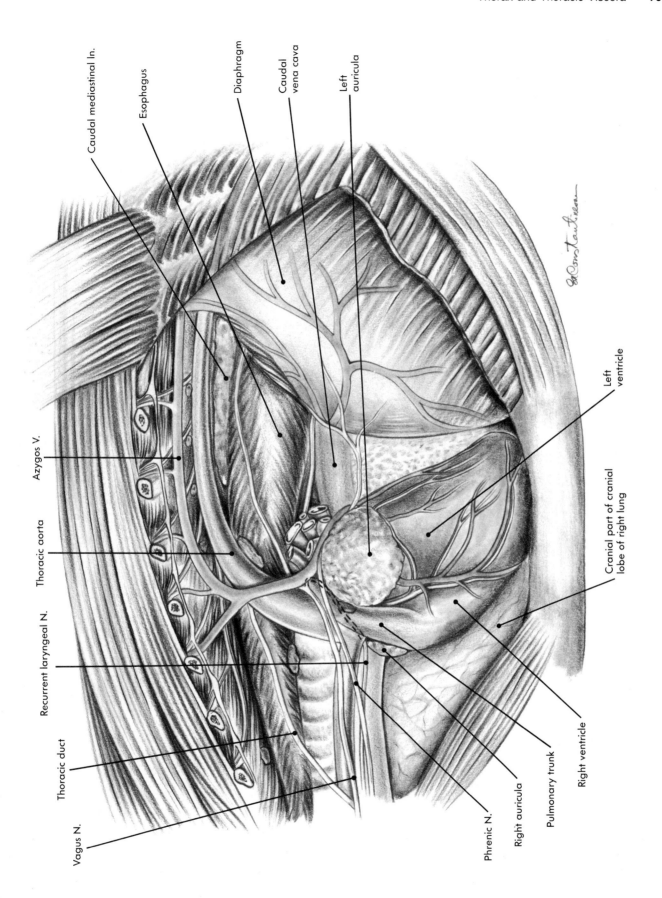

Caudal mediastinal ln.

Esophagus

Diaphragm

Caudal
vena cava

Left
auricula

Azygos V.

Thoracic aorta

Recurrent laryngeal N.

Thoracic duct

Vagus N.

Left
ventricle

Cranial part of cranial
lobe of right lung

Right ventricle

Pulmonary trunk

Right auricula

Phrenic N.

Fig. 2-30

Heart, vessels, and nerves (left thoracic cavity).

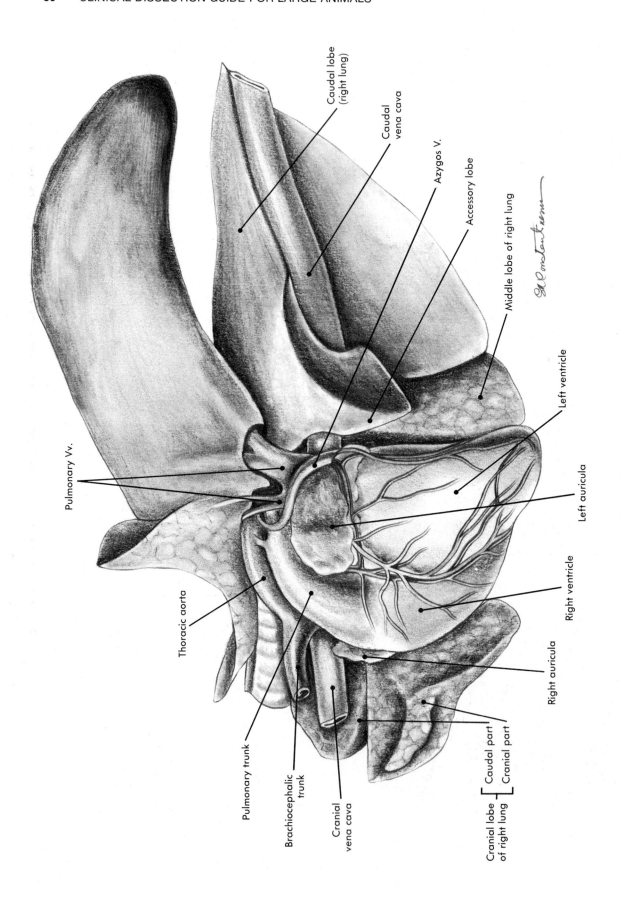

Caudal lobe (right lung)

Caudal vena cava

Azygos V.

Accessory lobe

Middle lobe of right lung

Left ventricle

Left auricula

Right ventricle

Right auricula

Caudal part
Cranial part
Cranial lobe of right lung

Cranial vena cava

Brachiocephalic trunk

Pulmonary trunk

Thoracic aorta

Pulmonary Vv.

Fig. 2-31   Heart (left aspect) and lungs (ventral aspect).

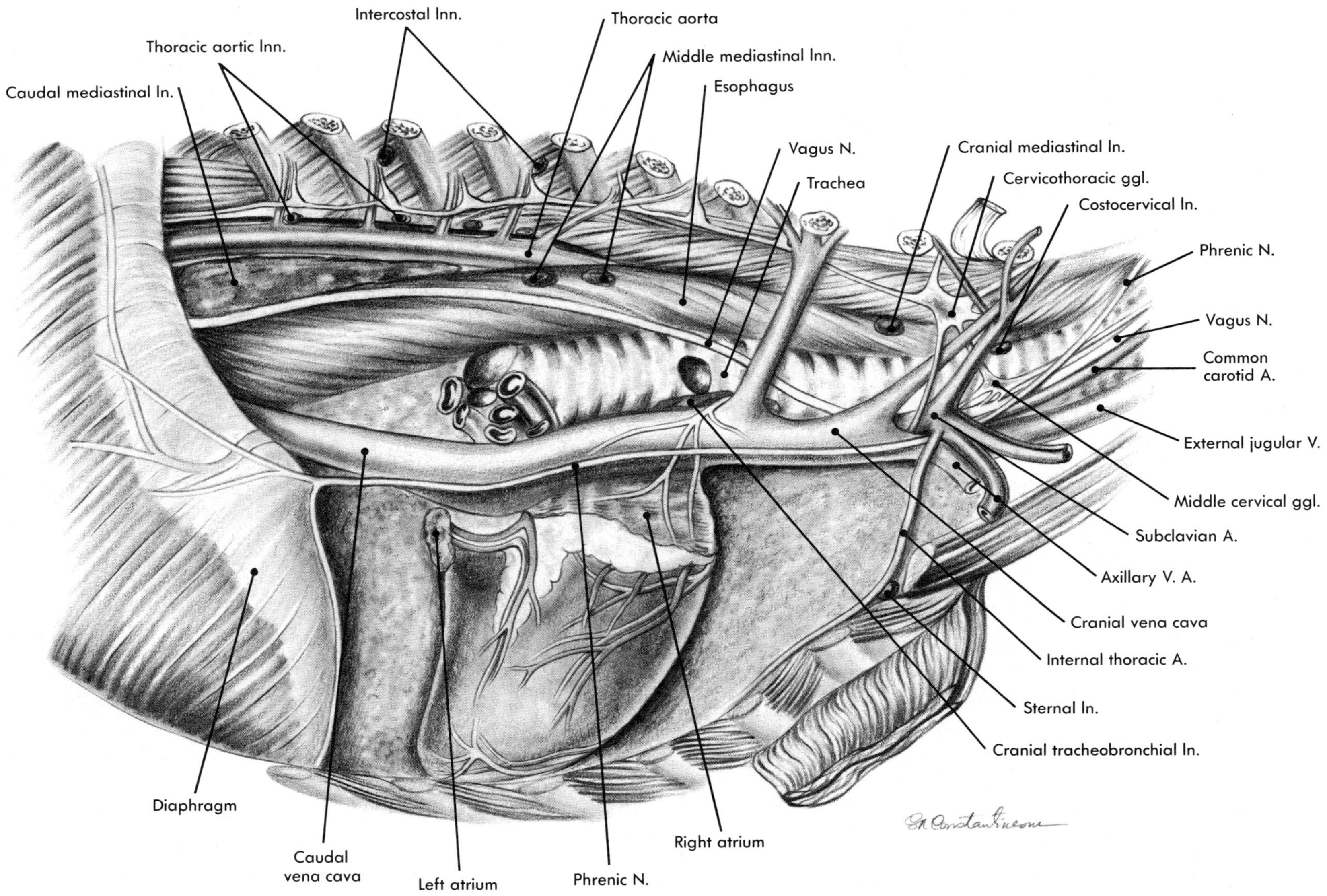

Phrenic N.

Vagus N.

Common carotid A.

External jugular V.

Middle cervical ggl.

Subclavian A.

Axillary V. A.

Cranial vena cava

Internal thoracic A.

Sternal ln.

Cranial tracheobronchial ln.

Cranial mediastinal ln.

Cervicothoracic ggl.

Costocervical ln.

Vagus N.

Trachea

Middle mediastinal lnn.

Esophagus

Thoracic aorta

Intercostal lnn.

Thoracic aortic lnn.

Caudal mediastinal ln.

Right atrium

Phrenic N.

Left atrium

Caudal vena cava

Diaphragm

Fig. 2-32

Heart, vessels, and nerves (right thoracic cavity).

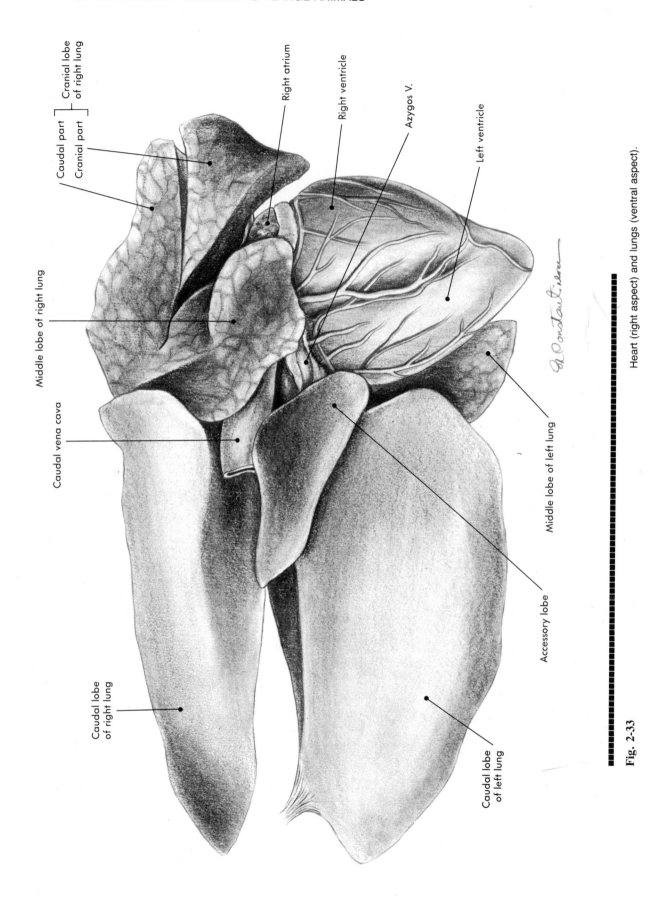

Caudal part ⎤
Cranial part ⎦ Cranial lobe of right lung

Right atrium

Right ventricle

Azygos V.

Left ventricle

Middle lobe of right lung

Caudal vena cava

Middle lobe of left lung

Accessory lobe

Caudal lobe of right lung

Caudal lobe of left lung

Heart (right aspect) and lungs (ventral aspect).

Fig. 2-33

Follow the directions for the other species and dissect the thorax; then remove all the ribs to expose, identify, dissect, remove, and examine the structures within the thoracic cavity (Figs. 2-34 to 2-37). Since the pig is lying in lateral recumbency you can remove all the ribs.

Pig
(Figs. 2-34 to 2-37)

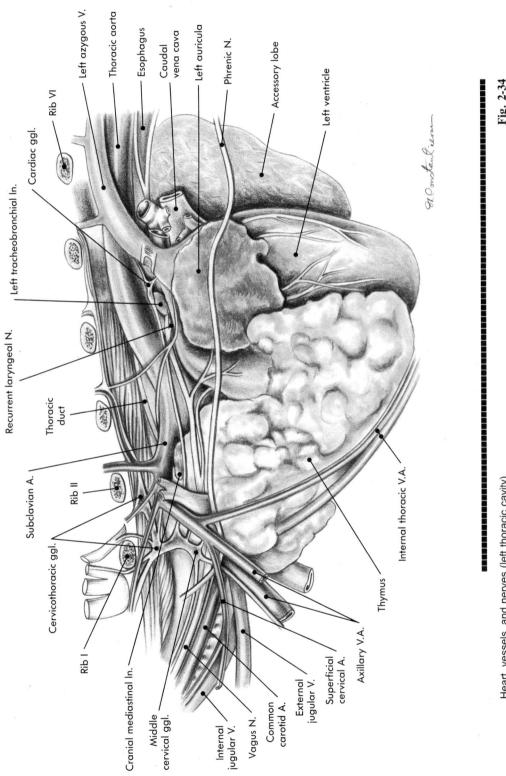

Fig. 2-34

Heart, vessels, and nerves (left thoracic cavity).

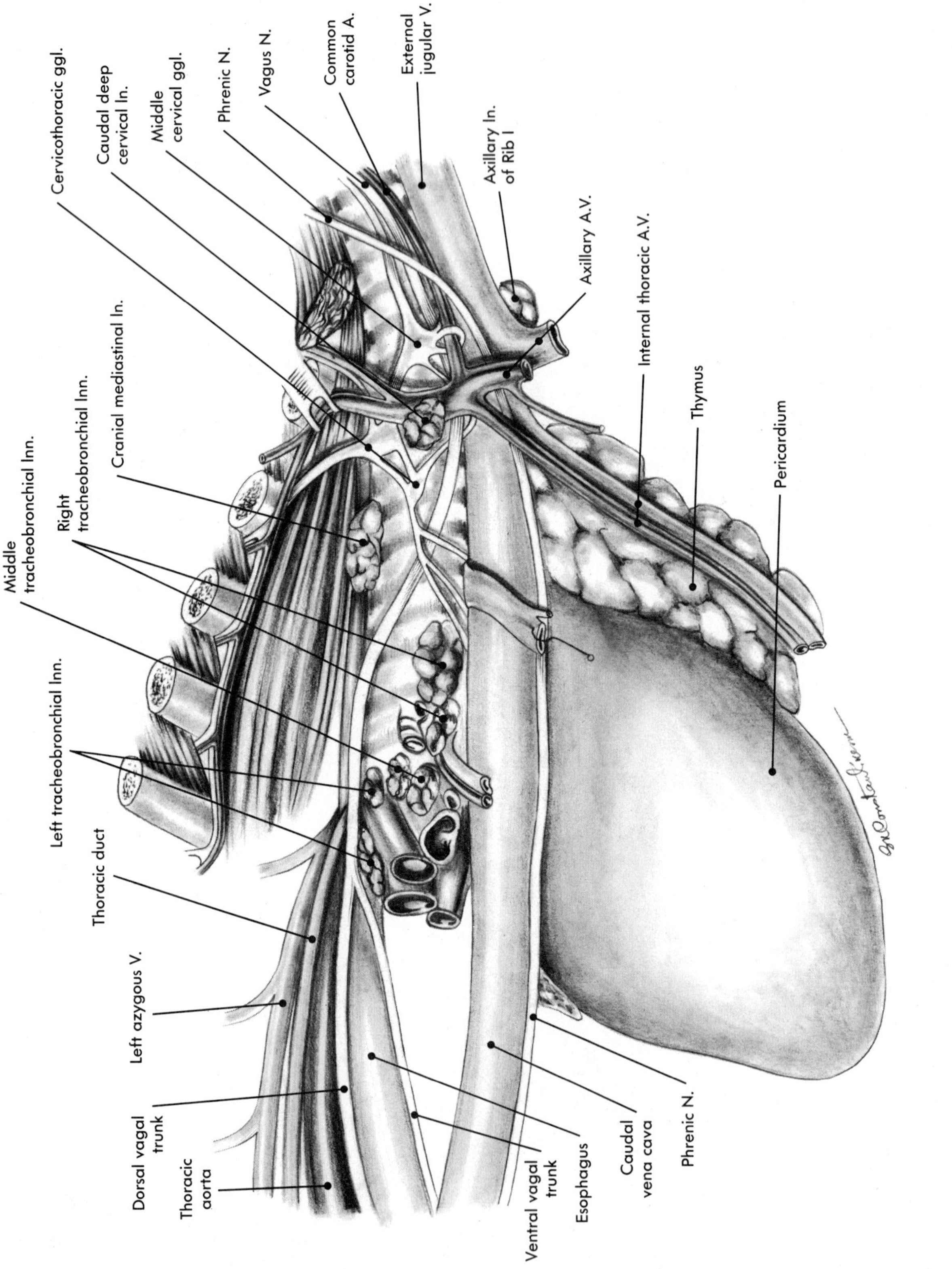

Cervicothoracic ggl.

Caudal deep cervical ln.

Middle cervical ggl.

Phrenic N.

Vagus N.

Common carotid A.

External jugular V.

Axillary ln. of Rib I

Axillary A.V.

Internal thoracic A.V.

Thymus

Pericardium

Middle tracheobronchial lnn.

Right tracheobronchial lnn.

Cranial mediastinal ln.

Left tracheobronchial lnn.

Thoracic duct

Left azygous V.

Dorsal vagal trunk

Thoracic aorta

Ventral vagal trunk

Esophagus

Caudal vena cava

Phrenic N.

Heart, vessels, and nerves (right thoracic cavity).

Fig. 2-35.

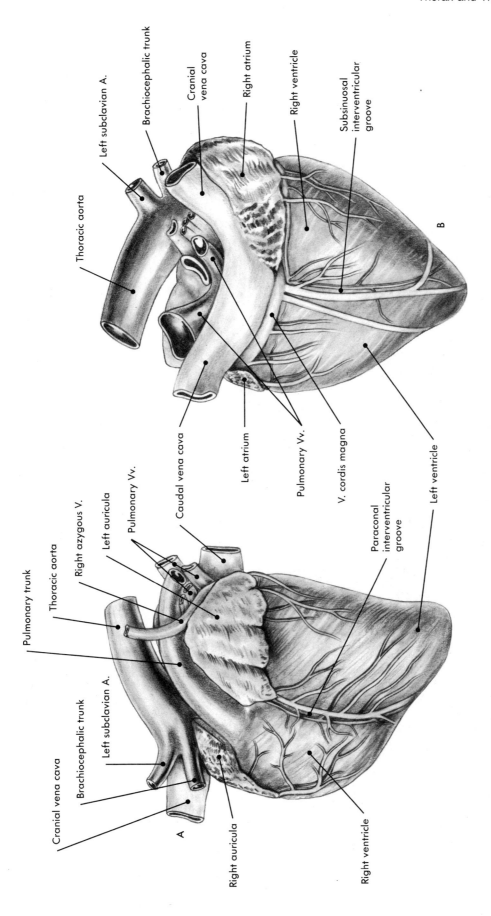

Left subclavian A.

Brachiocephalic trunk

Cranial vena cava

Right atrium

Right ventricle

Subsinuosal interventricular groove

Thoracic aorta

B

Caudal vena cava

Left atrium

Pulmonary Vv.

V. cordis magna

Left ventricle

Paraconal interventricular groove

Right ventricle

Pulmonary trunk

Thoracic aorta

Right azygous V.

Left auricula

Pulmonary Vv.

Cranial vena cava

Brachiocephalic trunk

Left subclavian A.

Right auricula

A

Fig. 2-36

A, Auricular (left) side of heart. **B,** Atrial (right) side of heart.

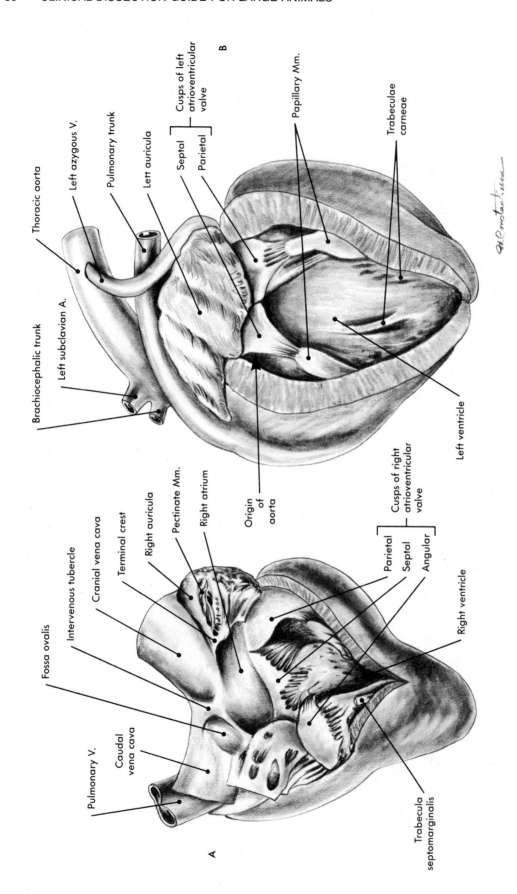

Fig. 2-37

A, Right side of heart (opened). **B,** Left ventricle (opened).

Chapter 3

Abdomen and abdominal viscera

ABDOMEN
Horse

Skin the abdomen up to the tuber coxae and the fold of the flank, and reflect it ventrally.

Caution! Skin carefully to dissect the dorsolateral cutaneous branches of the costoabdominal N. and the lumbar Nn. *Notice* that these branches become superficial when emerging from beneath the retractor costae M. (the costoabdominal N.) and from between the lumbar origin of the gluteus medius M. and the longissimus thoracolumborum M. (the lumbar Nn.). They can be identified at the ventral edge of the aponeurosis of the latissimus dorsi M., which is fused with the thoracolumbar fascia; the edge corresponds to a line from the tuber coxae to the tuberosity of the scapular spine.

Caution! When dissecting the ventrolateral cutaneous branches of the same nerves, use care at the level of a line from the tuber coxae to the olecranon. This line corresponds approximately to the end of the dorsolateral and to the origin of the ventrolateral cutaneous branches of these nerves (Fig. 3-1).

Continue the dissection of the cutaneus trunci M. and reflect it ventrally. The most caudoventral attachment (or the origin) of this muscle is located within the stifle fold (the fold of the flank). At the middistance between the tuber coxae and patella and at the cranial border of the tensor fasciae latae M., the subiliac ln., lateral cutaneous femoral N. (ventral branches of L_3 and L_4), and caudal branches of the deep circumflex iliac A.V. are exposed. Dissect all these structures carefully.

Dissect an area of the abdominal tunic and compare its thickness and elasticity to those of the cow. Dissect and reflect the muscular fibers of the external abdominal oblique M. from the last ribs, from the thoracolumbar fascia, and from the tuber coxae. Transect its aponeurosis from the tuber coxae to the patellar area and reflect the muscle ventrally. The muscle will be anchored by the perforating ventrolateral cutaneous branches of nerves T_{18}, L_1, and L_2; transect them to liberate the muscle and to better and more easily examine the internal abdominal oblique M. and the ventral muscular branches of the nerves.

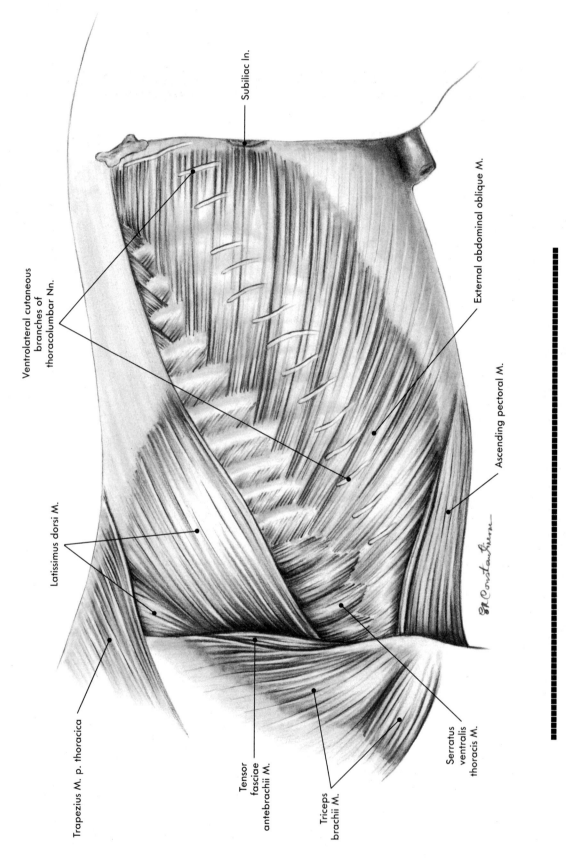

Subiliac ln.

Ventrolateral cutaneous
branches of
thoracolumbar Nn.

External abdominal oblique M.

Ascending pectoral M.

Latissimus dorsi M.

Trapezius M. p. thoracica

Tensor
fasciae
antebrachii M.

Triceps
brachii M.

Serratus
ventralis
thoracis M.

First level, abdomen (left side).

Fig. 3-1

The reflection of the external abdominal oblique M. reveals the retractor costae, transversus abdominis, and internal abdominal oblique Mm. Identify the ventral branches of the costoabdominal and first two lumbar Nn. Observe the prominent part of the internal abdominal oblique M., running from the tuber coxae to the last costochondral joint, which corresponds to the cord of the flank (Fig. 3-2, *A*). Transect the origin and insertions of this muscle from the tuber coxae, the last rib, and the costal arch; then transect the muscle from the tuber coxae to the patellar area and reflect it ventrally.

Close to the lateral border of the rectus abdominis M., the aponeurosis of the internal abdominal oblique M. fuses with the aponeurosis of the external abdominal oblique M. to form the external lamina of the rectus abdominis M. sheath.

On the lateral aspect of the transversus abdominis M., lie the ventral branches of the last thoracic (intercostal) Nn., of the costoabdominal N., and of the first two lumbar Nn. Between the internal abdominal oblique and transversus abdominis Mm., the cranial branches of the deep circumflex iliac A.V. are visible, positioned ventral to the tuber coxae (Fig. 3-2, *B*).

Transect the transversus abdominis M. from the transverse processes of the lumbar vertebrae and from the medial aspect of the last rib and costal arch, and reflect it ventrally. On the medial aspect of the rectus abdominis M., the aponeurosis of the transversus abdominis M. represents the cranial part of the internal lamina of the rectus abdominis M. sheath. The caudal part of the sheath is represented by the transverse fascia.

Caution! The transversus abdominis M. is lined by the endoabdominal (transverse) fascia and parietal peritoneum. To penetrate the peritoneal cavity, transect both together, starting at the tuber coxae and continuing cranially on the transverse processes of the lumbar vertebrae, the last rib, and the costal arch and ventrally toward the patellar area. Cranial to the last rib, the endoabdominal fascia and parietal peritoneum line the diaphragm; do not displace them from this position.

Remove the intercostal Mm. from the last intercostal space (right and left).

A short summary of the topographic anatomy of the projection of the viscera on lateral and ventral walls of the abdomen is necessary here, before removing the viscera from the abdominal cavity.

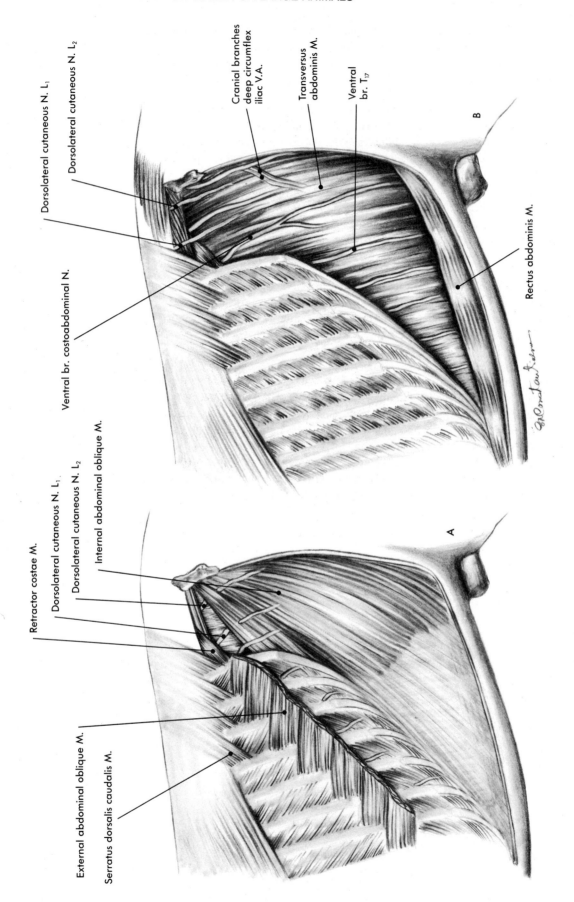

Dorsolateral cutaneous N. L_1

Dorsolateral cutaneous N. L_2

Cranial branches deep circumflex iliac V.A.

Transversus abdominis M.

Ventral br. T_{17}

Ventral br. costoabdominal N.

Rectus abdominis M.

Retractor costae M.

Dorsolateral cutaneous N. L_1

Dorsolateral cutaneous N. L_2

Internal abdominal oblique M.

External abdominal oblique M.

Serratus dorsalis caudalis M.

B

A

Fig. 3-2

A, Second level, abdomen (left side). **B,** Third level, abdomen (left side).

The following viscera may be identified on the *right lateral wall* of the abdomen: the cecum, liver, right kidney, right ventral and right dorsal colon, pancreas, and duodenum.

The *cecum* is positioned with its base in the hole of the flank, exceeding the last rib cranially to the third intercostal space (from caudal to cranial) and ventrally to a horizontal line between the shoulder and hip joints. Its body continues ventrally in the cord and slope of the flank, one-hand caudal to the costal arch.

The *right lobe of the liver* is located between the centrum tendineum of the diaphragm, the third intercostal space (from caudal to cranial) to the base of the cecum, and along a line almost horizontal to the shoulder joint.

The *right kidney* is located at the level of the last two to three intercostal spaces (the last four ribs), between the caudate lobe of the liver and the base of the cecum and on the roof of the abdominal cavity (sublumbar area).

The *right ventral* and the *right dorsal colon* are located between the body of the cecum, liver, centrum tendineum of the diaphragm, and floor of the abdominal cavity. The costal arch separates their area of projection.

The *pancreas* lies in contact with the liver, duodenum, and kidney.

The *descending duodenum* is located between the liver and the base of the cecum, surrounding the lateral border of the kidney (Fig. 3-3, *A*).

On *the left side* of the abdomen, the spleen, liver, stomach, left kidney, left ventral and left dorsal colon, descending colon, jejunum, and pancreas can be identified.

The *spleen* is located with its base on the dorsal wall of the abdominal cavity, at the level of the last four ribs. The concave cranial border extends from the cranial extremity of the base, whereas the convex caudal border runs from the caudal extremity of the base. They meet in the apex toward the costal arch between the ninth and eleventh intercostal spaces.

The *liver* is the structure closest to the diaphragm. Its caudal border starts at the tenth rib (RX) dorsally and ends at the seventh rib (RVII) ventrally, at the level of the costal arch.

The *stomach* is located between the liver and spleen.

The *left kidney* is supported by the base of the spleen at the level of the last three ribs; it exceeds the last rib caudally.

The *left ventral colon* lies on the floor of the abdominal cavity, 5 to 10 cm dorsal to the hypochondrium, and in the slope of the flank.

The *left dorsal colon* is located dorsal to the left ventral colon, at the level of the costal arch and cord of the flank.

The *descending colon* fills the hole of the flank.

The *jejunum* is located between the descending colon, spleen, and left dorsal colon. Compare the arteries supplying the jejunum (the jejunal Aa.) and the arteries of the descending colon (the left colic A.). Be careful to *notice* the distance between the arterial arcades and the lesser curvature of each of the intestinal segments.

The *pancreas* follows the cranial extremity of the kidney, lying between the base of the spleen and the stomach (Fig. 3-3, *B*).

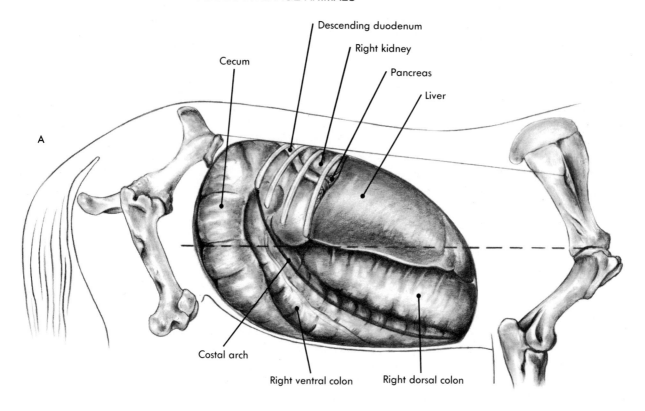

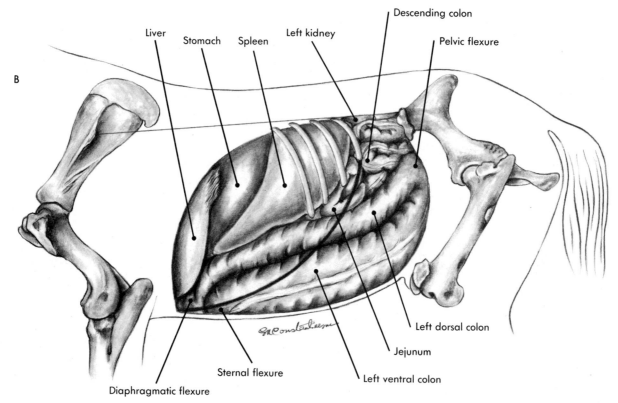

Fig. 3-3   **A,** Abdominal viscera (right aspect). **B,** Abdominal viscera (left aspect).

The *ventral wall* (or floor) of the abdomen is divided into the following areas: cranial abdominal region: hypochondrial and xyphoid regions; middle abdominal region: the flank and the fold of the flank; caudal abdominal region: inguinal and pubic regions, preputial (in male) or the udder region (in female). On the floor of the abdominal cavity, the following viscera and structures can be identified: the right ventral colon, sternal flexure, left ventral colon, cecum, and occasionally the diaphragmatic flexure and jejunal ansae.

The *right ventral colon* is located on the right side of the abdominal cavity and extends up to the corresponding hypochondrium.

The *sternal flexure* lies on the xiphoid cartilage.

The *left ventral colon* is located on the left side of the abdominal cavity and extends up to the corresponding costal arch.

The *body of the cecum,* in continuation from the base of the cecum, is located between the right and left ventral colon, with its apex positioned toward the sternal flexure.

The *diaphragmatic flexure* is sometimes located cranial to the sternal flexure.

Jejunal ansae can be observed on the left caudal side, either between the left ventral colon and the left wall of the abdominal cavity or between the left ventral colon and the cecum (Fig. 3-4).

An important characteristic of the horse is the lack of any contact between the stomach and liver with the floor of the abdominal cavity, as in other species, because of the enormous size and development of the cecum and ascending colon. The apex of the spleen also does not extend to the floor of the abdominal cavity.

Palpate the triangular ligg. of the liver. The right lig. connects the dorsal border of the right lobe of the liver to the costal part of the diaphragm, whereas the left lig. connects the dorsal border of the left lateral lobe of the liver to the tendinous center of the diaphragm. Palpate the coronary lig., which is attached to the right side of the tendinous center of the diaphragm and the cranial aspect (diaphragmatic surface) of the liver, accompanying the caudal vena cava. On the right side, palpate the hepatorenal lig. (between the caudate process of the liver and the right kidney), the intercolic fold (between the right dorsal and right ventral colon), the mesoduodenum (attached to the lesser curvature of the descending duodenum and continuing under the ventral aspect of the right kidney to form the duodenorenal lig.), and the adherence of the base of the cecum to the dorsal wall of the abdominal cavity (Fig. 3-5).

Palpate the caudal flexure of the duodenum and continue to palpate the transverse and ascending duodenum up to the duodenocolic fold, which attaches the last segment of the duodenum to the transverse colon at its limit with the descending colon. The duodenum continues with the duodenojejunal flexure, and from here the jejunum starts. Palpate the continuation between the mesoduodenum with the mesojejunum.

Reflect the liver as far cranially as possible and follow the descending duodenum in a cranial direction, continuing with the cranial flexure and the cranial duodenum toward and along the visceral aspect of the liver. Locate the pancreas, which is attached to the base of the cecum, and *notice* its relationships with the surrounding viscera. Identify the portal V., which lies between the pancreas and the liver, and the caudal vena cava; *note* their relationships to the liver (Fig. 3-6). Attempt to introduce a finger between the caudate process of the liver (cranially), the right lobe of the pancreas (caudally), the caudal vena cava (dorsally), and the portal V. (ventrally). This narrow foramen is the epiploic foramen, which is the passage between the peritoneal cavity and the omental bursa.

Ventral to this area, reflect the liver cranially and palpate the lesser omentum, which is attached to the visceral (caudal) aspect of the liver and to the lesser curvature of the stomach (the hepatogastric lig.). The attachment of the lesser omentum to the cranial duodenum is the hepatoduodenal lig.

Remember! The lesser omentum is represented by two structures: the hepatogastric and the hepatoduodenal ligg.

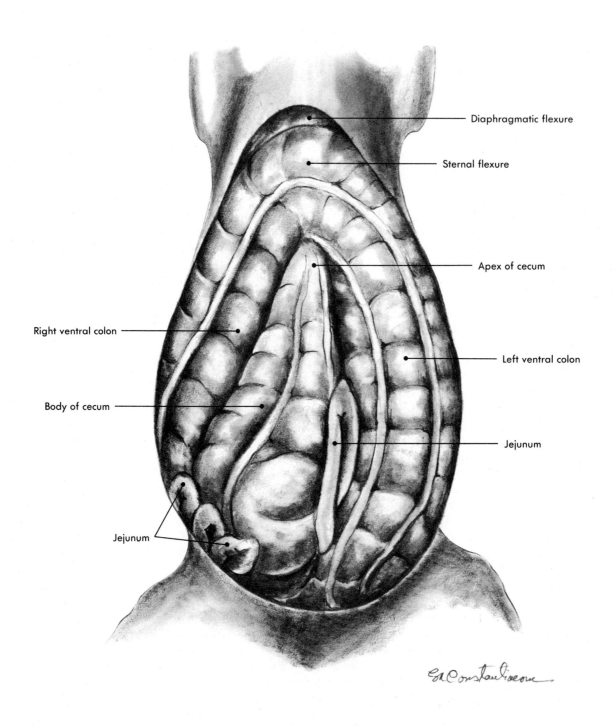

Diaphragmatic flexure

Sternal flexure

Apex of cecum

Right ventral colon

Left ventral colon

Body of cecum

Jejunum

Jejunum

Fig. 3-4

Abdominal viscera (ventral aspect).

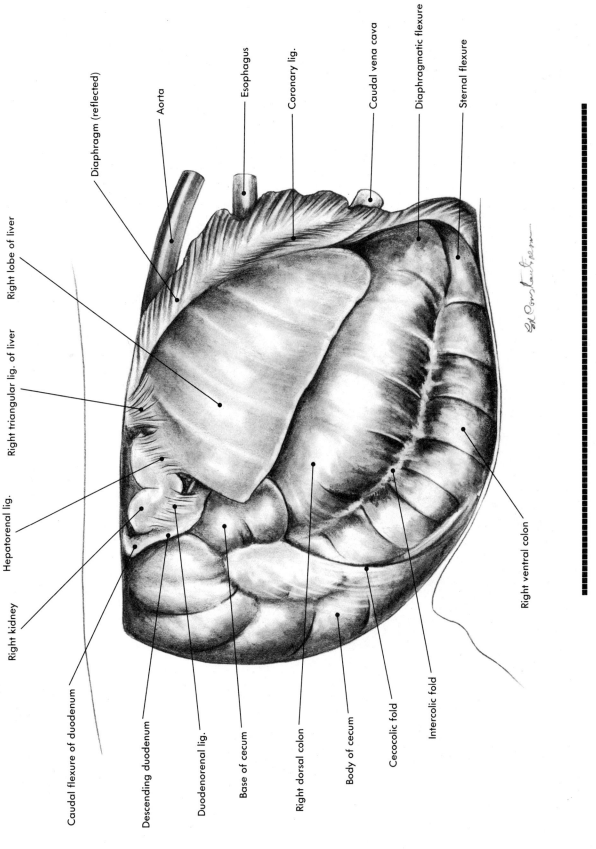

Caudal flexure of duodenum

Descending duodenum

Duodenorenal lig.

Base of cecum

Right dorsal colon

Body of cecum

Cecocolic fold

Intercolic fold

Right ventral colon

Right kidney

Hepatorenal lig.

Right triangular lig. of liver

Right lobe of liver

Diaphragm (reflected)

Aorta

Esophagus

Coronary lig.

Caudal vena cava

Diaphragmatic flexure

Sternal flexure

Fig. 3-5

Abdominal viscera (right aspect).

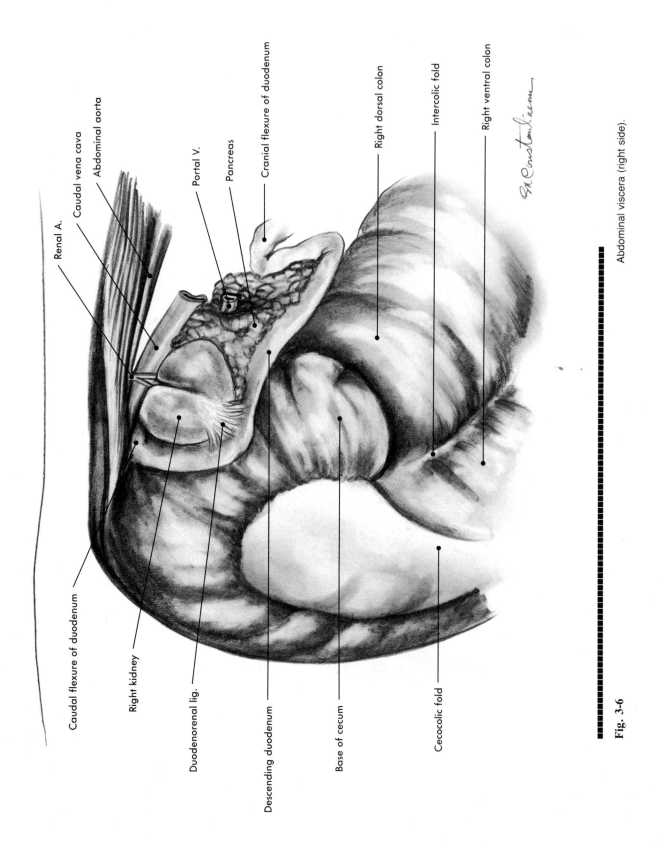

Renal A.

Caudal vena cava

Abdominal aorta

Portal V.

Pancreas

Cranial flexure of duodenum

Right dorsal colon

Intercolic fold

Right ventral colon

Caudal flexure of duodenum

Right kidney

Duodenorenal lig.

Descending duodenum

Base of cecum

Cecocolic fold

Abdominal viscera (right side).

Fig. 3-6

On the left side, palpate the base of the spleen and the splenorenal lig. (between the spleen and the left kidney), which is part of the greater omentum. Palpate the intercolic fold (between the left ventral colon and the left dorsal colon), which broadens toward the pelvic flexure, and palpate the pelvic flexure. Reflect the spleen caudally to liberate the greater curvature of the stomach and look at the origin of the greater omentum and at its first attachment on the medial aspect of the spleen (the gastrosplenic lig.). *Notice* the presence of the hilus on the visceral aspect of the spleen, which is the place of attachment of the gastrosplenic lig. The greater omentum continues toward the pelvic inlet as the superficial wall. It changes its direction, returning to the stomach as the deep wall. Palpate it between the caudal aspect of the stomach and the intestines, up to where it ends on the transverse colon and pancreas. Introduce a hand dorsal to the base of the spleen, and palpate the phrenicosplenic lig., which is also part of the greater omentum and which is located between the spleen and the left crus of the diaphragm. The last component of the greater omentum, the gastrophrenic lig., can be palpated between the dorsal end of the greater curvature of the stomach and the crura of the diaphragm.

Remember! The greater omentum is represented by the gastrophrenic, gastrosplenic, phrenicosplenic, and splenorenal ligg. and the superficial and deep walls.

Push a hand deeper toward the dorsal wall of the abdominal cavity to palpate the origin of the mesentery (the root of the mesentery), which is on the left side of the aorta. From the root of the mesentery, palpate the mesojejunum, which is attached to the lesser curvature of the jejunum. Caudal to the mesojejunum and still on the dorsal wall of the abdominal cavity, palpate the origin of the descending mesocolon, attached to the lesser curvature of the descending colon (Fig. 3-7).

The following dissection steps describe the removal of the abdominal viscera. Two techniques are suggested. In the first technique, all of the large intestine (except the rectum—that is, the cecum and ascending, transverse, and descending colon), the ileum, the jejunum, and the caudal segment of the duodenum are removed as a unit, followed by the stomach and spleen together, and finally by the liver with the cranial segment of the duodenum and the pancreas or, if possible, all of the viscera starting with the stomach and ending with the pancreas together. The second technique starts with the removal of the liver, continues with the stomach, the duodenum, the spleen, the rest of the small intestine (jejunum and ileum), the descending colon, and finally the cecum, the ascending colon, and the transverse colon. A detailed description of each technique follows.

The author suggests the first technique as the method of choice.

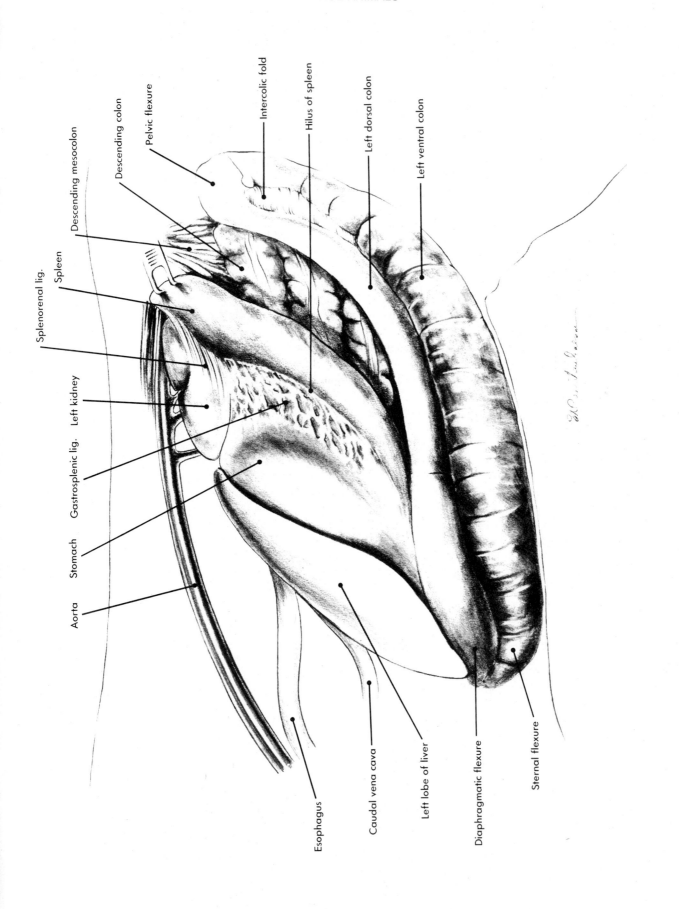

Intercolic fold

Hilus of spleen

Left dorsal colon

Left ventral colon

Pelvic flexure

Descending colon

Descending mesocolon

Spleen

Splenorenal lig.

Left kidney

Gastrosplenic lig.

Stomach

Aorta

Esophagus

Caudal vena cava

Left lobe of liver

Diaphragmatic flexure

Sternal flexure

Abdominal viscera (left side).

Fig. 3-7

The first technique. Liberate the cecum from its attachment to the dorsal wall of the abdominal cavity. Isolate the descending duodenum from the base of the cecum. Double ligate the duodenum caudal to the pancreas, and transect it between the two ligatures. Double ligate the descending colon caudal to the kidneys, after identifying the caudal mesenteric A. and the testicular or the ovarian A., and transect it in a manner similar to that for the duodenum. On the left side, introduce a hand and palpate the root of the mesentery (between the spleen, stomach, and kidney). Transect it as far as possible from the abdominal aorta (which is easily palpable), including the cranial mesenteric A., to preserve the cranial mesenteric ganglion and plexus. Separate the greater omentum from its attachments to the transverse colon, pancreas, and ascending duodenum and from its continuation with the lesser omentum. The entire intestinal mass, except the cranial part of the duodenum, can then be removed for later examination.

A variant of this technique may be utilized, leaving the transverse colon and the entire duodenum attached to the greater omentum and the mesoduodenum. To accomplish this, transect between double ligatures the transverse colon between the right dorsal colon and the descending colon, and the duodenum at the duodenojejunal flexure.

Double ligate the esophagus either in the thoracic or in the abdominal cavity, and transect it between the two ligatures. Identify and transect the celiac trunk (A.) far enough from the aorta to not damage the celiac ganglia and plexus. Retain the cranial duodenum, cranial flexure, and part of the descending duodenum with the stomach, pancreas, and liver for examination of the bile duct and pancreatic ducts. Transect the splenorenal, phrenicorenal, and gastrophrenic ligg.

Transect the triangular ligg., the coronary and falciform ligg., and the caudal vena cava as close as possible to the foramen venae cavae (of the diaphragm) and 5 cm caudal to the dorsal border of the liver. Remove the spleen, stomach, liver with the duodenum, and pancreas, and place together on the table.

Examine the continuity of the falciform lig. on the floor of the abdominal cavity toward the umbilicus, where it meets the median vesical lig. (Fig. 3-8, *A*).

Examine the two adrenal glands, the two kidneys surrounded by the retroperitoneal fat (capsula adiposa) and the visceral peritoneum covering only their ventral aspect. Because of the uncomfortable examining position of the sublumbar structures, transect the abdominal aorta caudal to the aortic hiatus (of the diaphragm). Transect both aorta and caudal vena cava caudal to the kidneys and cranial to the caudal mesenteric A.; transect the lumbar Aa., freeing the great vessels from any other attachment to the sublumbar area; and liberate the kidneys and adrenal glands from the perirenal fat, fibrous capsule, and peritoneum. Transect the ureters a short distance from where they leave the kidneys, and place all these structures on a tray.

Examine the relationship between the kidneys, adrenal glands, renal lnn., vessels supplying the kidneys (the renal A.V.), and origin of each ureter. Identify the hilus of the kidneys, and *notice* the position of the hilus and the shape of each kidney (Fig. 3-9, *A* to *C*).

Make a longitudinal section in each kidney from the lateral border toward the hilus, and remove the renal capsule. The renal sinus (with the two terminal recesses lined by the renal pelvis) and the renal crest are surrounded by the cortex and the medulla. Identify the macroscopic structures as they are shown in Fig. 3-9, *B*. Make a longitudinal section in the adrenal glands, and examine the capsule, cortex, and medulla.

Notice that the horse has a smooth unipyramidal type of kidney.

Identify and dissect the celiac ganglia and plexus (around the origin of the celiac A.), the cranial mesenteric ganglion and plexus (around the origin of the cranial mesenteric A.), and the renal ganglia and plexuses (between the renal Aa. and the hilus of the kidneys). All these sympathetic structures are visible on the ventral aspect of the aorta (Fig. 3-10, *A* and *B*).

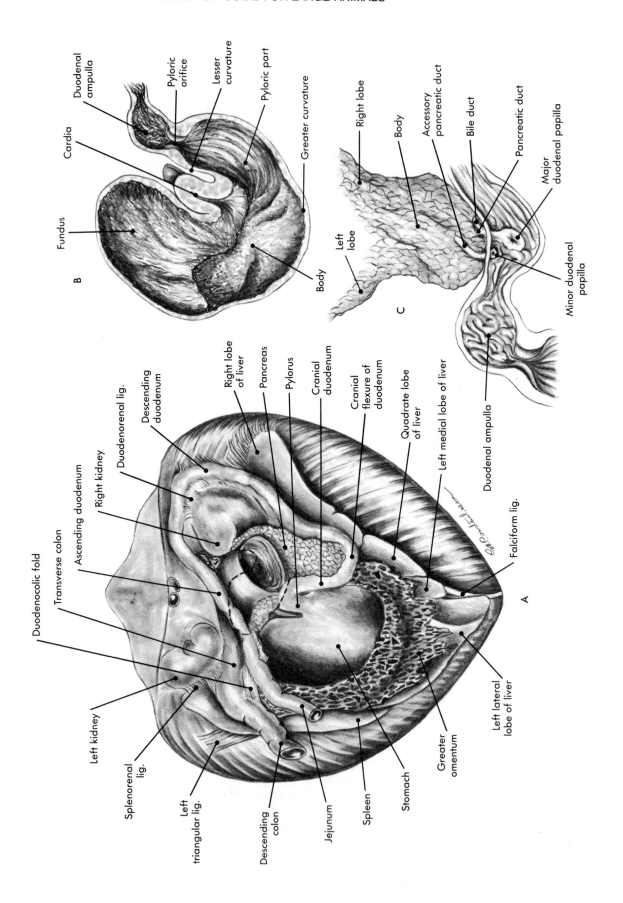

Duodenal ampulla
Pyloric orifice
Lesser curvature
Pyloric part
Cardia
Greater curvature
Fundus
Body
B

Right lobe
Body
Accessory pancreatic duct
Bile duct
Pancreatic duct
Major duodenal papilla
Left lobe
Minor duodenal papilla
C

Right lobe of liver
Pancreas
Pylorus
Cranial duodenum
Cranial flexure of duodenum
Quadrate lobe of liver
Left medial lobe of liver
Duodenal ampulla
Descending duodenum
Duodenorenal lig.
Right kidney
Ascending duodenum
Transverse colon
Duodenocolic fold
Left kidney
Splenorenal lig.
Left triangular lig.
Descending colon
Jejunum
Spleen
Stomach
Greater omentum
Left lateral lobe of liver
Faliciform lig.
A

Fig. 3-8   **A,** Stomach and associated structures. Dotted line indicates the boundaries of the pancreas. **B,** Stomach, opened. **C,** Pancreas.

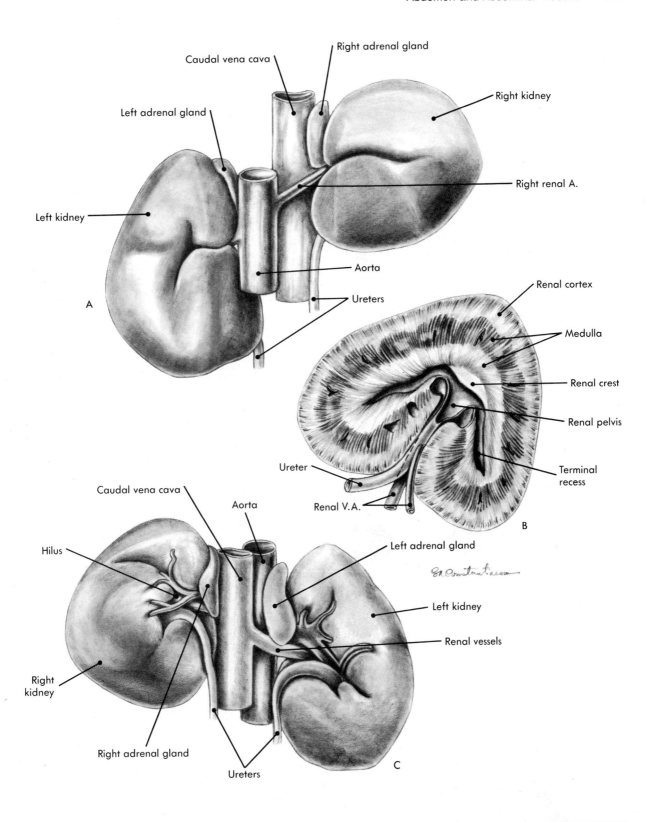

A, Dorsal aspect of kidneys in place. **B,** Longitudinal section of right kidney. **C,** Ventral aspect of kidneys in place.

Fig. 3-9

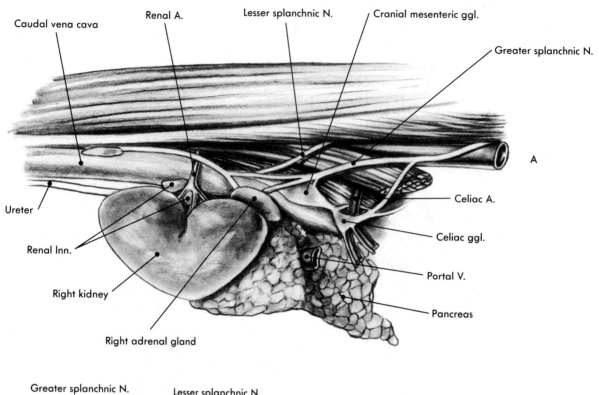

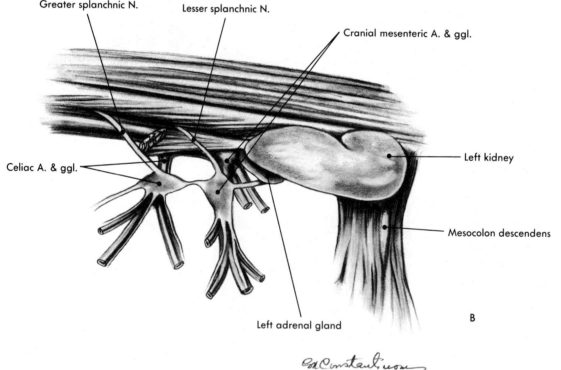

Fig. 3-10   **A,** Topography of right kidney. **B,** Topography of left kidney.

The second technique. Leaving the triangular ligg. of the liver still attached to the diaphragm, carefully remove the rest of the diaphragm from its costal insertions. Leave the crura of the diaphragm in place, and retain a strip of diaphragm approximately 10 cm wide, corresponding to the aortic hiatus (dorsally), the esophageal hiatus (in the middle), and the foramen venae cavae (ventrally), up to the sternal attachment of the diaphragm.

Double ligate the esophagus in the thoracic cavity close to the esophageal hiatus, and transect it between the two ligatures. Liberate the esophagus from the diaphragm and the esophageal impression of the liver by pulling the stomach backward. At the same time the gastrophrenic lig. will be broken.

Transect the left crus of the diaphragm at the level of the phrenicosplenic lig. Transect the diaphragm around the aortic hiatus, leaving the abdominal aorta in place. Transect the two triangular ligg. after examining their relationship to the diaphragm. Complete the technique on the right side.

Transect the lesser omentum between the liver and the stomach (the hepatogastric lig.) and between the liver and the cranial part of duodenum (the hepatoduodenal lig.). Transect the hepatorenal lig. (between the caudate process of the liver and the right kidney), and at this level, transect the right crus of the diaphragm. Palpate the caudal vena cava, which travels toward the dorsal border of the liver, and transect it there. Palpate the visceral aspect of the liver, and identify the hilus with the portal V., hepatic A. and common hepatic duct (the bile duct). Transect the vessels and the duct as far as possible from the liver. Remove the liver with the median strip of the diaphragm from the abdominal cavity, and place it on a tray.

Transect the renosplenic and the phrenicosplenic ligg. Double ligate the duodenum between the cranial flexure and the descending part, and transect it between the two ligatures. Liberate the stomach and the attached duodenum from the pancreas and the right dorsal colon. Transect the celiac A. far enough from the aorta to preserve the celiac ggl., so that the stomach, cranial duodenum, and spleen are free and easy to remove.

Free the duodenum from all its attachments up to the duodenojejunal flexure. The ileum can be identified by the ileocecal fold attached to its greater curvature, whereas the mesoileum (a segment of the mesentery) is attached to its lesser curvature. Transect the small intestine at the end of the ileocecal fold and between two ligatures. Notice that the mesoileum is the continuation of the mesojejunum. To remove the jejunoileum, it is necessary to palpate the root of the mesentery and to transect it far enough from the aorta to preserve the cranial mesenteric ggl. Remove and place the small intestine on a table.

Identify the transverse colon and the junction with the descending colon. Double ligate the latter, and transect it between the ligatures. Follow the course of the descending colon toward the pelvic inlet, double ligate it, and transect it as performed above. Identify the dorsal (lumbar) attachment of the descending mesocolon and transect it there. Remove the descending colon, and compare it to the small intestine.

The last viscera to be removed are the cecum, the ascending colon with its four segments, and the transverse colon. Isolate them from their attachments on the dorsal wall of the abdominal cavity and on the right kidney, and remove and place them on a low table or on the floor.

Regardless of which technique is utilized, all the viscera removed from the abdominal cavity is subjected to external and internal examination. The blood supply of the viscera is described on pp. 109-111. The sympathetic ggl. and plexuses are dissected with the kidneys and adrenal glands.

Begin with the liver. Dissect the portal V., the hepatic A., and the bile duct, and identify their major branches. *Notice* that the bile duct is still connected to the duodenum, if the first technique of exenteration was used. The bile duct fuses with the pancreatic duct before opening inside the hepatopancreatic ampulla within the major duodenal papilla. Make an incision along the greater curvature of the cranial duodenum close to the pylorus to identify the major and the minor duodenal papilla (see Figs. 3-8, *B* and *C*). Identify the portal lnn., which are scattered along the hilus. Examine the visceral and the diaphragmatic aspects of the liver (Fig. 3-11, *A* and *B*), with the hepatogastric lig. still in place or transected (depending on the technique of exenteration), and notice the "impressions" of other viscera on the embalmed liver. Try to discern the relationship between the liver and these viscera. Look at the fissures and at the limits between the lobes. *Notice* that the horse is the only species dealt with that does not have a gall bladder. The number of lobes is similar to the rabbit (left lateral, left medial, quadrate, right, and caudate). Examine the diaphragmatic aspect of the liver, and identify the caudal vena cava, the coronary lig. (dorsal to the vein), the falciform lig. (ventral to the vein), and the esophageal impression.

Remember! The round lig. of the liver lies in the free edge of the falciform lig.

Notice the branches of the celiac A., which supplies the stomach, spleen, and liver. Do not transect them (if the first technique of exenteration was used); if the second technique was used, the hepatic A. was previously transected. Examine the gastric lnn. located on the lesser curvature of the stomach.

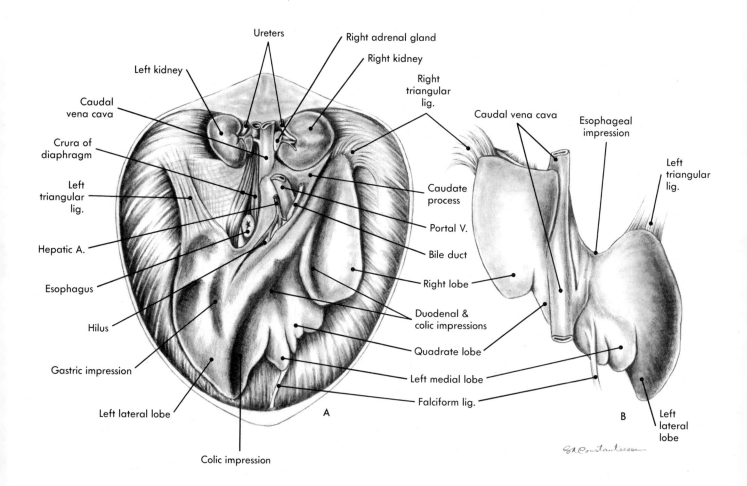

Fig. 3-11

A, Visceral aspect of liver. **B,** Diaphragmatic aspect of liver.

Unfold and spread out the jejunum to examine the jejunal lnn. (close to the root of the mesentery) and the arterial and venous arcades, which are 2 to 4 cm away from the lesser curvature of the jejunum. *Notice* its smooth surface. Near the cecum, on the greater curvature of the small intestine, a small peritoneal fold is observed. This is the ileocecal fold, which separates the jejunum from the ileum. The latter is attached by the mesoileum from the lesser curvature and has no delimitation between it and the mesojejunum.

Examine the cecum and its four longitudinal bands (Fig. 3-12). Incise the base of it along the greater curvature. Reflect the lateral slip and look at the ileocecal and cecocolic communications: both are provided with mucosal folds, and in addition, an ileal papilla is present (Fig. 3-13, *A* to *C*). Examine the four longitudinal bands and the sacculations of the ventral colon, as well as the sternal flexure between the right and left colon. Examine the pelvic flexure and the dorsal colon. The left dorsal colon has only one band on its lesser curvature (the line of attachment of the intercolic fold) and no sacculations, whereas the right dorsal colon has three longitudinal bands and sacculations. *Notice* the diaphragmatic flexure between the left and right dorsal colon. The cecocolic fold was previously examined. Examine the cecal and the colic lnn.

A sudden reduction in lumen size of the right dorsal colon is observed at the medial aspect of the base of the cecum. The following segment is the transverse colon, without sacculations or bands (sometimes one band is outlined). Unfold and spread the descending colon on the table; it has sacculations and two bands. Examine the arterial and venous arcades of the left colic A.V., lying on the lesser curvature of the descending colon. Compare with the similar arcades of the jejunum (Fig. 3-14). Observe also the colic lnn. Examine the spleen with the base, the two borders, and the apex. Inspect its medial aspect, including the hilus with the splenic A.V. and the splenic lnn.

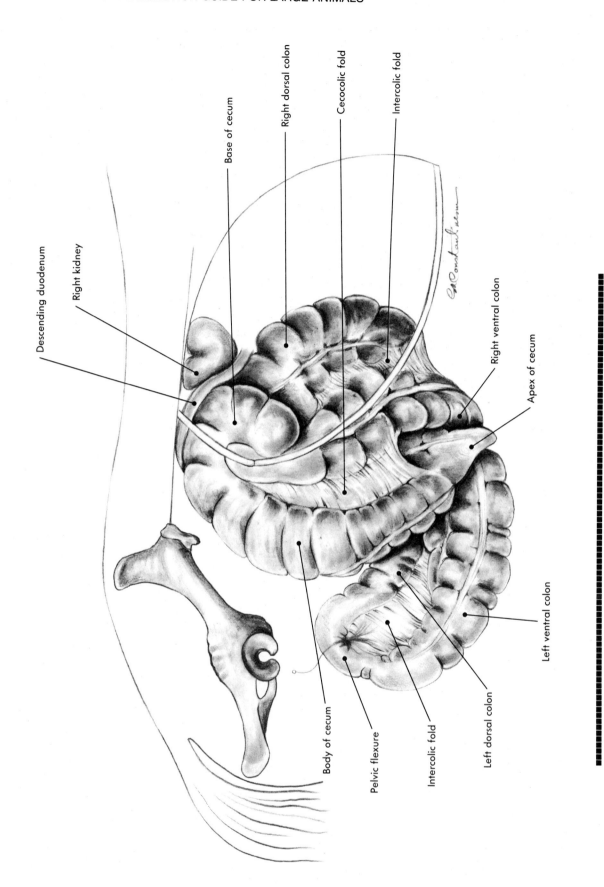

Fig. 3-12   Most of the large intestine (right side).

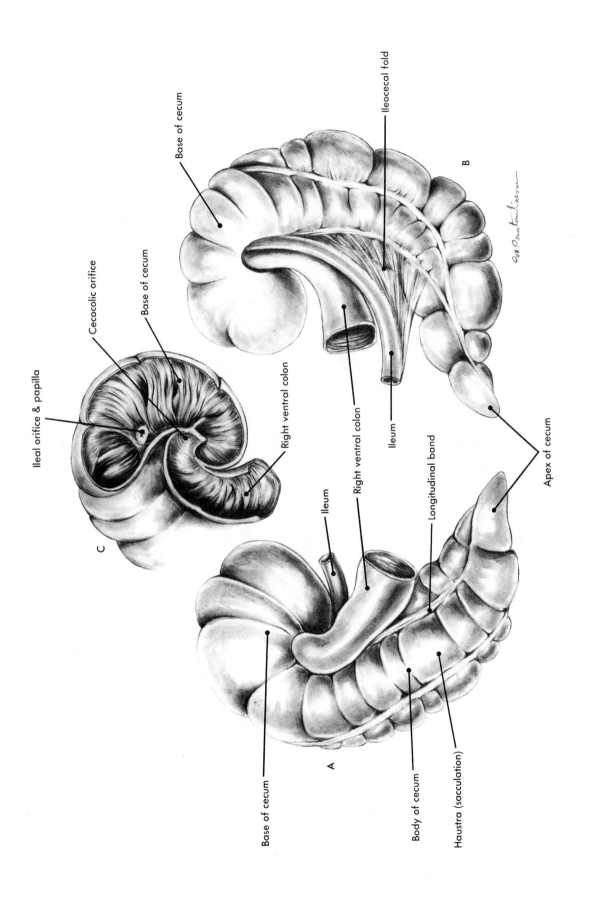

A, Cecum (right aspect). **B**, Ileum and cecum (left aspect). **C**, The communications of the cecum.

Fig. 3-13

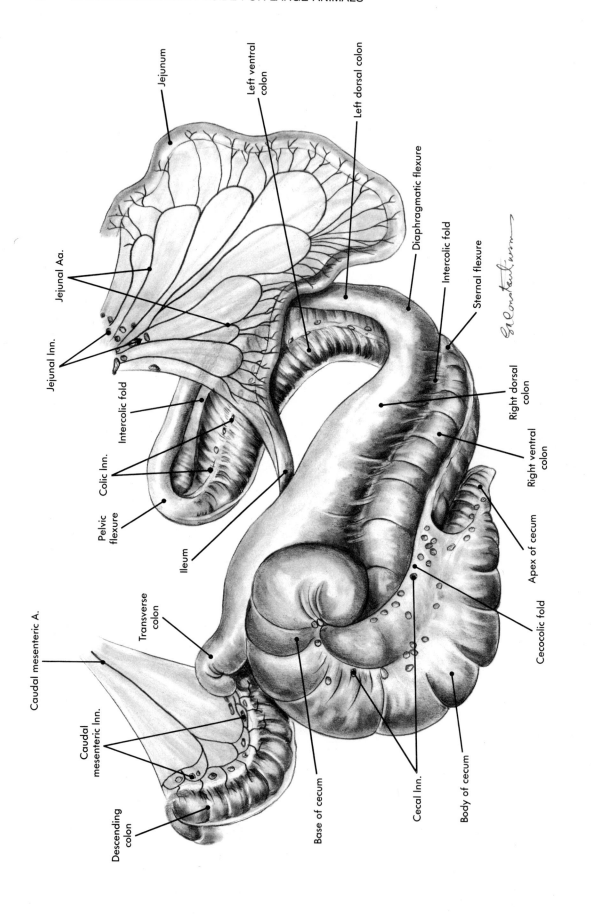

Jejunum

Left ventral colon

Left dorsal colon

Diaphragmatic flexure

Intercolic fold

Sternal flexure

Jejunal Aa.

Jejunal Inn.

Right dorsal colon

Intercolic fold

Colic Inn.

Right ventral colon

Pelvic flexure

Apex of cecum

Ileum

Cecocolic fold

Caudal mesenteric A.

Transverse colon

Caudal mesenteric Inn.

Descending colon

Base of cecum

Cecal Inn.

Body of cecum

Intestine of the horse.

Fig. 3-14

Dissect the celiac A. following its branches and *notice* the anastomoses that are indicated in Fig. 3-15 by circles.

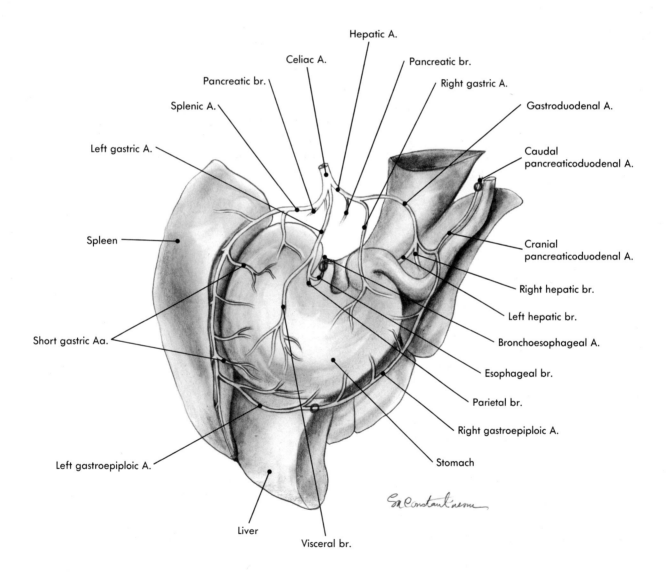

Celiac A. in the horse.

Fig. 3-15

Make an incision along the greater curvature of the stomach from the cardia to the pylorus, including the latter. Examine the margo plicatus (which separates the glandular from the nonglandular mucous membrane of the stomach), the cardiac and the pyloric sphincters, and especially the mucosal folds. The cardiac fold allows movement of ingesta *only* from the esophagus to the stomach and *never* back, whereas the pyloric folds allow transit from the stomach to the duodenum (Fig. 3-8, *B*).

From the cranial mesenteric A. come branches that supply the whole intestine up to the descending colon. Because of the complexity of the arteries supplying the intestines, certain correlations between the viscera and the name of the arteries are suggested. The "colic branch" supplies the first half of the ascending colon (the ventral right and left colon), whereas the "right colic A." supplies the second half of the ascending colon (the dorsal left and right colon). The middle colic A. supplies the middle portion of the colon, or the transverse colon.

The first branch of the cranial mesenteric A. is the caudal pancreaticoduodenal A., which anastomoses with the cranial pancreaticoduodenal A. (from the hepatic A.). The jejunal Aa. supply the jejunum, whereas the ileal Aa. supply the ileum. The ileocolic A. supplies the cecum and the ventral colon (notice the lateral and medial cecal Aa.). The mesenteric ileal br. supplies the ileum on its lesser curvature (where the mesentery is attached). The descending colon is supplied by the left colic A. from the caudal mesenteric A.

The following anastomoses occur: between the caudal pancreaticoduodenal A. and the cranial pancreaticoduodenal A. (from the hepatic A.); between the last jejunal A. and the ileal Aa.; between the ileal Aa. and the mesenteric ileal br. (A.); between the colic br. and the right colic A.; and between the middle colic A. and the left colic A. (from the caudal mesenteric A.). The anastomoses are indicated in Fig. 3-16 by circles.

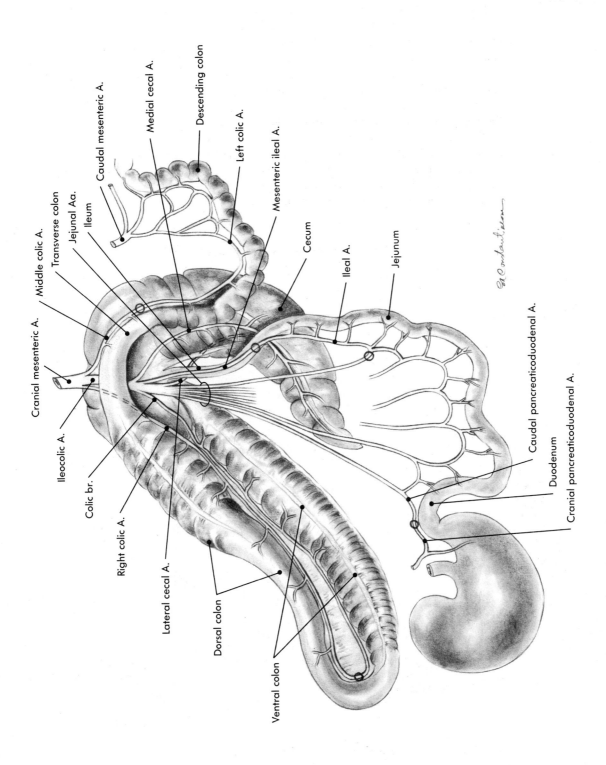

Fig. 3-16

Cranial and caudal mesenteric Aa. in the horse.

Large ruminants

Reflect the skin from the abdomen ventrally as in the horse.

Cautiously dissect the dorsolateral and the ventrolateral (both cutaneous) branches of the costoabdominal and first two lumbar spinal nerves.

Dissect the caudal extent of the cutaneus trunci M. and reflect it ventrally. Palpate and dissect the subiliac ln., which is in a similar location and position as in the horse; then identify the lateral cutaneous femoral N. and the caudal branches of the deep circumflex iliac A.V.

Similar to the dissection of the horse, dissect a small area of the abdominal tunic and sever the muscular attachments and the aponeurosis of the external abdominal oblique M., reflecting the muscle ventrally. The retractor costae and the internal abdominal oblique Mm. are now exposed. Identify the nerves lying on the muscle (Fig. 3-17).

Transect the origin and insertions of this muscle, which are similar to that in the horse, and reflect it ventrally. Identify the nerves and the vessels within the area.

Transect the transversus abdominis M. and reflect it ventrally, taking the same precautions as in horse with regard to the close relationship to the transverse fascia and the parietal peritoneum (Fig. 3-18, A and B).

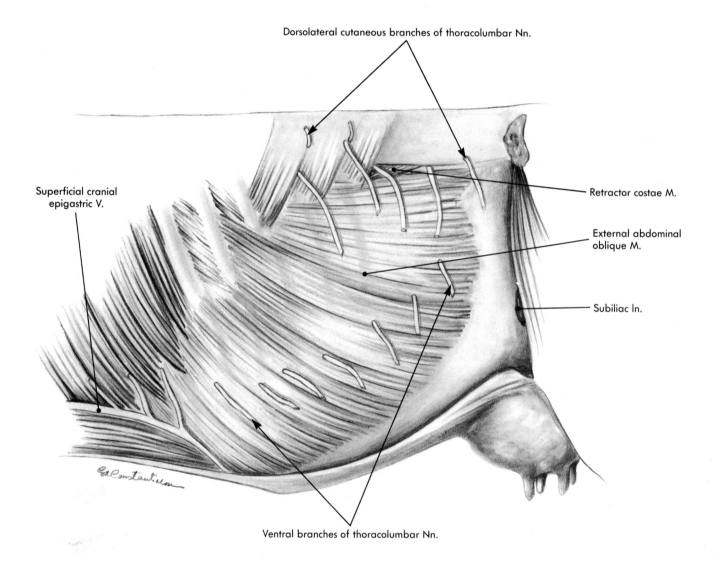

Dorsolateral cutaneous branches of thoracolumbar Nn.

Superficial cranial epigastric V.

Retractor costae M.

External abdominal oblique M.

Subiliac ln.

Ventral branches of thoracolumbar Nn.

Fig. 3-17

First level, abdomen (left side).

On the ventral area of the transverse processes of the lumbar vertebrae, gently reflect the psoas major and minor Mm. ventrally and expose part of the proximal segments of the ventral branches of nerves L_1 and L_2 lying between them, and the quadratus lumborum and intertransversarii lumborum Mm. Transect the intertransversarii lumborum Mm., the intertransverse ligg., and the quadratus lumborum M. up to the intervertebral foramina to expose the ventral branch joining the dorsal branch of each lumbar nerve.

The abdominal topography (the projection of viscera on the lateral walls of the abdomen) is as follows:

I. **Left side** (Fig. 3-19)
  A. Rumen
    • Almost the whole wall of abdominal cavity
  B. Reticulum
    • On the ventral wall of the abdominal cavity, between ribs VI and VII or VIII
  C. Spleen
    • Between the tendinous center of the diaphragm and rumen up to the reticulum and 10 cm in breadth
  D. Abomasum
    • Between the reticulum and the ventral and cranial sacs of the rumen, on the floor of the abdominal cavity
II. **Right side** (Fig. 3-20)
  A. Liver
    • Between the diaphragm and a convex line joining the last rib dorsally (the caudate process of the caudate lobe) to rib VI ventrally
  B. Gallbladder
    • Between ribs X and XI, ventral to the liver
  C. Reticulum
    • At the cranioventral end of the liver, between ribs VI and VII
  D. Omasum
    • Between ribs VII and X in their ventral half
  E. Abomasum
    • Behind the omasum in the next two intercostal spaces, limited by the costal arch and the ventral wall of the abdominal cavity
  F. Duodenum
    • Cranial part in the 10th intercostal space
    • With the descending part lying horizontally in the flank, from the caudate process of the liver toward the tuber coxae
    • Caudal flexure close to tuber coxae
  G. Pylorus
    • Between the abomasum and cranial part of the duodenum, ventral to the gallbladder
  H. Ascending colon
    • Proximal loop lying in the flank, ventral to the duodenum
    • Distal loop lying in the flank, dorsal to the duodenum
  I. Pancreas (right lobe)
    • Between the distal loop of the ascending colon, the descending colon, the descending duodenum, the kidney, and the caudate process of the liver
  J. Descending colon (Fig. 3-21, *A*)
    • Dorsal to the pancreas and distal loop of the ascending colon
  K. Cecum (Fig. 3-21, *B*)
    • In the flank and ventral to the proximal (sometimes to the distal) loop of the ascending colon and dorsal to the level of the last costochondral joint
  L. Jejunum
    • Lying in the remainder of the flank
  M. Right kidney
    • On the dorsal wall of the abdominal cavity, between the last rib and the transverse process of L_2 vertebra

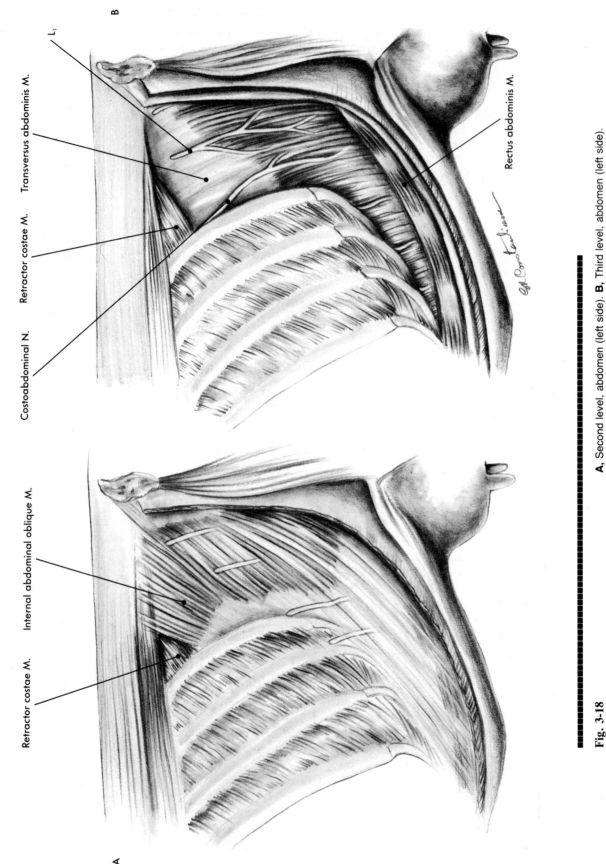

Fig. 3-18   **A,** Second level, abdomen (left side). **B,** Third level, abdomen (left side).

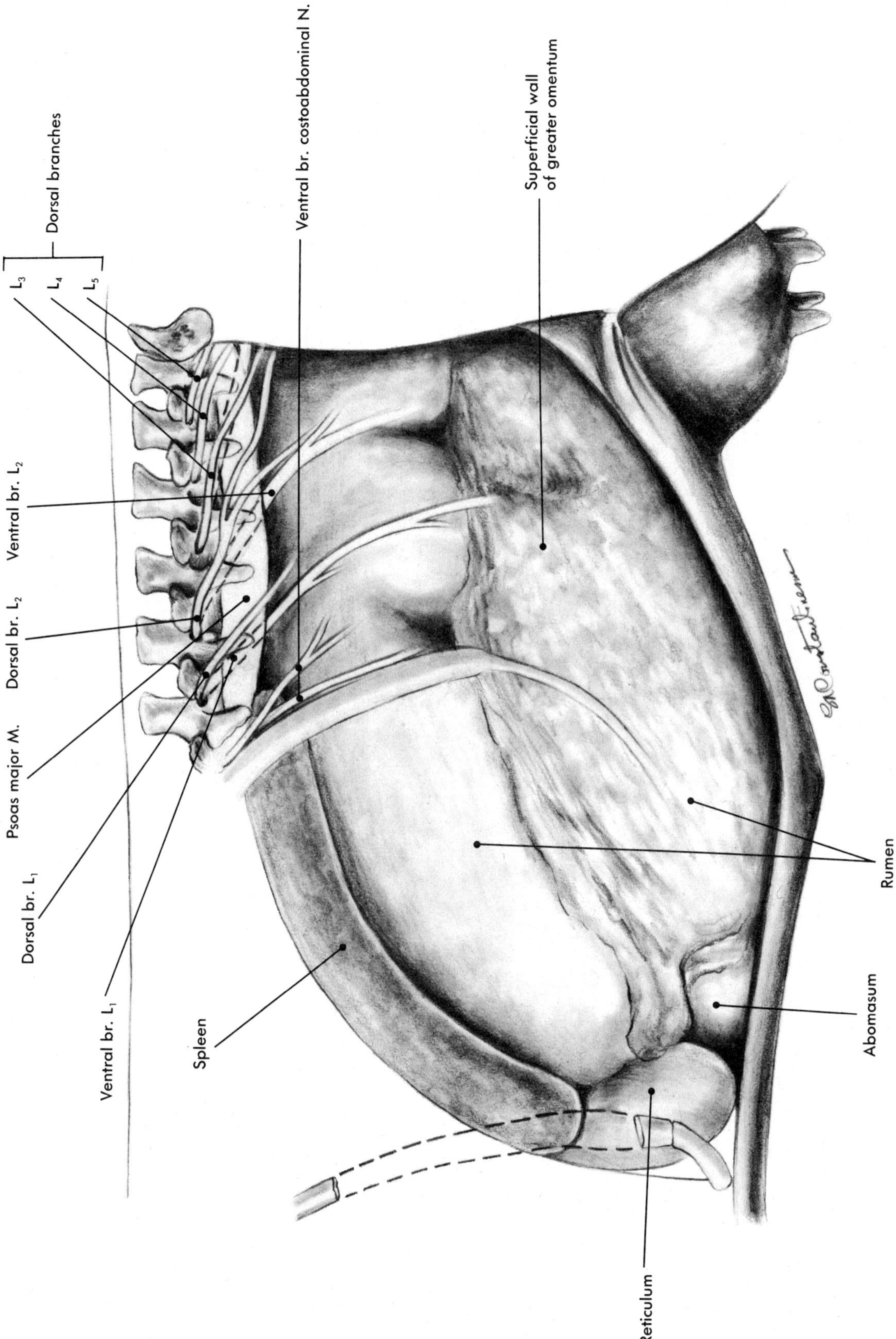

Dorsal branches

L₃
L₄
L₅

Ventral br. costoabdominal N.

Superficial wall
of greater omentum

Ventral br. L₂

Dorsal br. L₂

Psoas major M.

Dorsal br. L₁

Ventral br. L₁

Spleen

Rumen

Abomasum

Reticulum

Abdominal viscera (left side).

Fig. 3-19

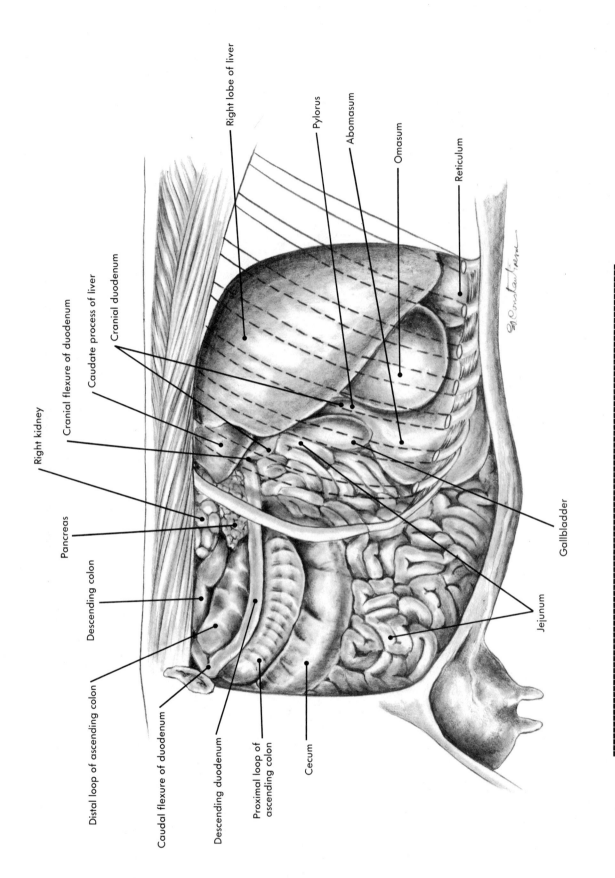

Right lobe of liver

Pylorus

Abomasum

Omasum

Reticulum

Right kidney

Cranial flexure of duodenum

Caudate process of liver

Cranial duodenum

Pancreas

Descending colon

Distal loop of ascending colon

Caudal flexure of duodenum

Descending duodenum

Proximal loop of ascending colon

Cecum

Jejunum

Gallbladder

Abdominal viscera (right side).

Fig. 3-20

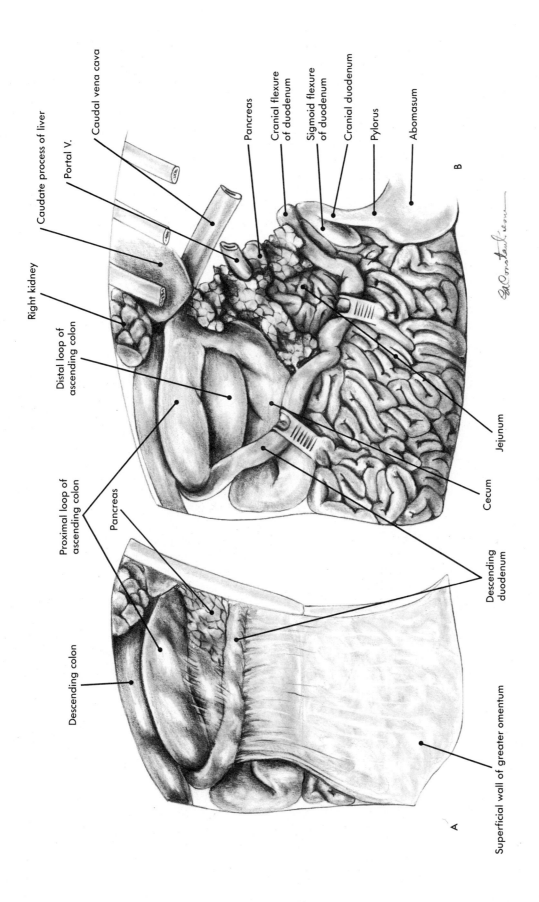

Caudal vena cava

Caudate process of liver

Portal V.

Pancreas

Cranial flexure of duodenum

Sigmoid flexure of duodenum

Cranial duodenum

Pylorus

Abomasum

Right kidney

Distal loop of ascending colon

Jejunum

Cecum

B

Proximal loop of ascending colon

Pancreas

Descending duodenum

Descending colon

Superficial wall of greater omentum

A

A, Abdominal viscera in flank (left aspect). **B,** Abdominal viscera in flank after removing the superficial wall of greater omentum (left aspect).

Fig. 3-21

On the left side, cut off ribs IX and XII and remove the intercostal Mm. from the last intercostal space. Remove the diaphragm from its costal insertions, and leave the left portion of a median strip extending over the cranial border of the spleen. To avoid damaging the phrenicosplenic lig., palpate it between the diaphragm and the dorsocaudal extremity of the spleen.

Identify the dorsal sac and the caudodorsal blind sac of the rumen, the left longitudinal groove and the superficial wall of the greater omentum attached to it, the reticulum, the ruminoreticular groove, and the fundus of the abomasum. Introduce your right hand between the caudal extent of the greater omentum and the pelvic inlet, and palpate the continuity of the greater omentum toward the right side.

On the right side, remove the same ribs and the intercostal muscles. Before removing the diaphragm, locate the right triangular lig. of the liver (which travels from the caudodorsal extent of the right lobe to the dorsolateral abdominal wall). Preserve the ligament and remove the rest of the diaphragm, leaving the right portion of a 2 to 3 cm wide median strip to the right of the caudal vena cava. The vein courses toward the foramen venae cavae, between the liver and the diaphragm. Palpate the right fold of the coronary lig. on the diaphragmatic aspect of the liver up to the foramen venae cavae, and then palpate the falciform lig. to the right, up to the notch for the round lig. This notch is not always distinct, but it is visible ventral to the gall bladder, on the ventral (right) border of the liver.

The superficial wall of the greater omentum lies between the pelvic inlet, the descending part of the duodenum, the gall bladder, the cranial part of the duodenum, the pylorus, and the greater curvature of the abomasum. The mesoduodenum extends dorsally to the descending duodenum, toward the left side, and passes under the distal loop of the ascending colon, the descending colon, and the right kidney and cranially up to the caudate process of the liver. It includes the right lobe of the pancreas.

Cautiously reflect as much as possible of the right lobe of the liver. Identify the cranial flexure of the duodenum, the cranial duodenum with the sigmoid flexure, and the pylorus. The ligament attached between the liver and the duodenum is the hepatoduodenal lig., which is continuous with the hepatogastric lig. and lies between the liver, the omasum, and the lesser curvature of the abomasum.

Remember! Both the hepatogastric and the hepatoduodenal ligg. are parts of the lesser omentum. The lesser omentum continues with the superficial wall of the greater omentum over the cranial duodenum, the pylorus, and the abomasum. Introduce your left hand over the caudal extent of the greater omentum toward the left side, and palpate the continuity of this structure.

Reflect the caudate process and continue to reflect the right lobe of the liver, identifying the portal V. between the body of the pancreas and the hilus of the liver. Introduce a finger over the mesoduodenum and the portal V. through the epiploic foramen into the omental bursa. Notice the transition between the mesoduodenum and the hepatoduodenal lig.

Transect the superficial wall of the greater omentum 10 cm ventral and parallel to the descending duodenum, and identify its two walls. This is the omental bursa. The deep wall of the bursa is the deep wall of the greater omentum. Introduce your hand into the omental bursa and explore the caudal recess, then palpate the fusion between the two walls, which closes the omental bursa caudally and dorsally. Pass a hand in a cranial direction up to the cranial duodenum. This is still the omental bursa. Cranial to the duodenum, your hand will slip between the lesser omentum (on the right side) and the rumen into the vestibule of the omental bursa. Within the most cranial extent of the vestibule, the omasum can be palpated. Notice that the epiploic foramen opens into the vestibule. Palpate the attachment of the deep wall to the superficial wall of the greater omentum before reaching the descending duodenum. Transect the deep wall in a manner similar to the transection of the superficial wall, and enter the supraomental recess, which corresponds to the peritoneal cavity. Identify the two leaves of the deep wall. The proximal loop of the ascending colon, the cecum, and the jejunum are now exposed. Sometimes the caudal extent of the proximal loop of the ascending colon, of the cecum, and of the jejunum exceed the supraomental recess caudally and are visible between the greater omentum and the pelvic inlet.

Palpate, isolate, and transect the hepatorenal lig. Bluntly isolate the kidneys and the adrenal glands (with the corresponding vessels), the ureters, and the renal lnn. by removing the perirenal fat. Identify the structures and their topographic relationships. Examine the renal lobes, which are characteristic for the large ruminants.

There are several different techniques for exenteration. In one technique, the liver is removed first, followed by the intestinal mass, and finally the stomach with the pancreas, the duodenum, and the spleen. In another technique, the intestinal mass is first removed, followed by the stomach, duodenum, pancreas, spleen, and liver as a unit; there are two choices: to remove first the stomach with the duodenum, pancreas and spleen, and then the liver, or vice versa. Removal of the intestinal mass first is the choice suggested by the author.

Transect the mesoduodenum dorsal to the descending duodenum, between the descending duodenum and the distal loop of the ascending colon. Double ligate the duodenum at the caudal flexure, and transect it between the ligatures. Identify the sigmoid colon, double ligate it, and transect it between the ligatures. Transect the descending mesocolon; identify and transect the root of the mesentery and the cranial mesenteric A. Cranial to the artery and surrounding it is the transverse colon. Transect the transverse mesocolon and the attachment of the deep wall of the greater omentum on the transverse colon. Transect the cranial extent of the superficial wall of the greater omentum on the cranial duodenum and the greater curvature of the abomasum. Transect the portal V. as close as possible to the pancreatic ring and remove the intestinal mass (Fig. 3-22, *A* and *B*).

For the first time, the spiral loop of the ascending colon will be exposed. Attempt to reconstitute the natural position of the intestines lying on a table. Then spread out the jejunum and mesojejunum and identify the centripetal and centrifugal coils (gyri) and the central flexure of the spiral loop; they are close to one another. The jejunum surrounds the spiral loop and continues on to the ileum, whose ileocecal fold connects its greater curvature to the cecum. Identify the jejunal lnn., which are close to the lesser curvature of the jejunum. Examine the distal loop of the ascending colon, the transverse colon, and the whole descending colon. The spiral colon and its relationships to the jejunum, mesojejunum, and left kidney can be examined in situ much better from the left side, after removing the rumen. The disadvantage of this technique is the difficulty in exposing the arteries.

Return to the previous position (with the right side of the intestines exposed), and properly identify each segment of the small and large intestine, the major branches of the cranial mesenteric A., and their divisions, characteristics, and anastomoses. There are some correlations between the names of the arteries and the structures supplied. The *branches* that are named "colic" supply the first half of the ascending colon (the proximal loop and the centripetal coils), whereas the *arteries* that are named "right colic" supply the second half of the ascending colon (the centrifugal coils and the distal loop). The middle colic A. supplies the middle segment of the colon, or the transverse colon. The mesenteric ileal br. supplies the ileum at the lesser curvature (corresponding to the attachment of the mesentery), whereas the antimesenteric ileal br. supplies the ileum at the greater curvature (the opposite side of the mesentery).

Also very important are the anastomoses between the caudal pancreaticoduodenal A. and the cranial pancreaticoduodenal A. (from the hepatic A.), between the last jejunal A. and the ileal Aa., between the ileal Aa. and the mesenteric ileal br., and between the middle colic A. and the left colic A. (form the caudal mesenteric A.). The anastomoses are highlighted in Fig. 3-23 by circles.

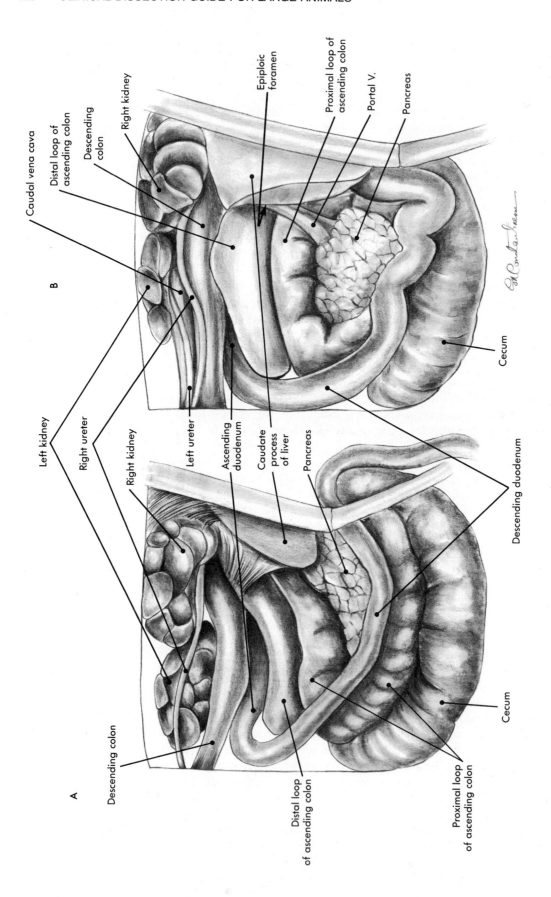

Fig. 3-22   **A,** Abdominal viscera in left flank with pancreas exposed. **B,** Abdominal viscera in left flank with wider exposure of the pancreas.

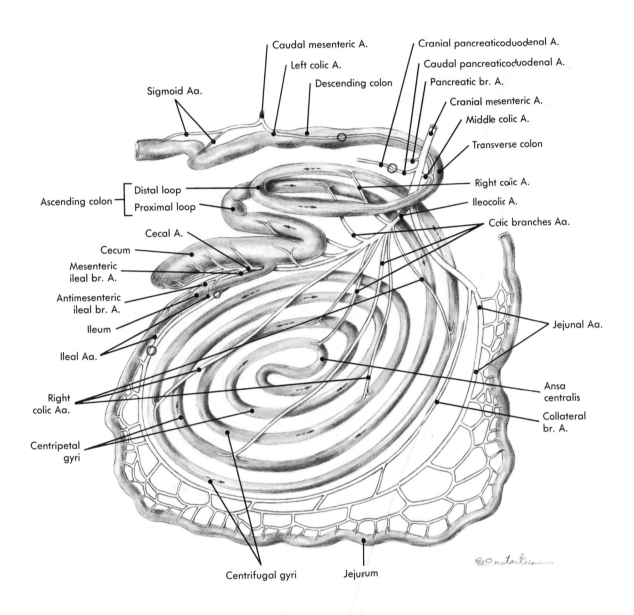

Cranial and caudal mesenteric Aa. in the large ruminants.

Fig. 3-23

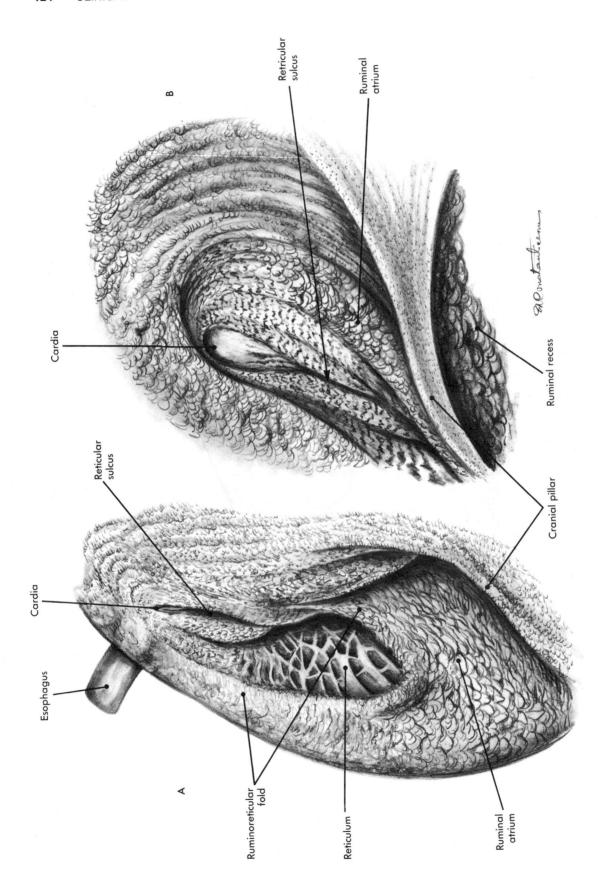

A, Reticular sulcus and adjacent structures. B, Reticular sulcus.

Fig. 3-25

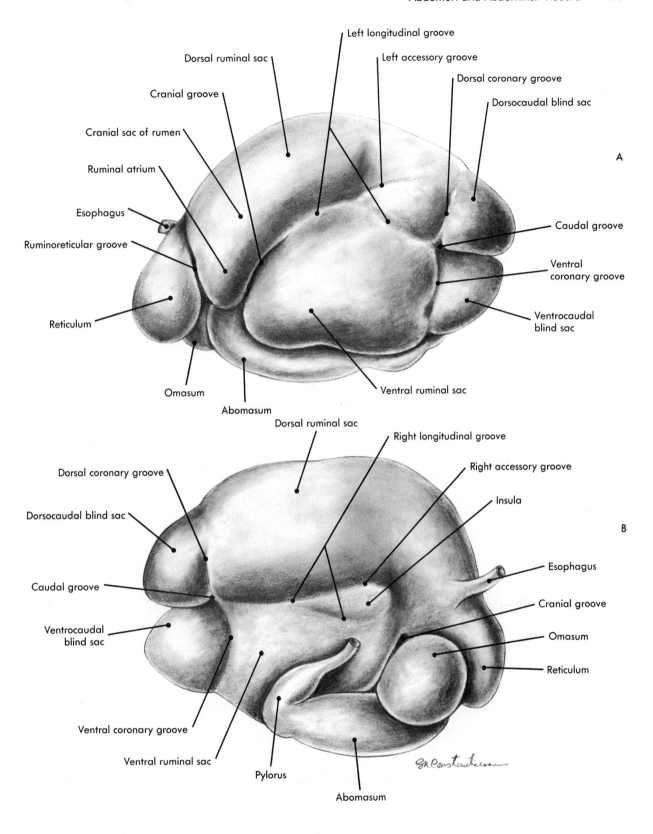

Left longitudinal groove

Left accessory groove

Dorsal ruminal sac

Cranial groove

Dorsal coronary groove

Dorsocaudal blind sac

Cranial sac of rumen

Ruminal atrium

A

Esophagus

Caudal groove

Ruminoreticular groove

Ventral coronary groove

Reticulum

Ventrocaudal blind sac

Omasum

Abomasum

Ventral ruminal sac

Dorsal ruminal sac

Right longitudinal groove

Dorsal coronary groove

Right accessory groove

Insula

B

Dorsocaudal blind sac

Esophagus

Caudal groove

Cranial groove

Ventrocaudal blind sac

Omasum

Reticulum

Ventral coronary groove

Ventral ruminal sac

Pylorus

Abomasum

A, Stomach (left aspect). **B,** Stomach (right aspect).

Fig. 3-26

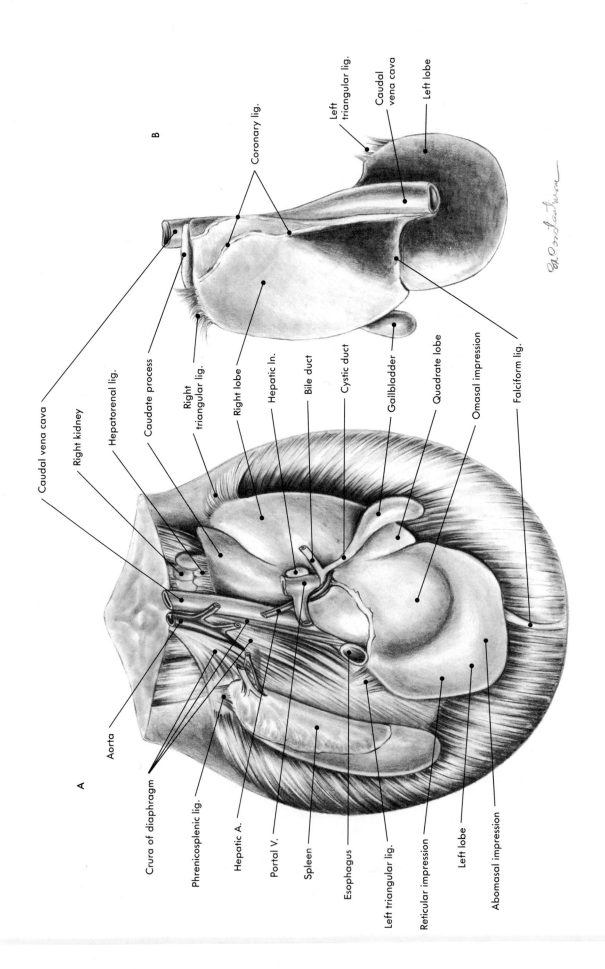

B

Coronary lig.

Left triangular lig.

Caudal vena cava

Left lobe

Caudal vena cava

Right kidney

Hepatorenal lig.

Caudate process

Right triangular lig.

Right lobe

Hepatic ln.

Bile duct

Cystic duct

Gallbladder

Quadrate lobe

Omasal impression

Falciform lig.

Aorta

A

Crura of diaphragm

Phrenicosplenic lig.

Hepatic A.

Portal V.

Spleen

Esophagus

Left triangular lig.

Reticular impression

Left lobe

Abomasal impression

Fig. 3-27

A, Visceral aspects of liver and spleen in place. B, Diaphragmatic aspect of liver.

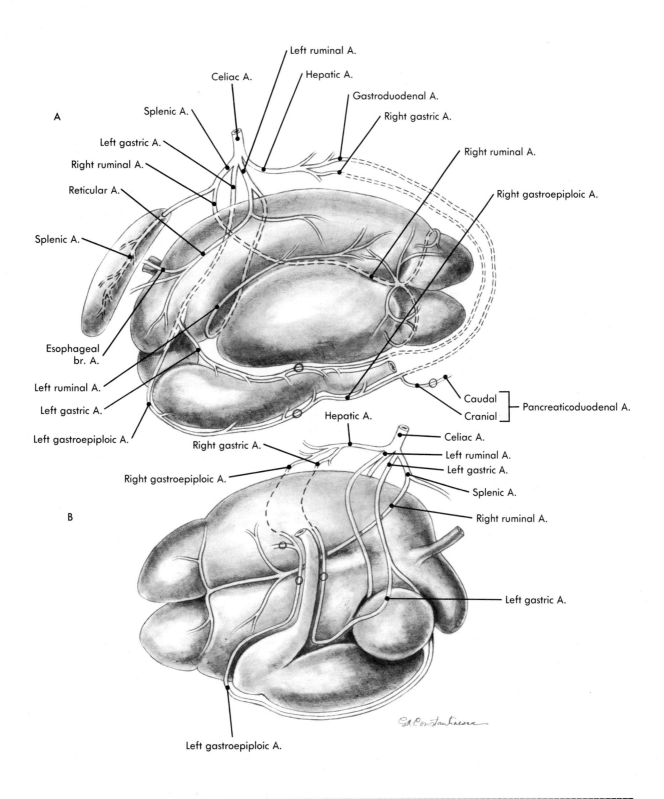

A, Arterial supply of large ruminants stomach (left aspect). **B,** Arterial supply of large ruminants stomach (right aspect).

Fig. 3-28

Make a continuous incision on the greater curvature of the reticulum, omasum, and abomasum including the pylorus; take out the ingesta, and wash the interior of these viscera. Examine the reticular, omasal, and abomasal grooves, which are parts of the gastric groove (see Figs. 3-24, 3-25, and 3-29). Examine the characteristics of the mucosa of these three compartments of the stomach, and *notice* the cellular aspect of the reticulum, the four categories of omasal folds (laminae) provided with marginal thickenings, the omasal pillar, the spiral folds of the abomasum, the pyloric sphincter, and the torus pyloricus (Fig. 3-29, *A* to *D*).

Pull the descending duodenum ventrally, together with the right lobe of the pancreas, and identify and isolate the accessory pancreatic duct. Make a longitudinal incision on the greater curvature of the duodenum, and expose the minor duodenal papilla (Fig. 3-30, *A* and *B*).

Examine the kidneys and the adrenal glands, and *notice* the twisted shape of the left kidney (as a consequence of the pressure of the rumen from the left to the right). Examine the relationship between the hilus and the three structures entering and exiting the kidneys (the renal A.V. and the ureter, respectively), as well as the direction of the renal vessels. Make a longitudinal section through each kidney at the level of the hilus, and examine the minor and major calices and the cortical and medullary parts of each lobe (Fig. 3-31, *A* to *C*).

Notice that the large ruminants have a lobated multipyramidal type of kidney.

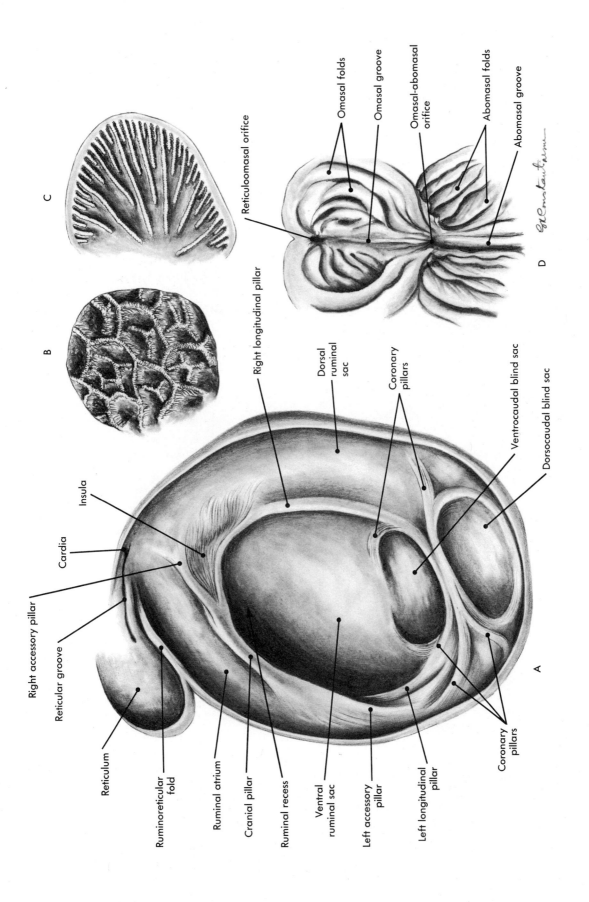

Right accessory pillar

Reticular groove

Reticulum

Ruminoreticular fold

Ruminal atrium

Cranial pillar

Ruminal recess

Ventral ruminal sac

Left accessory pillar

Left longitudinal pillar

Coronary pillars

Cardia

Insula

Right longitudinal pillar

Dorsal ruminal sac

Coronary pillars

Ventrocaudal blind sac

Dorsocaudal blind sac

Reticuloomasal orifice

Omasal folds

Omasal groove

Omasal-abomasal orifice

Abomasal folds

Abomasal groove

A

B

C

D

Fig. 3-29

A, Opened stomach (dorsal view). **B,** Reticulum (internal aspect). **C,** Omasum (internal aspect, transverse section). **D,** Omasum and abomasum (internal aspect).

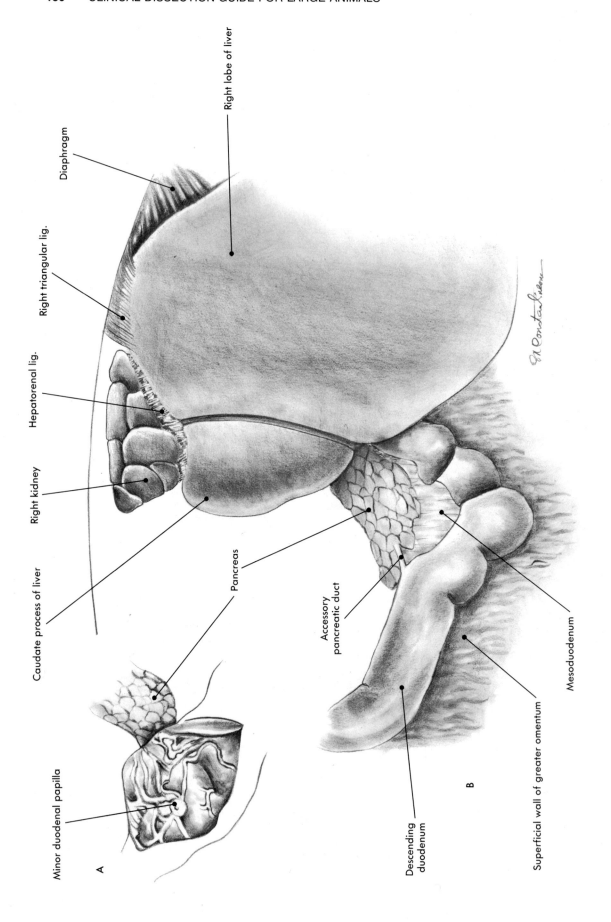

Diaphragm

Right lobe of liver

Right triangular lig.

Hepatorenal lig.

Right kidney

Caudate process of liver

Pancreas

Accessory pancreatic duct

Minor duodenal papilla

Descending duodenum

Mesoduodenum

Superficial wall of greater omentum

A

B

Fig. 3-30 **A,** Minor duodenal papilla. **B,** Accessory pancreatic duct.

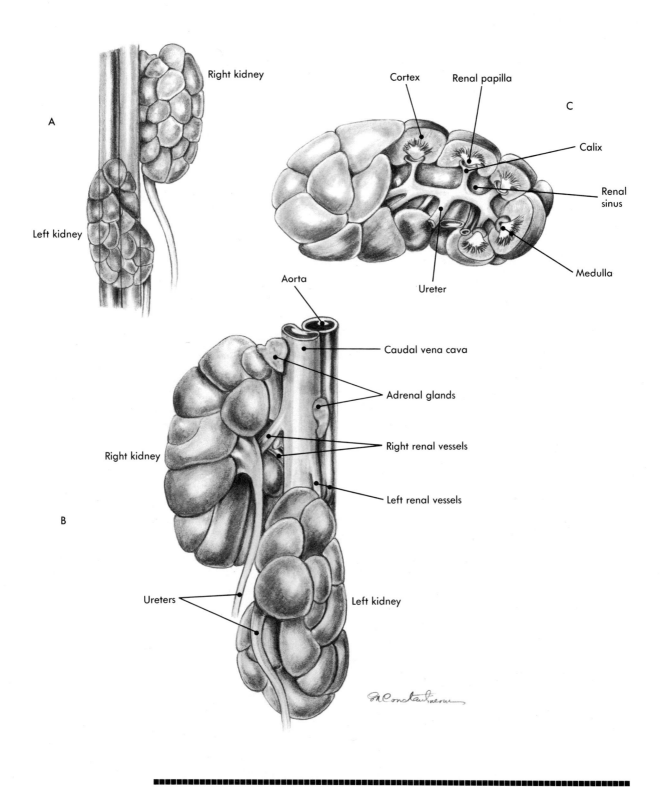

A, Kidneys in the large ruminants (dorsal view). **B,** Kidneys in the large ruminants (ventral view). **C,** Internal aspect of right kidney.

Fig. 3-31

Sheep
■■■■■■■■■
(Figs. 3-32
to 3-35)

The directions for dissection of the abdomen and abdominal viscera in the large ruminants are the same for the small ruminants.

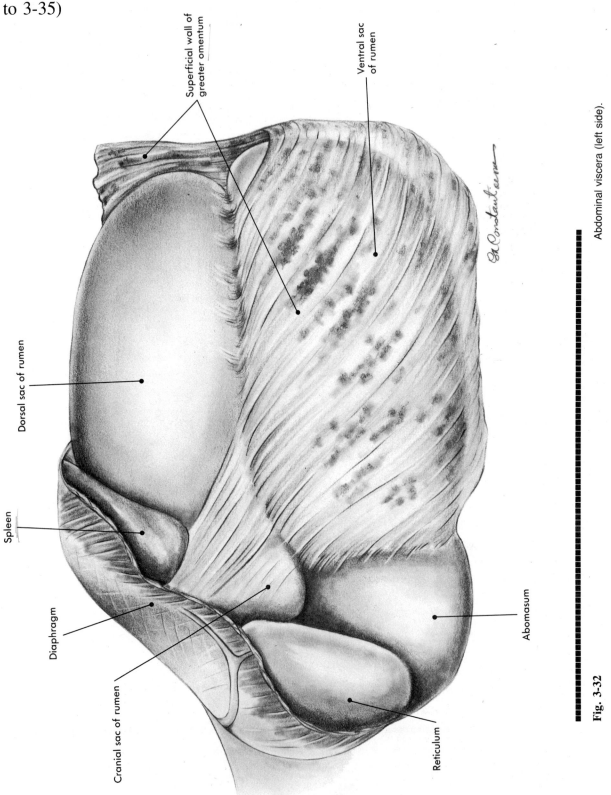

Fig. 3-32
Abdominal viscera (left side).

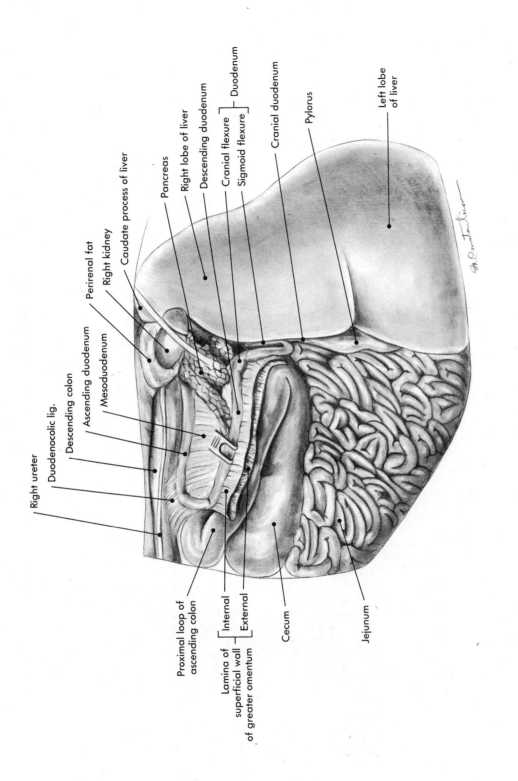

Right ureter

Duodenocolic lig.

Descending colon

Ascending duodenum

Mesoduodenum

Perirenal fat

Right kidney

Caudate process of liver

Pancreas

Right lobe of liver

Descending duodenum

Cranial flexure ⎤
Sigmoid flexure ⎦ Duodenum

Cranial duodenum

Pylorus

Left lobe of liver

Proximal loop of ascending colon

Internal ⎤
External ⎦ Lamina of superficial wall of greater omentum

Cecum

Jejunum

Superficial level of abdominal viscera after removing superficial wall of greater omentum (right aspect).

Fig. 3-33

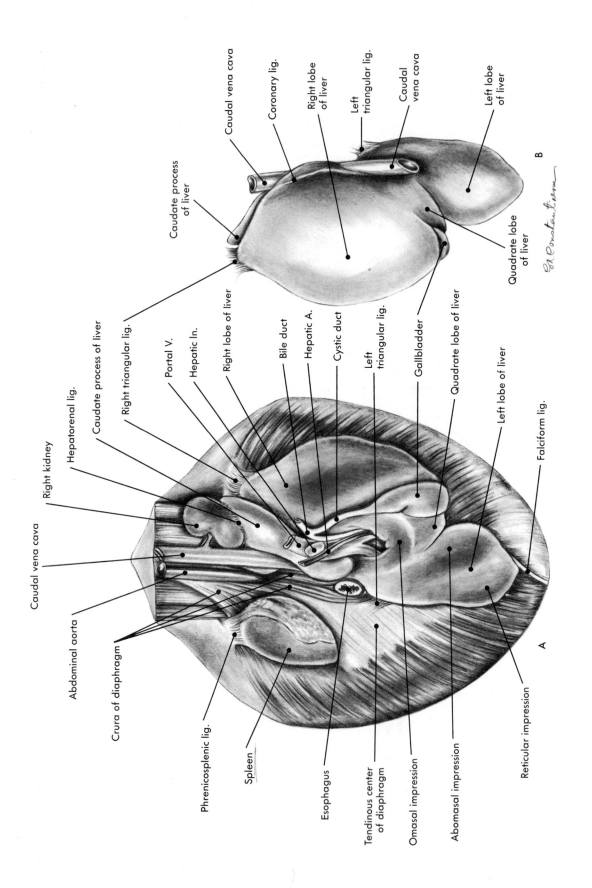

Fig. 3-34

A, Visceral aspect of liver and spleen in place. **B,** Diaphragmatic aspect of liver.

Caudal vena cava

Coronary lig.

Right lobe of liver

Left triangular lig.

Caudal vena cava

Left lobe of liver

Caudate process of liver

Quadrate lobe of liver

B

Caudate process of liver

Right triangular lig.

Caudate process of liver

Hepatorenal lig.

Right kidney

Portal V.

Hepatic ln.

Right lobe of liver

Bile duct

Hepatic A.

Cystic duct

Left triangular lig.

Gallbladder

Quadrate lobe of liver

Left lobe of liver

Falciform lig.

Caudal vena cava

Abdominal aorta

Crura of diaphragm

Phrenicosplenic lig.

Spleen

Esophagus

Tendinous center of diaphragm

Omasal impression

Abomasal impression

Reticular impression

A

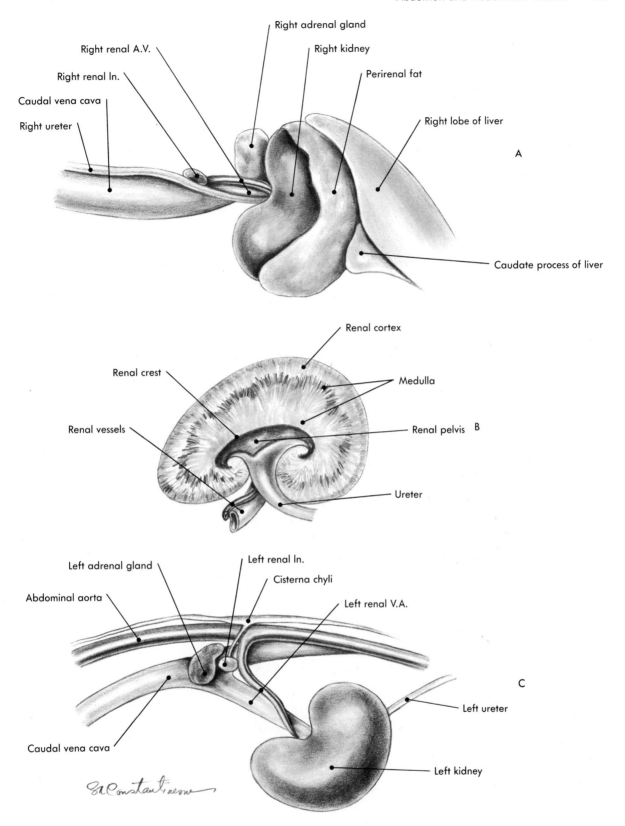

A, Right kidney in place. **B,** Longitudinal section through kidney. **C,** Left kidney in place.

Fig. 3-35

Goat
■■■■■■■
(Figs. 3-36
to 3-43)

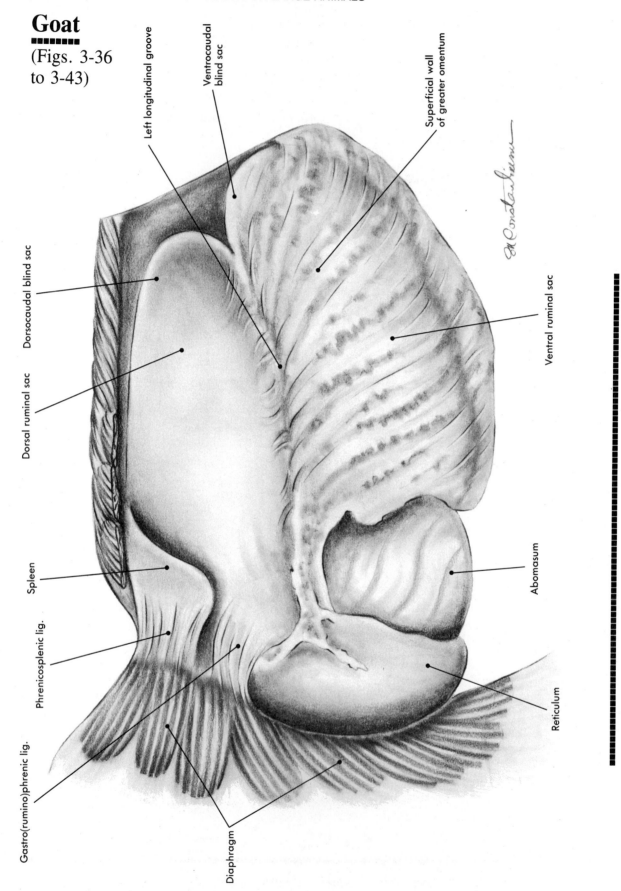

Stomach and spleen (left aspect).

Fig. 3-36

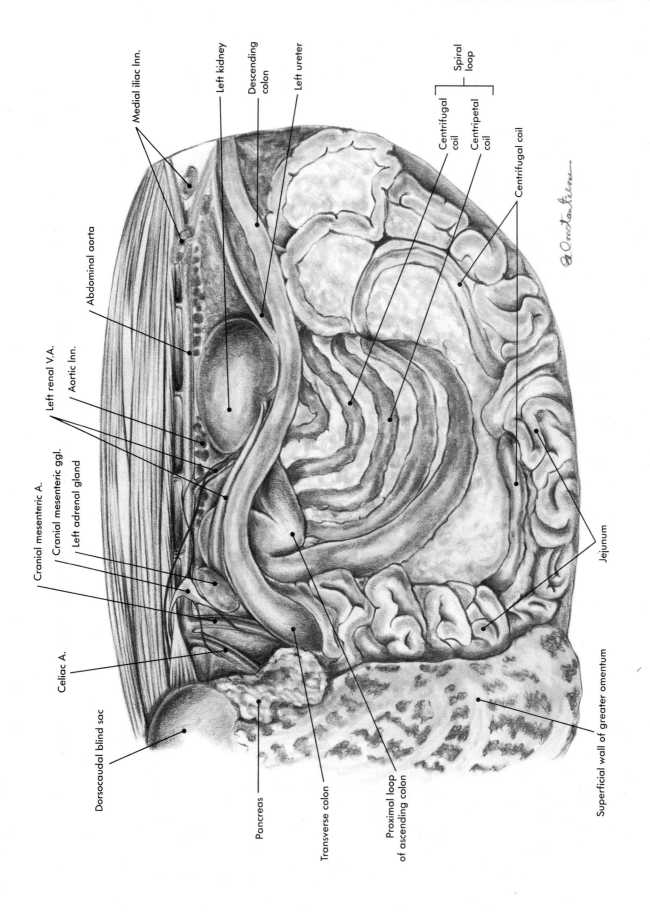

Medial iliac lnn.

Left kidney

Descending colon

Left ureter

Spiral loop

Centrifugal coil

Centripetal coil

Centrifugal coil

Abdominal aorta

Left renal V.A.

Aortic lnn.

Cranial mesenteric A.

Cranial mesenteric ggl.

Left adrenal gland

Celiac A.

Dorsocaudal blind sac

Jejunum

Pancreas

Transverse colon

Proximal loop of ascending colon

Superficial wall of greater omentum

Abdominal viscera after reflecting stomach cranially (left aspect).

Fig. 3-37

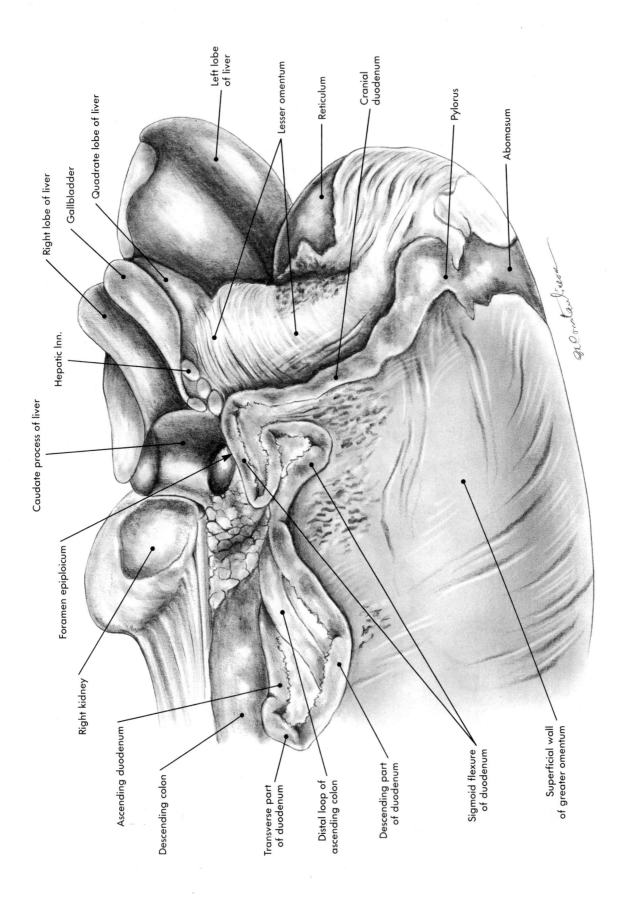

Left lobe of liver

Lesser omentum

Reticulum

Cranial duodenum

Pylorus

Abomasum

Quadrate lobe of liver

Gallbladder

Right lobe of liver

Hepatic lnn.

Caudate process of liver

Foramen epiploicum

Right kidney

Ascending duodenum

Descending colon

Transverse part of duodenum

Distal loop of ascending colon

Descending part of duodenum

Sigmoid flexure of duodenum

Superficial wall of greater omentum

Fig. 3-38   Abdominal viscera in place (right side). Liver and right kidney are reflected.

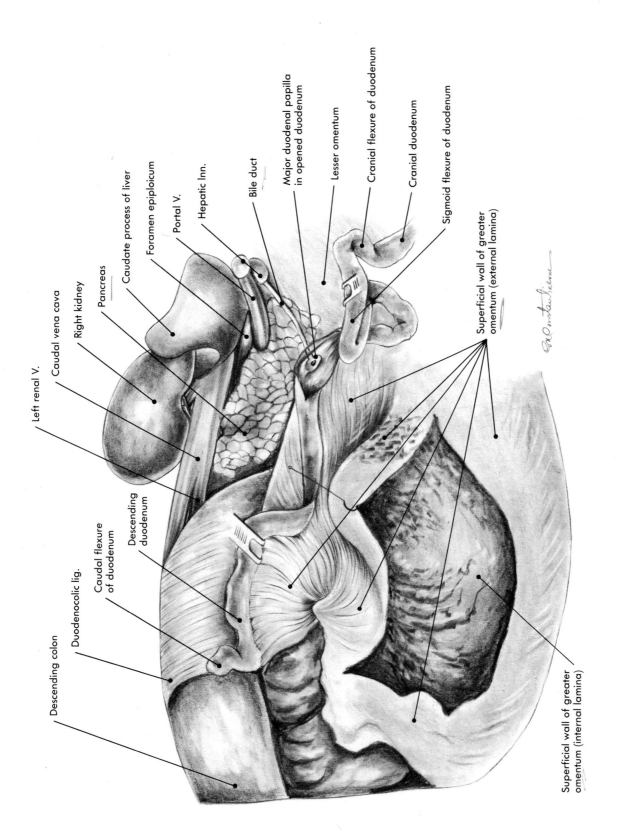

Left renal V.

Caudal vena cava

Right kidney

Pancreas

Caudate process of liver

Foramen epiploicum

Portal V.

Hepatic lnn.

Bile duct

Major duodenal papilla in opened duodenum

Lesser omentum

Cranial flexure of duodenum

Cranial duodenum

Sigmoid flexure of duodenum

Superficial wall of greater omentum (external lamina)

Descending colon

Duodenocolic lig.

Caudal flexure of duodenum

Descending duodenum

Superficial wall of greater omentum (internal lamina)

External lamina of superficial wall of greater omentum fenestrated.

Fig. 3-39

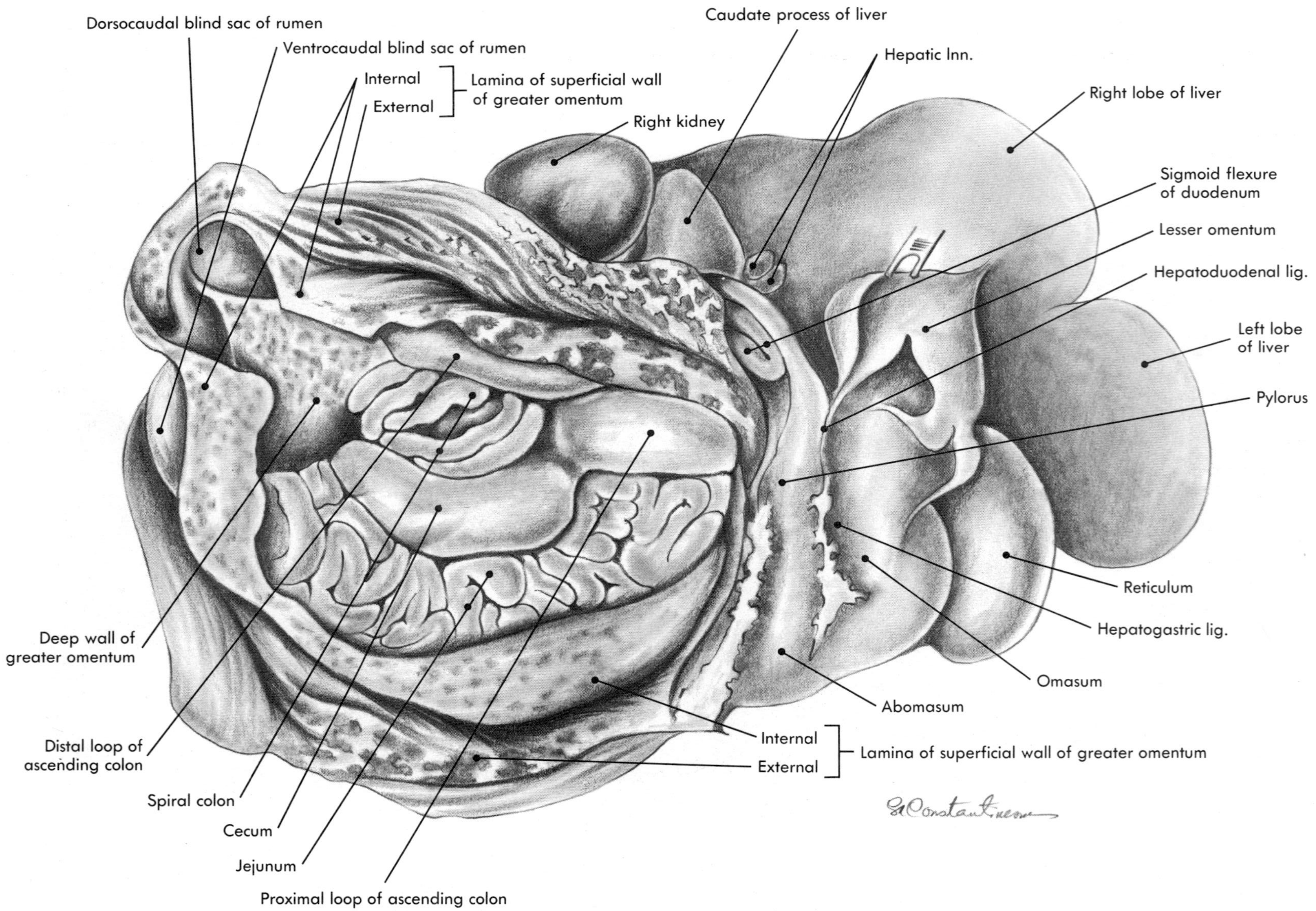

Right lobe of liver

Sigmoid flexure
of duodenum

Lesser omentum

Hepatoduodenal lig.

Left lobe
of liver

Pylorus

Reticulum

Hepatogastric lig.

Omasum

Lamina of superficial wall of greater omentum

Abomasum

Internal

External

Hepatic lnn.

Caudate process of liver

Right kidney

Lamina of superficial wall
of greater omentum

Ventrocaudal blind sac of rumen

Internal

External

Dorsocaudal blind sac of rumen

Deep wall of
greater omentum

Distal loop of
ascending colon

Spiral colon

Cecum

Jejunum

Proximal loop of ascending colon

G. Constantinescu

Omental bursa opened.

Fig. 3-40

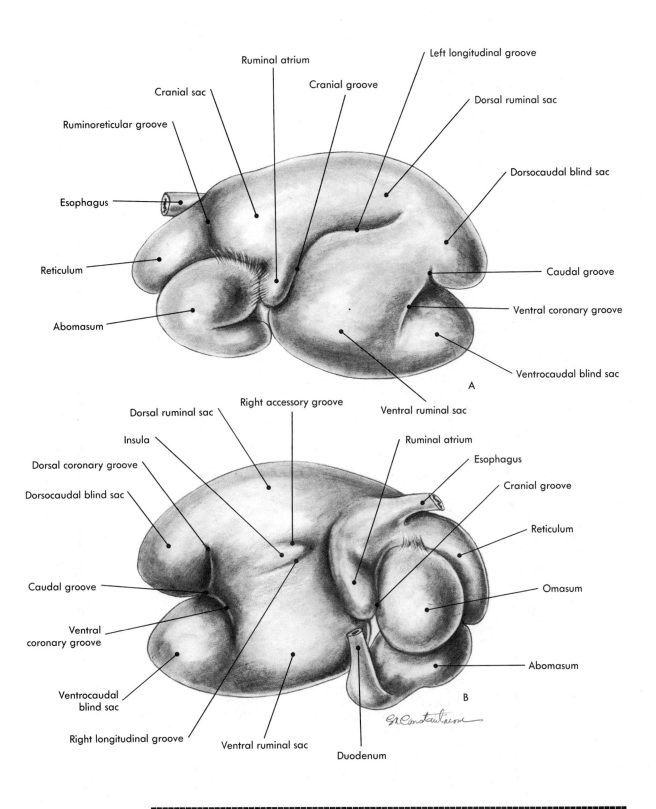

Ruminal atrium

Cranial sac

Cranial groove

Left longitudinal groove

Ruminoreticular groove

Dorsal ruminal sac

Esophagus

Dorsocaudal blind sac

Reticulum

Caudal groove

Abomasum

Ventral coronary groove

Ventrocaudal blind sac

Ventral ruminal sac

A

Right accessory groove

Dorsal ruminal sac

Insula

Ruminal atrium

Dorsal coronary groove

Esophagus

Dorsocaudal blind sac

Cranial groove

Reticulum

Caudal groove

Omasum

Ventral coronary groove

Ventrocaudal blind sac

Abomasum

Right longitudinal groove

Ventral ruminal sac

B

Duodenum

A, Stomach (left aspect). **B,** Stomach (right aspect).

Fig. 3-41

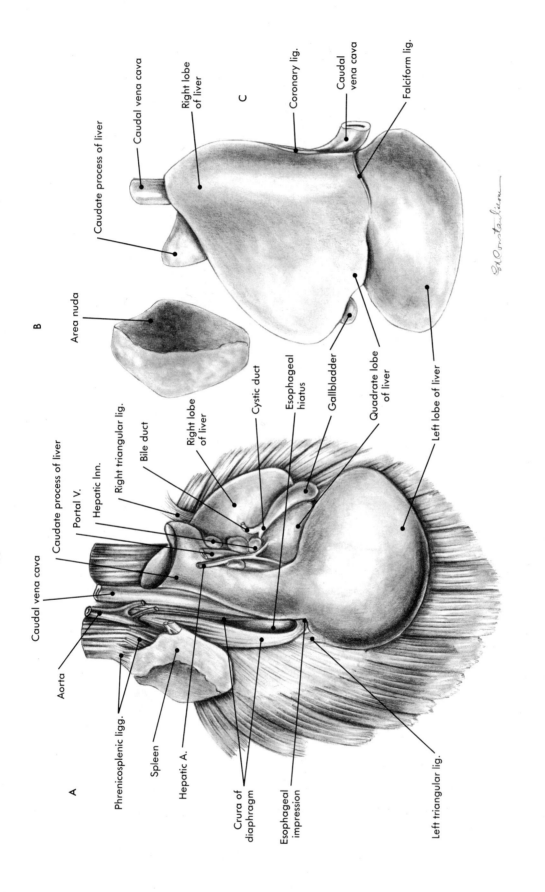

Fig. 3-42   **A,** Visceral aspect of liver and spleen in place. **B,** Diaphragmatic aspect of spleen. **C,** Diaphragmatic aspect of liver.

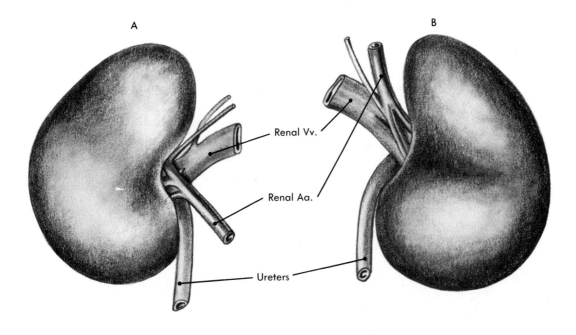

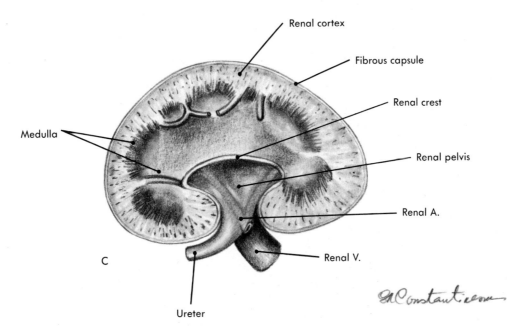

A, Right kidney (ventral aspect). **B,** Left kidney (ventral aspect). **C,** Longitudinal section through kidney.

Fig. 3-43

Pig
▪▪▪▪▪
(Figs. 3-44
to 3-53)

There are two ways to dissect the abdomen to expose and examine the abdominal viscera: through a lateral approach (Figs. 3-44 and 3-45) or through the ventral approach (Figs. 3-46 and 3-47).

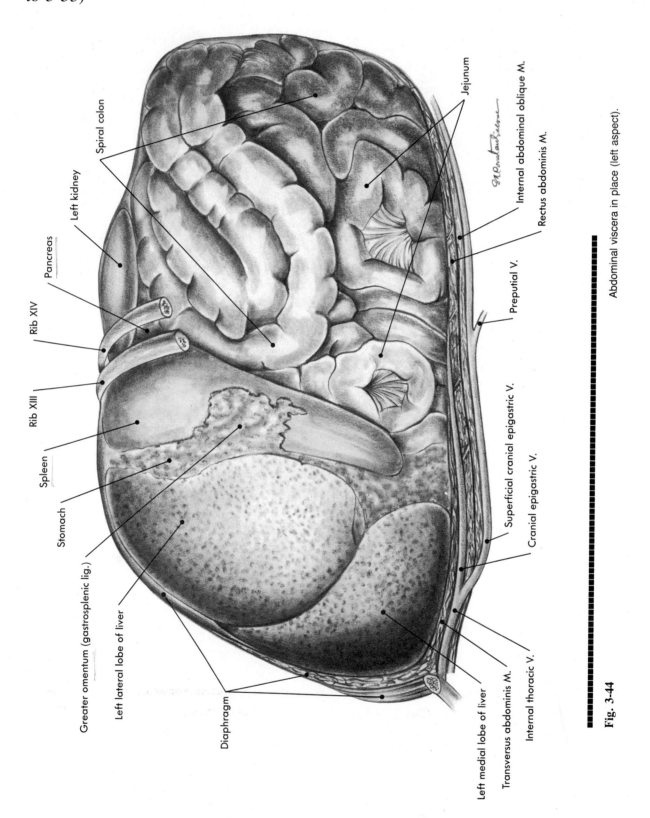

Spiral colon

Left kidney

Pancreas

Rib XIV

Rib XIII

Spleen

Stomach

Greater omentum (gastrosplenic lig.)

Left lateral lobe of liver

Diaphragm

Jejunum

Internal abdominal oblique M.

Rectus abdominis M.

Preputial V.

Superficial cranial epigastric V.

Cranial epigastric V.

Left medial lobe of liver

Transversus abdominis M.

Internal thoracic V.

Abdominal viscera in place (left aspect).

Fig. 3-44

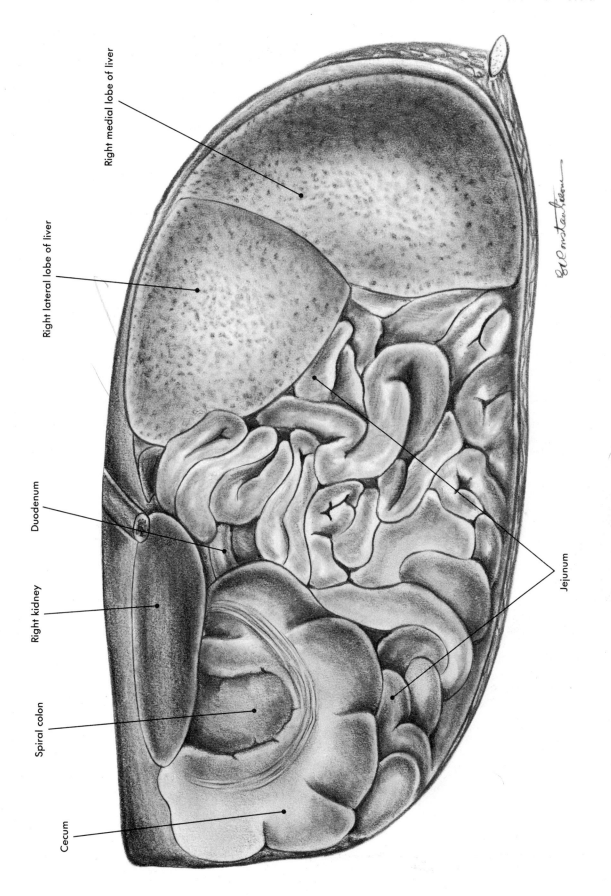

Right medial lobe of liver

Right lateral lobe of liver

Duodenum

Right kidney

Spiral colon

Cecum

Jejunum

Fig. 3-45

Abdominal viscera in place (right aspect).

When choosing the ventral approach, first identify the sex of the specimen. If the specimen is female, make a midventral incision in the abdominal wall, followed by two symmetrical perpendicular additional incisions, one in front of the tensor fasciae latae M. and the other caudal to the last rib and cartilage. If the specimen is male, make a U-shaped incision in the ventral abdominal wall, starting from the area cranial to one superficial inguinal ring and parallel to the penis, continuing cranially 5 to 10 cm around the prepuce, and coming back parallel to the penis on the other side to the area cranial to the opposite superficial inguinal ring. Then perform the same incisions as for the female specimen.

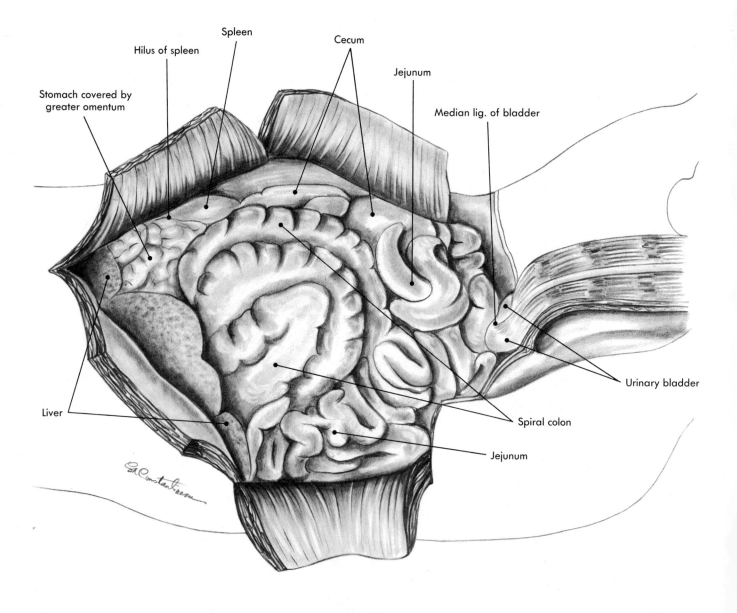

Fig. 3-46   Abdominal viscera in place (ventral aspect).

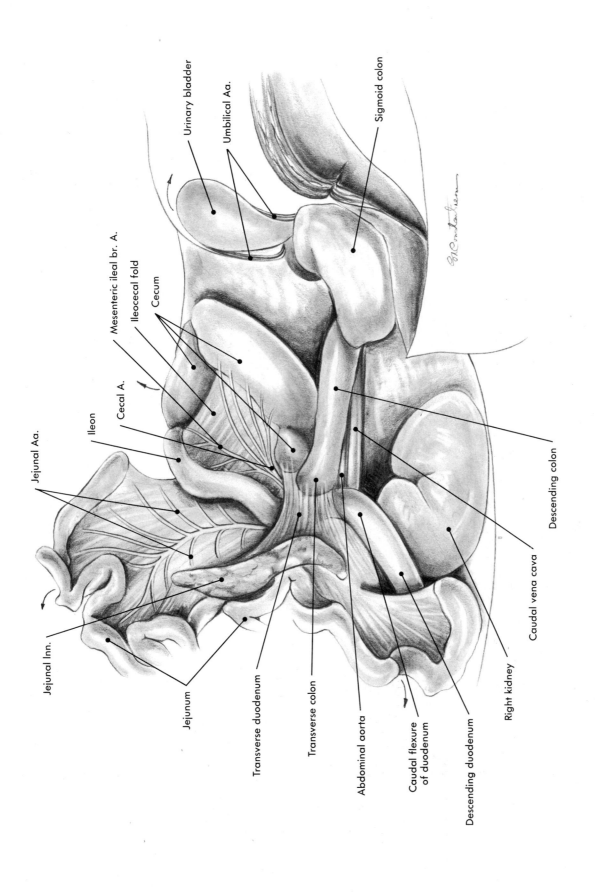

Urinary bladder

Umbilical Aa.

Sigmoid colon

Mesenteric ileal br. A.

Ileocecal fold

Cecum

Cecal A.

Ileon

Jejunal Aa.

Jejunal Inn.

Jejunum

Transverse duodenum

Transverse colon

Abdominal aorta

Caudal flexure
of duodenum

Descending duodenum

Right kidney

Caudal vena cava

Descending colon

Fig. 3-47

Abdominal viscera in the dorsal compartment of abdomen.

Carefully pull the spiral colon out of the abdominal cavity. Identify the centripetal and the centrifugal coils and the central flexure of the spiral loop (Fig. 3-48).

Notice that, in comparison with the ruminants, not all the centripetal coils are smooth in the pig; they are provided with haustra. Some of the centrifugal coils also have haustra. In young pigs, the centrifugal coils are almost smooth.

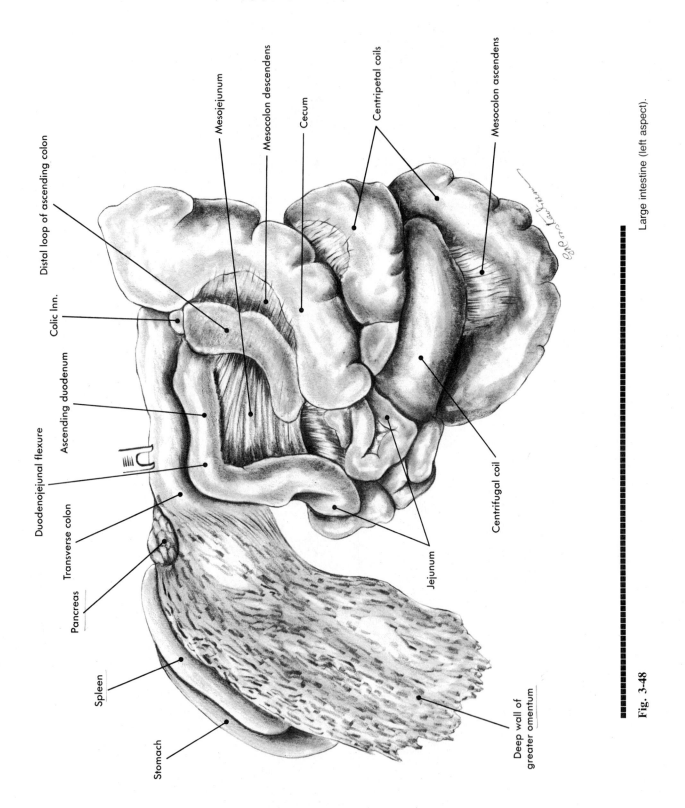

Fig. 3-48

On the right side, palpate the caudal vena cava from the dorsal wall of the abdominal cavity toward the diaphragmatic aspect of the liver. The caudal vena cava is surrounded and protected by the coronary lig., which continues with the falciform lig. ventral to the foramen venae cavae of the diaphragm. In some specimens, especially older specimens, the falciform lig. is very short and does not exceed the area of the liver.

Reflect the liver cranially and the right kidney dorsally to expose the deep structures. These include the pylorus, the cranial duodenum, the cranial flexure of the duodenum, a part of the descending duodenum, the caudal flexure of the duodenum, the lesser omentum (with both hepatogastric and hepatoduodenal ligg.), the right (deep) wall of the greater omentum, the portal V. and caudal vena cava, the right lobe of the pancreas, the renoduodenal lig., the mesentery, and the right adrenal gland. Attempt to palpate the foramen epiploicum by introducing a finger between the liver, pancreas, portal V., and caudal vena cava (Fig. 3-49).

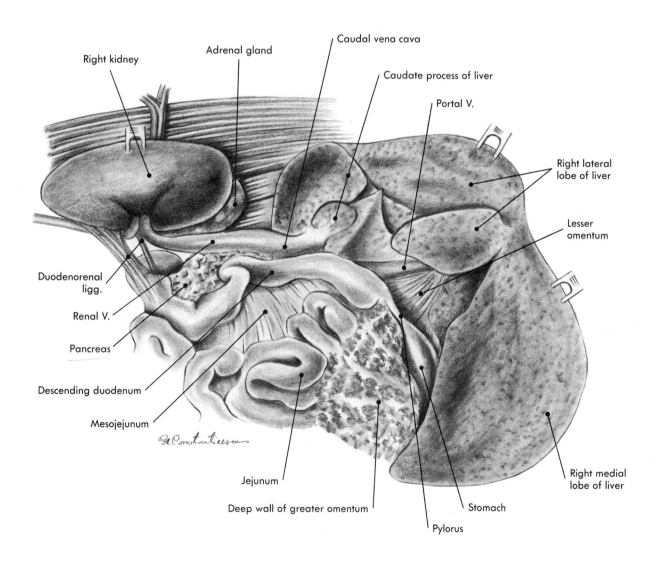

Abdominal viscera (deep right aspect):

Fig. 3-49

Identify the gastric diverticulum, which is a dilation located on the top of the fundus. Make an incision on the greater curvature of the stomach from the diverticulum toward the cranial duodenum and passing through the pylorus. Examine the internal aspect of the stomach, in which there are four distinct mucosal areas: the diverticulum, cardia, pyloric sphincter, and torus pyloricus (an adipose and muscular prominence within the pyloric canal). A few centimeters from the pylorus, within the duodenum, is the major duodenal papilla, which contains the opening of the bile duct. The accessory pancreatic duct opens from the right lobe of the pancreas into the descending duodenum, about 10 to 12 cm from the pylorus (Fig. 3-50).

Notice that, in the pig, the pancreatic duct is not developed.

Dissect the celiac A. and the arterial anastomoses shown in Fig. 3-51 by circles.

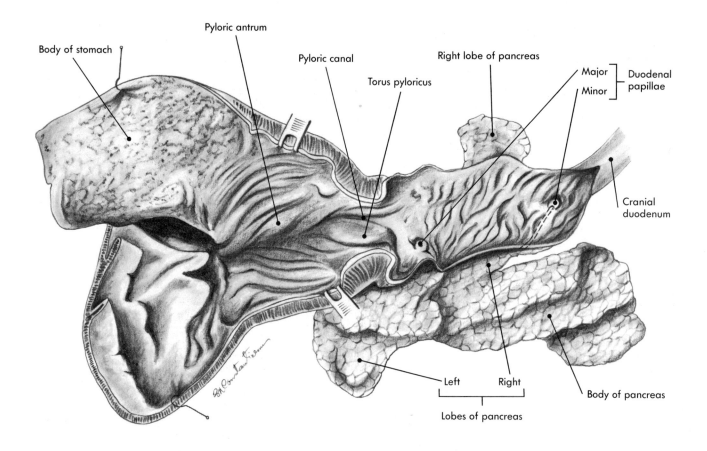

Fig. 3-50   Stomach and duodenum opened on greater curvature.

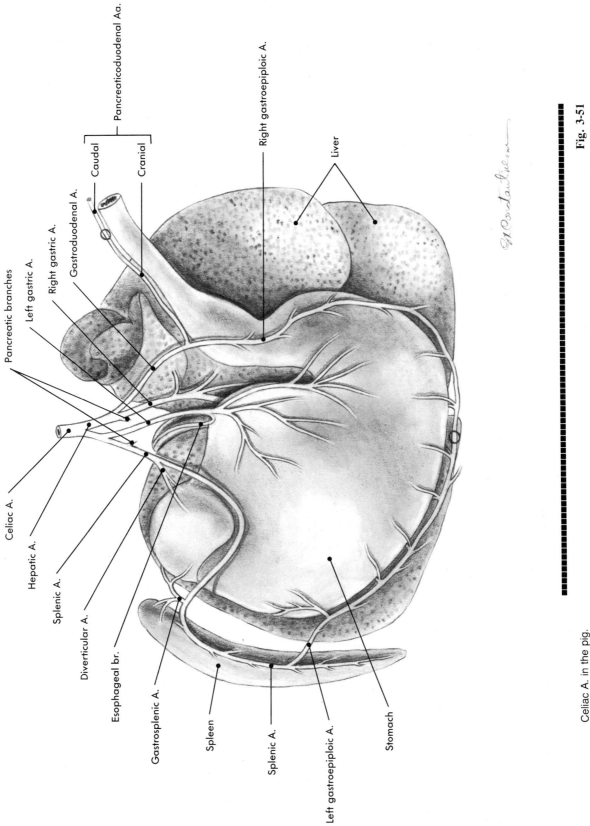

Pancreaticoduodenal Aa.

Caudal

Cranial

Right gastroepiploic A.

Liver

Gastroduodenal A.

Right gastric A.

Left gastric A.

Pancreatic branches

Celiac A.

Hepatic A.

Splenic A.

Diverticular A.

Esophageal br.

Gastrosplenic A.

Spleen

Splenic A.

Left gastroepiploic A.

Stomach

Fig. 3-51

Celiac A. in the pig.

Dissect the branches of the cranial mesenteric A., whose names have a similar significance to those of the previous species. The anastomoses are highlighted by circles in Fig. 3-52).

Examine the kidneys and the adrenal glands.

Notice that the pig is the only species (of the large animals that this guide describes) with a hilus located on the dorsal side of the kidneys. In addition, the shape of the hilum is characteristic (Fig. 3-53, *A* and *B*).

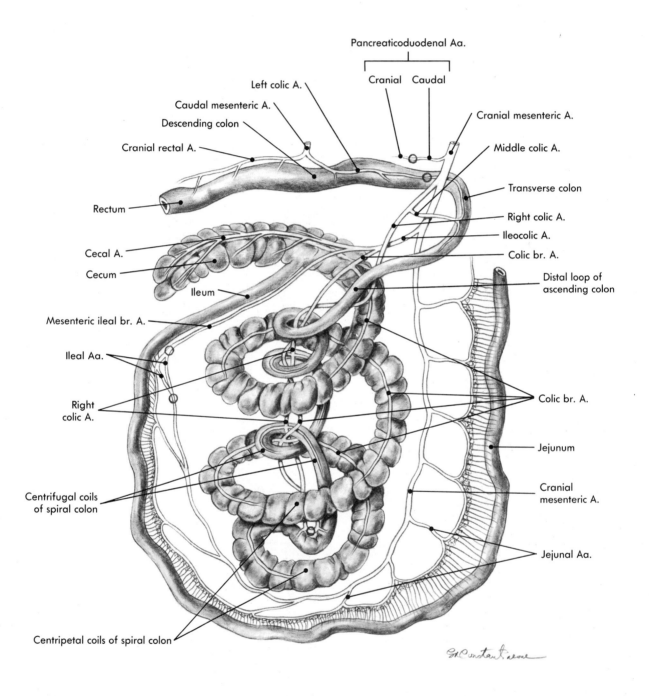

Fig. 3-52   Cranial and caudal mesenteric Aa. in the pig.

Dissect the renal lnn. and the sympathetic ganglia of the area. Make a horizontal midsection in one of the kidneys, and examine the pelvis, calices, and renal pyramids (Fig. 3-53, *C* and *D*).

Notice that the pig has a smooth, multipyramidal type of kidney.

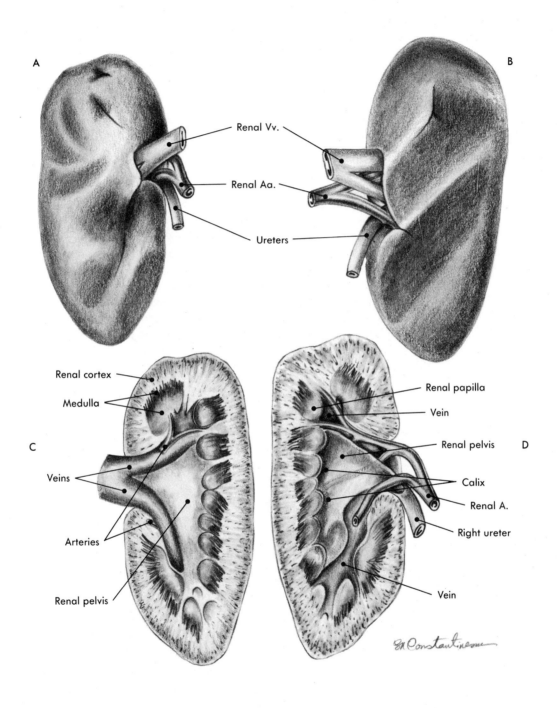

A, Right kidney (ventral aspect). B, Left kidney (ventral aspect). C, Ventral half of the right kidney (median aspect). D, Dorsal half of right kidney (median aspect).

Fig. 3-53

Chapter 4

Pelvis, pelvic viscera, tail, and external genitalia

PELVIS: PELVIC INLET

Horse

The pelvic inlet (cranial pelvic aperture) (Fig. 4-1) is the communication between the abdominal and pelvic cavities. Transect the lumbar muscles transversely with a knife and the vertebrae with a saw at approximately the level of L_2 or L_3 (cranial to the origin of the caudal mesenteric A.). Transect (transversely) the ventral abdominal muscles with a knife at the level of the lumbar section.

Examine the common structures for a male or a female specimen: the symmetrical rectus abdominis Mm.; the linea alba; the transversus abdominis M.; the internal abdominal oblique M.; the arteries and veins; and the parietal peritoneum lining the transverse fascia.

Make a longitudinal incision in the parietal peritoneum and the transverse fascia parallel and 5 cm lateral to the linea alba, and reflect them. The specific fibrous intersections of the rectus abdominis M. are visible. Identify the psoas major and psoas minor Mm. and the origin of the genitofemoral N.

In both male and female, the descending and sigmoid colons, which are suspended by the descending mesocolon and supplied by the caudal mesenteric A., are exposed. Close to the aorta and around the origin of the caudal mesenteric A., find and dissect the caudal mesenteric ggl. and the hypogastric Nn., which are the sympathetic contribution to the pelvic (parasympathetic) plexuses. The urinary bladder, lying on the floor of the pelvic cavity, is attached to the lateral walls by two lateral ligaments and to the floor by the median vesical lig., which extends up to the umbilicus. Follow the most caudal extent of the aorta, and identify the two external iliac Aa. and then the two internal iliac Aa. The external iliac A. takes a caudoventral direction on the lateral side of the pelvic inlet branching into the deep circumflex iliac A. ventral to the tuber coxae. Identify the cremasteric, or uterine, A., which is a branch of the external iliac A., unique to the horse.

Identify the lumbar lnn. that are scattered along the ventral aspect of the aorta, the medial iliac lnn. that are located around the quadrifurcation of the aorta, and the lateral iliac lnn. that are found around the bifurcation of the deep circumflex iliac A.

In the stallion (or gelding), execute an incision of the parietal peritoneum and the transverse fascia at the level of the cranial br. of the deep circumflex iliac A. toward the main artery and its origin from the external iliac A. Then follow the course of the external iliac A. up to the deep femoral A., which is the second branch of the external iliac A. Make an elliptical incision around the deep inguinal ring; then reach the linea alba cranial to the pubic symphysis, and continue the incision cranially at least 20 to 25 cm. Carefully free the parietal peritoneum and the transverse fascia, and reflect them cranially. The caudal border of the internal abdominal oblique M. with the cremaster M. is visible. Caudally and deep to them (in fact, superficial to them), the arcus (lig.) inguinalis is also visible.

154

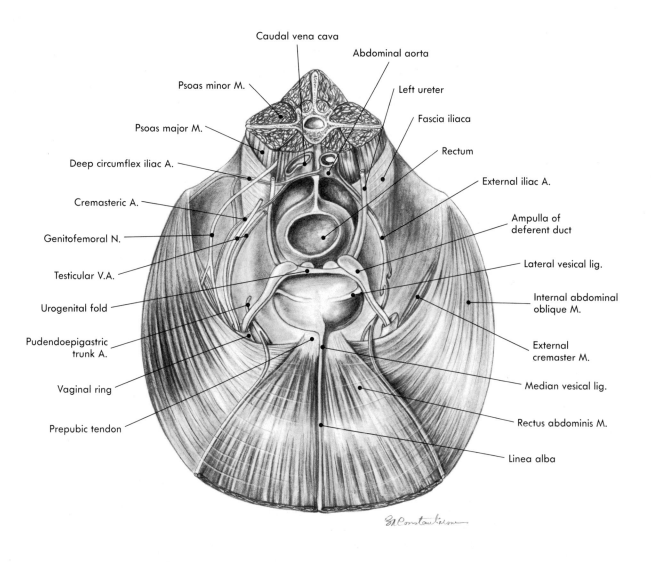

Pelvic inlet in the stallion.

Fig. 4-1

Remember! The internal abdominal oblique M. represents the craniolateral border of the deep inguinal ring and of the inguinal canal, whereas the arcus inguinalis represents the caudomedial border of the same structures (Fig. 4-2). The lateral commissure of the deep inguinal ring is located dorsally, in comparison to the medial commissure, which is located ventrally. The medial commissure of the deep inguinal ring is close to and overlaps the caudomedial commissure of the superficial inguinal ring. The distance between the lateral commissure of the deep inguinal ring and the craniolateral commissure of the superficial inguinal ring is much longer. Between the two inguinal rings the inguinal canal is found.

Remember! The superficial inguinal ring is a slit in the aponeurosis of the external abdominal oblique M. before its attachment to the pubic symphysis, which is outlined by two borders (crura) and two commissures. From the caudomedial commissure of the ring, a strong fibrous structure of 8 to 10 cm breadth (the arcus inguinalis) courses toward and attaches to the tuber coxae within the abdominal cavity. Another structure originating from the superficial inguinal ring, the lamina femoralis, travels toward the medial aspect of the thigh.

Remember! The reflection of the parietal peritoneum and the transverse fascia through the deep inguinal ring is named the vaginal ring.

Caution! The vaginal ring—surrounding the deferent duct, the vessels (the testicular A. and pampiniform plexus), and the nerves supplying the testicle and epididymis (all of which are covered with visceral peritoneum)—is much smaller than the deep inguinal ring.

Remember! The transverse fascia becomes the internal spermatic fascia and the peritoneum becomes the vaginal tunic. The parietal lamina (lamina parietalis) of the vaginal tunic is the continuation of the parietal peritoneum and the visceral lamina (lamina visceralis) is the continuation of the visceral peritoneum. Between the two laminae, the cavity of the vaginal tunic is enclosed; it communicates with the peritoneal cavity. The internal spermatic fascia and the parietal lamina of the vaginal tunic lining the inguinal canal are called the vaginal canal.

Remember! The spermatic cord is made up of all the structures surrounded by and including the internal spermatic fascia and the parietal lamina of the vaginal tunic. Consequently, in addition to the visceral lamina of the tunic, the spermatic cord includes the ductus deferens with the attached mesoductus deferens, the vessels and nerves supplying the testicle and the epididymis (surrounded by the proximal mesorchium), and the smooth muscular fibers (formerly the internal cremaster M.). The mesofuniculus is the narrow strip of mesorchium between the origin of the mesoductus deferens and the parietal lamina of the vaginal tunic.

Notice that the mesorchium (including the proximal mesorchium) and the structures belonging to it, such as the mesofuniculus and mesoductus deferens, represent the visceral lamina of the vaginal tunic.

Identify all these structures.

Transect the internal abdominal oblique M. in a cranial direction from the vaginal ring to expose the inguinal canal and the superficial inguinal ring. Return to the deep femoral A., which gives off the pudendoepigastric trunk. The external iliac and deep femoral Aa. (along with the saphenous N.) leave the abdominal cavity to supply the pelvic limb through the lacuna vasorum. Close to them, identify the lacuna musculorum; this space allows the iliopsoas and sartorius Mm. to exit the abdominal cavity toward the pelvic limb.

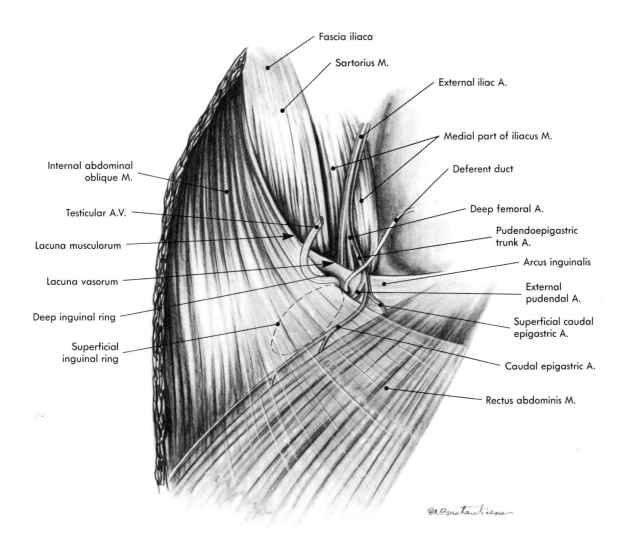

Inguinal rings and associated structures in the stallion.

Fig. 4-2

In the mare, follow the flexuous course of the ovarian A. and the corresponding venous plexus within the mesovarium up to the ovary.

Remember! The broad lig., which suspends the ovary, the uterine tube (or salpinx), and the uterus, has distinct segments: the mesovarium, mesosalpinx, and mesometrium, respectively. The mesovarium is divided into proximal and distal segments. The proximal mesovarium is considered the portion dorsal to the uterine tube, whereas the distal mesovarium is the portion dorsal to the proper lig. of the ovary (the medial wall of the ovarian bursa); the latter extends from the mesosalpinx to the ovary.

The mesosalpinx is attached to the ventral aspect of the uterine tube and continues with the proximal mesovarium (Fig. 4-3).

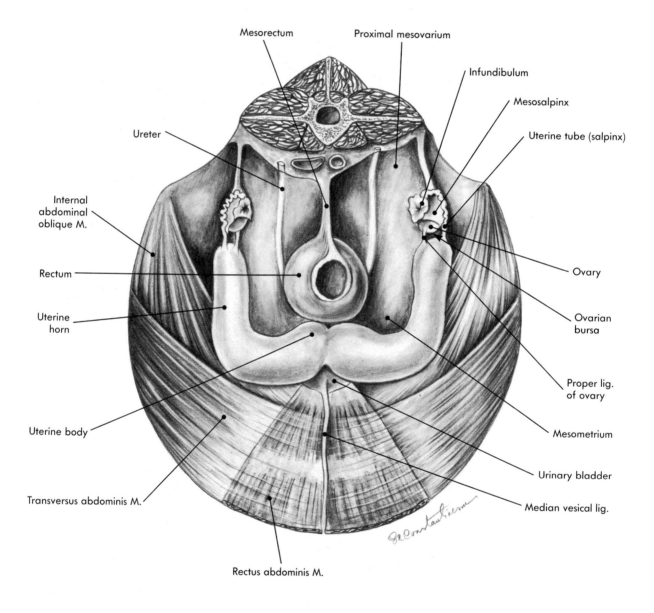

Pelvic inlet in the mare.

Remember! The ovarian bursa is the duplication of the broad lig. between the ovary and the uterine horn; the lateral wall is represented by the mesosalpinx with the uterine tube inside, whereas the medial wall is represented by the distal mesovarium, which is bordered ventrally by the proper lig. of the ovary. The opening of the ovarian bursa is ventrally oriented in the mare.

Examine the infundibulum of the uterine tube, its abdominal opening (ostium) and the fimbriae. Introduce a finger into the ovarian bursa, and palpate the ovarian fossa of the ovary (present only in the mare). The uterine horns converge toward the uterine body. The body lies on the floor of the pelvic cavity and the horns are directed toward the dorsal wall of the abdominal cavity, caudal to the kidneys. Try to identify and palpate the uterine A.V.

Isolate the external iliac A. and its branches, the deep circumflex iliac and deep femoral Aa. Following the course of the pudendoepigastric trunk after it branches off the caudal epigastric A., the external pudendal A. penetrates the deep inguinal ring. However, there is no vaginal ring in the mare. The round lig. of the uterus, which in the female dog is well developed and penetrates the inguinal canal, is very short in the mare, like an appendix pendulating on the lateral aspect of the mesometrium (Figs. 4-4 and 4-5, *A* and *B*).

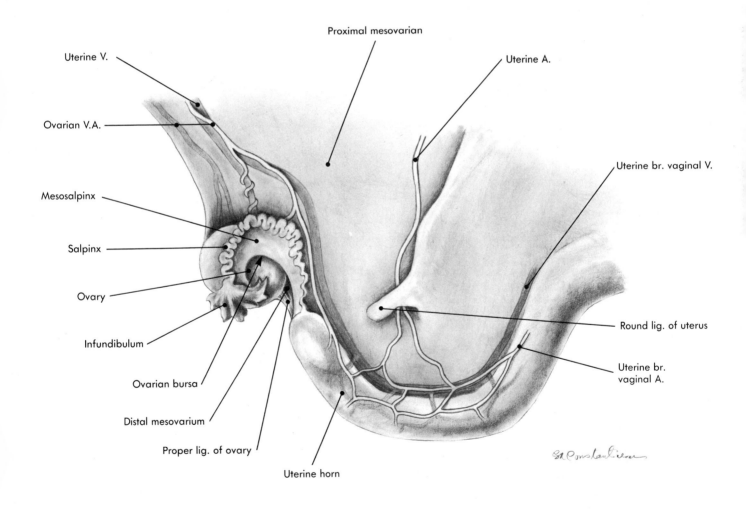

Ovary, salpinx, and uterus (left lateral view).

Fig. 4-4

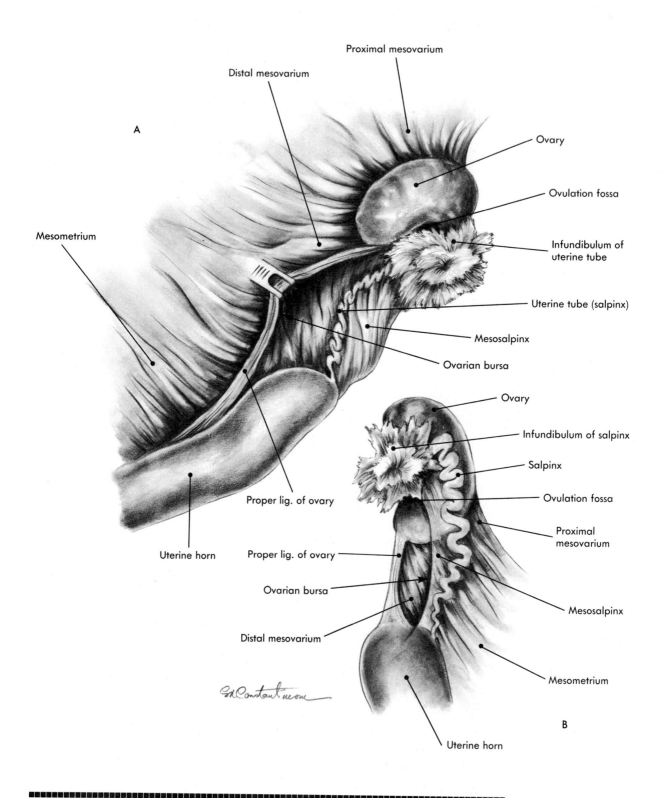

Fig. 4-5   **A,** Medial aspect of left ovary, salpinx and associated structures. **B,** Ventral aspect of left ovary, salpinx, and associated structures.

In the male, continue to dissect the external pudendal A. outside of the inguinal canal and identify the branch coming from the ventral aspect of the abdominal wall: the caudal superficial epigastric A. The continuation of the external pudendal A. that supplies the penis is the cranial A. of the penis, whereas the branch supplying the scrotum is the ventral scrotal br. (Fig. 4-6).

In the female, the branches of the external pudendal A. that supply the mammary glands are the cranial mammary A. (the caudal superficial epigastric A.) and the caudal mammary A. (the ventral labial br.).

In the male, the penis is protected within the prepuce. Examine the orifice of the prepuce and the two preputial folds that are specific for horse.* Pull the penis out of the prepuce, and identify the preputial ring. Examine the glans penis with the corona glandis and collum glandis, the urethral process with the external urethral orifice surrounded by the fossa glandis, and (dorsal to it) the urethral sinus (Figs. 4-7, *B* and 4-8, *A*). Make a midlongitudinal incision on the dorsal aspect of the free part of the penis; then continue the incision of the preputial ring, preputial fold, and external lamina of the prepuce (Figs. 4-7, *A* and 4-8, *A*). Reflect the two borders of the incision; the paired cranial Aa. and Vv. of the penis, the dorsal venous plexuses, and the dorsal Nn. of the penis will be exposed. Follow the cranial Aa. and Vv. of the penis toward the external pudendal Aa.Vv. and *notice* that the veins are very small. The main vein that drains the blood from the penis is the accessory external pudendal V. (not named in the N.A.V.). This vein perforates the origin of the gracilis M. (through a fibrous ring close to the pelvic symphysis) to join the deep femoral V. Transect the penis transversely to obtain two or three segments and examine each cross surface (Figs. 4-7, *C* to *E*). The common structures of all sections are the following: the corpus cavernosum (surrounded by the tunica albuginea) and the trabeculae (sent by the tunica albuginea to divide the corpus cavernosum into numerous chambers); the urethra (surrounded by the corpus spongiosum lying in the urethral groove on the ventral aspect of the corpus cavernosum); the bulbospongiosus M. (forming a hinge around the urethra); and the retractor penis M. (the most ventral structure of the penis). In the most cranial sections, the dorsal process of the glans penis overlaps the corpus cavernosum.

Next, in the testicular area, examine the scrotum and median raphe. Pull one testicle down into the scrotum, keeping your fingers clamped around the spermatic cord until the scrotum becomes tense. Make a 2 cm, sagittal skin incision lateral and parallel to the raphe. The so-called skin incision will pass through the scrotum (the scrotal skin and dartoic tunic). Bluntly dilacerate the external spermatic fascia from the internal spermatic fascia and the vaginal tunic and break the lig. of the tail of the epididymis (part of the embryonic gubernaculum testis) (Fig. 4-8, *B* and *C*).

Notice that the gubernaculum testis and the lig. of the tail of the epididymis connect the testicle to the dartoic tunic (of the scrotum). These same steps are utilized during the closed castration, the only procedure that is suggested.

---

*A disagreement exists between the 1983 *Nomina Anatomica Veterinaria* (N.A.V.) (page A65) and some anatomy books and guides for dissection as far as the prepuce is concerned. The present guide agrees with the N.A.V. and considers the prepuce as follows: In the horse, the prepuce has two laminae—the external (outer) lamina, which is a continuation of the skin, and the internal (inner) lamina, which is completely in intimate contact with the fully erect penis. When the penis is retracted within the prepuce, the internal lamina makes a fold called the preputial fold, which has a surface turned toward the external lamina and another turned toward the penis. The area of attachment of the internal lamina on the penis has the shape of a ring and is called the preputial ring. The preputial fold lies within the preputial cavity (whose communication with the external environment at the end of the penis is called the preputial orifice, or ostium) and is surrounded by the external lamina.

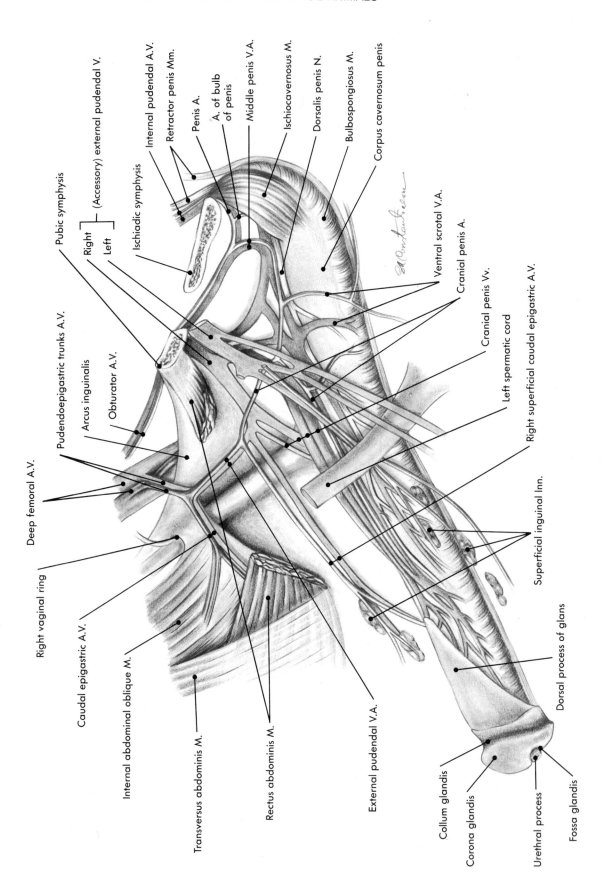

Pubic symphysis

(Accessory) external pudendal V.
{ Right
  Left

Ischiadic symphysis

Internal pudendal A.V.

Retractor penis Mm.

Penis A.

A. of bulb of penis

Middle penis V.A.

Ischiocavernosus M.

Dorsalis penis N.

Bulbospongiosus M.

Corpus cavernosum penis

Ventral scrotal V.A.

Cranial penis A.

Cranial penis Vv.

Left spermatic cord

Right superficial caudal epigastric A.V.

Superficial inguinal Inn.

Dorsal process of glans

Collum glandis

Corona glandis

Urethral process

Fossa glandis

External pudendal V.A.

Rectus abdominis M.

Transversus abdominis M.

Internal abdominal oblique M.

Caudal epigastric A.V.

Right vaginal ring

Deep femoral A.V.

Pudendoepigastric trunks A.V.

Arcus inguinalis

Obturator A.V.

Fig. 4-6   Penis and associated structures in the stallion.

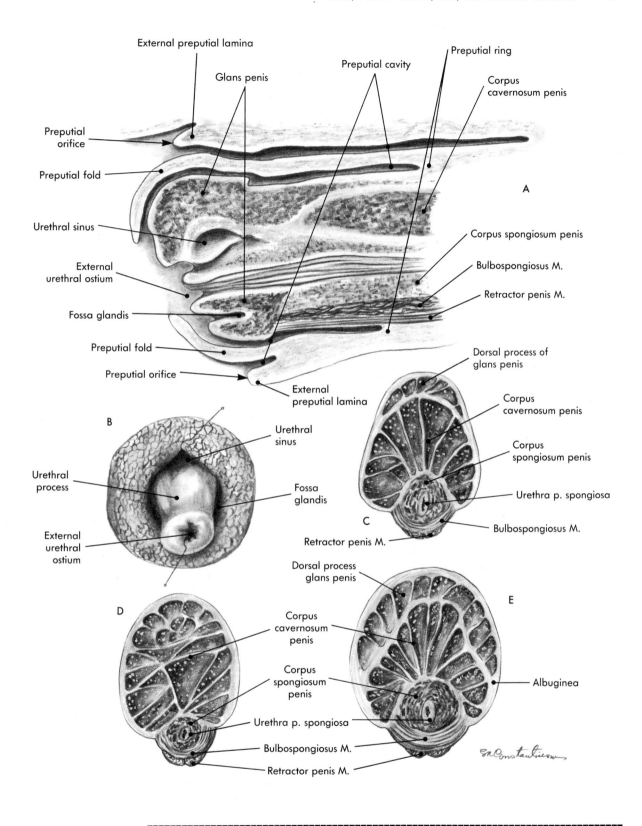

A, Median section of penis. **B,** Frontal view of glans penis. **C,** Cross section (middle third of penis). **D,** Cross section (caudal third of penis). **E,** Cross section through glans penis.

Fig. 4-7

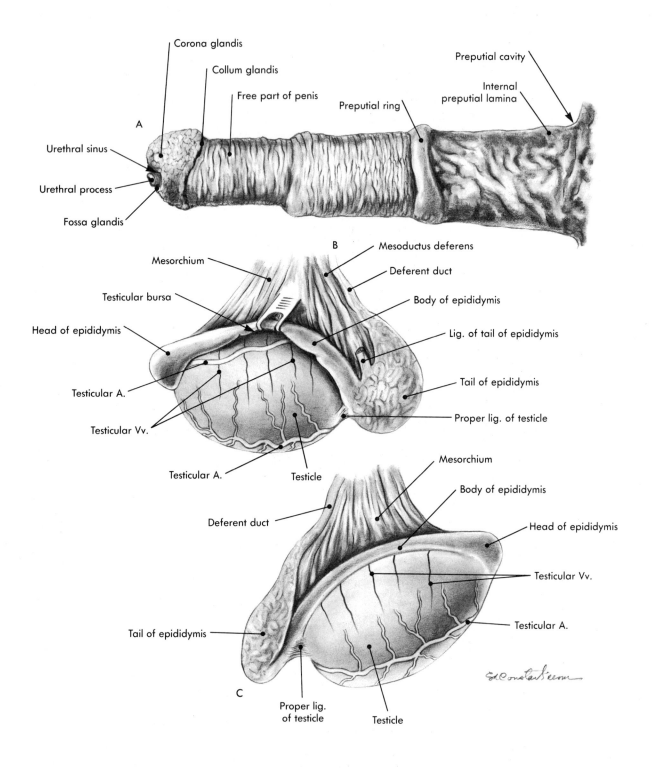

Fig. 4-8   **A,** Penis (lateral aspect). **B,** Left testicle (lateral aspect). **C,** Left testicle (medial aspect).

Make an incision through the internal spermatic fascia and the parietal lamina of the vaginal tunic to expose the cavity of the vaginal tunic. The testicle and epididymis are protected by the visceral lamina of the vaginal tunic. Examine the specific location and branches of the testicular A. Examine the testicular bursa, the narrow space on the lateral side of the testicle between the testicle and the body of the epididymis.

Notice that the scrotal skin and the dartoic tunic on one hand and the internal spermatic fascia and the parietal lamina of the vaginal tunic on the other are almost fused with each other and are very difficult to separate from one another. (Fig. 4-9).

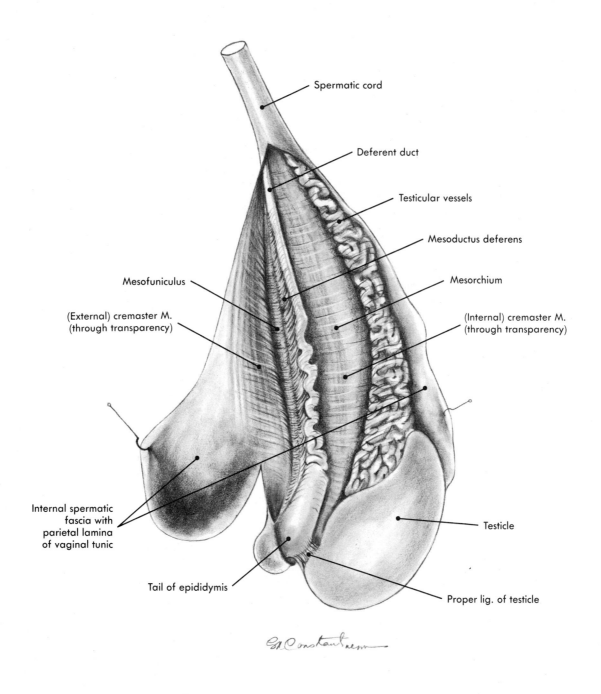

Spermatic cord

Deferent duct

Testicular vessels

Mesoductus deferens

Mesorchium

(Internal) cremaster M. (through transparency)

Testicle

Proper lig. of testicle

Mesofuniculus

(External) cremaster M. (through transparency)

Internal spermatic fascia with parietal lamina of vaginal tunic

Tail of epididymis

Spermatic cord in the stallion.

Fig. 4-9

Notice that the dartoic tunic makes two separate tunics, one for each testicle, and a tunnel for the penis to glide through.

In the mare, examine the mammary glands (Fig. 4-10, *A*) and the intermammary groove. Remove one gland from the midline and the ventral wall of the abdomen. In the removed gland, probe the two or three papillary ducts (teat canals) leading into the papillary part of the lactiferous sinus. Make an incision through the teat and the glandular tissue, exposing the glandular part of the lactiferous sinus; this part is connected to the glandular tissue by numerous lactiferous ducts (Fig. 4-10, *B*).

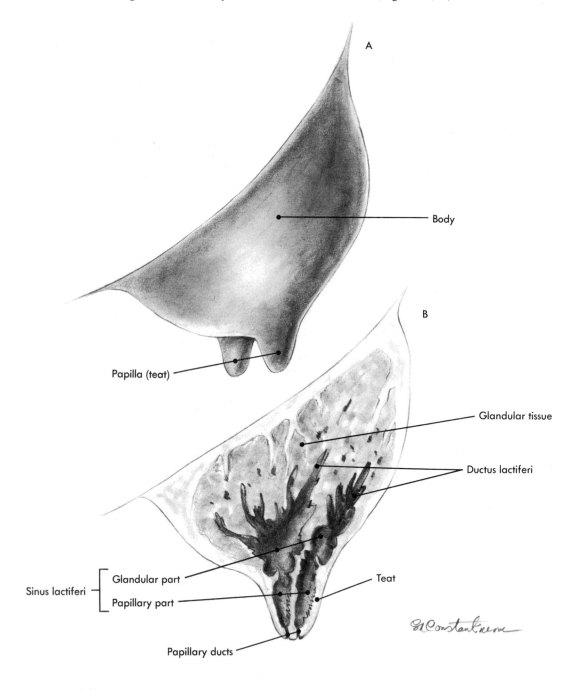

Fig. 4-10   **A,** Mammary gland of mare (left lateral aspect). **B,** Sagittal section through a half of mammary gland.

In both the male and female, identify the internal pudendal A., as a branch of the internal iliac A. The umbilical A. branches off close to the origin of the internal pudendal A. and courses toward the urinary bladder in the free edge of the lateral lig. of the bladder. From it come the deferential A. in the male and the ureteric br. in the female; it then becomes fibrous with an obliterated lumen (a nonpatent artery). However, a small artery (the cranial vesical A.) does branch off of it, and this artery supplies the apex of the bladder. The umbilical A. becomes the round lig. of the urinary bladder.

Identify the following structures as landmarks for the dissection of the pelvic walls: the tuber coxae, tuber sacrale, and tuber ischiadicum; the hip joint; and the greater and third trochanters. Skin the croup (or pelvic wall) between the abdominal and perineal areas and the base of the tail, and skin the thigh distal to the stifle joint. Reflect the skin over the crus. The dorsolateral (cutaneous) branches of the lumbar nerves (the so-called Nn. clunium craniales), the dorsolateral (cutaneous) branches of the sacral nerves (the so-called Nn. clunium medii), the caudal cutaneous femoral N. (or the Nn. clunium caudales), and the caudal Nn. are now exposed.

Between the tuber coxae and the third trochanter identify (through the gluteal fascia) a light white-yellow line separating the tensor fasciae latae M. from the cranial part of the superficial gluteal M. Make an incision along this line and separate the two muscles. From the third trochanter toward the sacral spine try to identify (through the gluteal fascia) the light-colored line separating the caudal border of the superficial gluteal M. from the biceps femoris M., and make an incision along this line. At a distance measured by three fingers cranial and parallel to the last line of separation is the cranial border of the caudal part of the superficial gluteal M. Incise the gluteal fascia carefully up to the tuber sacrale, and separate the superficial gluteal M. from the middle gluteal M.

Notice that the superficial gluteal M. is attached to the tuber sacrale by the gluteal fascia and not by its muscular portion. The cranial part of the superficial gluteal M. gets smaller and thinner from the third trochanter to the tuber coxae.

Remove the gluteal fascia from the middle gluteal M. Incise the superficial lamina of the fascia lata parallel and 2 cm caudal to the cranial border of the biceps femoris M. between the third trochanter and the patella. Reflect the fascia cranially up to the cranial border of the biceps femoris M., and examine the common origin of the superficial and deep laminae of the fascia lata. The deep lamina lies deep to the biceps femoris M. in a caudal direction. Cranial to the biceps femoris M., only the fascia lata is present as a unique structure attached to the tensor fasciae latae M. Examine the duplication of the superficial lamina toward the patella (Fig. 4-11).

Return to the croup, carefully separate the superficial gluteal M. from the biceps femoris M., and identify the aponeurosis of the superficial gluteal M., which lies deep to the biceps femoris and semitendinosus Mm. up to the tuber ischiadicum.

Notice the continuity between this aponeurosis and the deep fascia lata. Examine the gluteal fascia, which continues as the caudal fascia and the superficial lamina of the fascia lata.

Transect the sacral origin of the biceps femoris M. and isolate it from the sacrosciatic lig. Transect the ischiadic attachment of the muscle, and reflect it ventrally.

Caution! The biceps femoris M. has an attachment on the caudal aspect of the femur (the tuberositas m. bicipitis). Do not transect it yet.

Transect the sacrocaudal origin of the semitendinosus M., isolate it from the origin of the sacrococcygei Mm. and from the sacrosciatic lig., and reflect both the biceps femoris and semitendinosus Mm. ventrally. Reflecting the semitendinosus M., look at the ischiadic bursa of this muscle. Transect the origin of the superficial gluteal M. on the tuber sacrale, free its caudal border from the aponeurosis, and reflect the muscle ventrally.

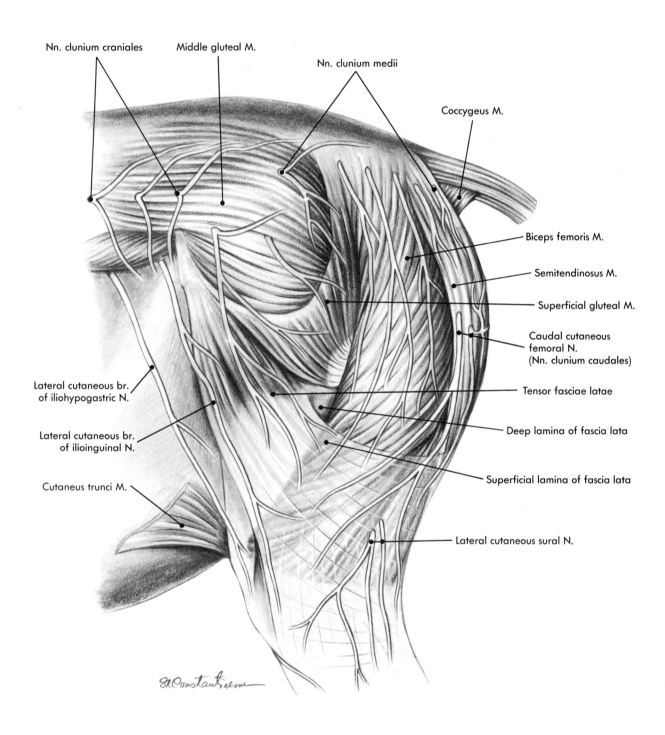

Nn. clunium craniales

Middle gluteal M.

Nn. clunium medii

Coccygeus M.

Biceps femoris M.

Semitendinosus M.

Superficial gluteal M.

Caudal cutaneous femoral N. (Nn. clunium caudales)

Lateral cutaneous br. of iliohypogastric N.

Tensor fasciae latae

Deep lamina of fascia lata

Lateral cutaneous br. of ilioinguinal N.

Superficial lamina of fascia lata

Cutaneus trunci M.

Lateral cutaneous sural N.

Fig. 4-11

Pelvic wall and thigh (left lateral aspect).

Identify the caudal part of the greater trochanter, one of the points of attachment of the middle gluteal M. At a point one-hand cranial to the greater trochanter, make a transverse incision through the middle gluteal M.; the accessory gluteal M. will be exposed.

Notice the glittery tendon of the accessory gluteal M., which differentiates it from the middle gluteal M. Identify the cranial part of the greater trochanter and its lateral crest, where the accessory gluteal M. inserts. Between the tendon and the cranial part of the trochanter, the trochanteric bursa of the accessory gluteal M. is located; probe it. Reflect the stumps of the middle gluteal M., and toward its caudal insertion, another muscle will be exposed: the deep gluteal M., with fibers connecting the ischiadic spine (of the coxal bone) to the cranial part of the greater trochanter. Probe the trochanteric bursa of the middle gluteal M. and identify the piriformis M., which is part of the middle gluteal M. and which is inserted on the intertrochanteric crest. Reflect the cranial stump of the middle gluteal M. over the accessory gluteal M. and up to the cranial border of the wing of the ilium. The gluteal line separates the territories of attachment of the middle (medially) and accessory (laterally) gluteal Mm. (Fig. 4-12).

Transect the tensor fasciae latae M. at a point 5 cm ventral to its insertion on the tuber coxae. Close to the tuber coxae, the iliolumbar A. and the ventral br. of L_4 are visible after surrounding the ilium medially to laterally around the lateral border of the wing of the ilium. In a deeper location, ventral to the tuber coxae and between the lateral part of the iliacus and internal abdominal oblique Mm., branches of the deep iliac circumflex A.V. and lateral cutaneous femoral N. are found. Deep to the middle gluteal M. between the wing of the ilium and the lateral border of the sacrum and between the tuber sacrale and the spinous processes of the sacrum, the two parts of the dorsal sacroiliac lig. are also exposed. From between these two parts of the dorsal sacroiliac lig., the dorsal sacrococcygei Mm. emerge. Deep to the middle gluteal, biceps femoris, and semitendinosus Mm., the sacrosciatic lig. is widely exposed. Identify the two ischiadic foramina: the greater and lesser. Identify and dissect the nerves and vessels passing through the greater ischiadic foramen from the pelvic cavity to the croup: the cranial gluteal vessels and the cranial gluteal, sciatic, caudal gluteal, and caudal cutaneous femoral Nn. Dissect them.

Identify the caudal gluteal vessels giving off the cranial gluteal and caudal Aa. and Vv. and supplying muscles in the corresponding area.

Transect the vertebral attachment of the semimembranosus M., isolate it from the sacrococcygei Mm., free its attachments on the sacrosciatic lig. and tuber ischiadicum, and reflect it ventrally. Transect the sacrosciatic lig. from its attachments on the lateral sacral crest (border) and on the transverse processes of the first two caudal vertebrae; reflect the ligament ventrally.

Caution! Some nerves and vessels can travel on either the lateral or the medial aspect of the sacrosciatic lig. and can perforate the ligament to change sides; they can even be located within the thickness of the ligament. To avoid damaging the nerves or vessels, identify first the origin of the nerves as they issue from the ventral sacral foramina and the main arteries and veins. At the medial aspect of the sacrosciatic lig., the following vessels and nerves are exposed: the caudal gluteal vessels; the cranial gluteal vessels and nerve; the obturator vessels and nerve; the internal pudendal vessels with the umbilical, prostatic (vaginal) vessels and their corresponding branches; the pudendal N.; the caudal rectal N. (or Nn.); and the pelvic Nn. (parasympathetic). From the transverse processes of the first 3 to 5 caudal vertebrae to the medial aspect of the sacrosciatic lig. and dorsal to the ischiadic spine is the territory of the coccygeus M. (in an oblique position). Medial to it and from the ischiadic spine to the anus, the levator ani M. with its three portions (that is, coccygeal, anal, and perineal) is exposed. Carefully transect the coccygeus M. to allow the dissection of the vessels and nerves toward the anogenital area and to widely expose the rectum (Fig. 4-13).

Internal abdominal oblique M.

Deep circumflex iliac A.V. & lateral cutaneous femoral N.

Tuber coxae

Middle gluteal M.

Lateral part of iliacus M.

Accessory gluteal M.

Iliacofemoral A.V. & cranial gluteal N.

Cranial gluteal V.A.

Deep gluteal M.

Sacrosciatic lig.

Caudal gluteal N.

Sciatic (ischiadicus) N.

Caudal cutaneous femoral N.

Biceps femoris M.

Ischiadic ln.

Caudal gluteal V.A.

Sacrococcygei Mm.

Coccygeus M.

Internal pudendal A.V.

Piriformis M.

Gemelli Mm. & tendon of internal obturator M.

Quadratus femoris M.

Biceps femoris M.

Middle gluteal M.

Semitendinosus M.

Tibialis N.

Semimembranosus M.

Superficial gluteal M.

Adductor major M.

Common fibular N.

Lateral cutaneous sural N.

Lateral saphenous V. & caudal cutaneous sural N.

Lateral head of gastrocnemius M.

Rectus femoris M.

Vastus lateralis M.

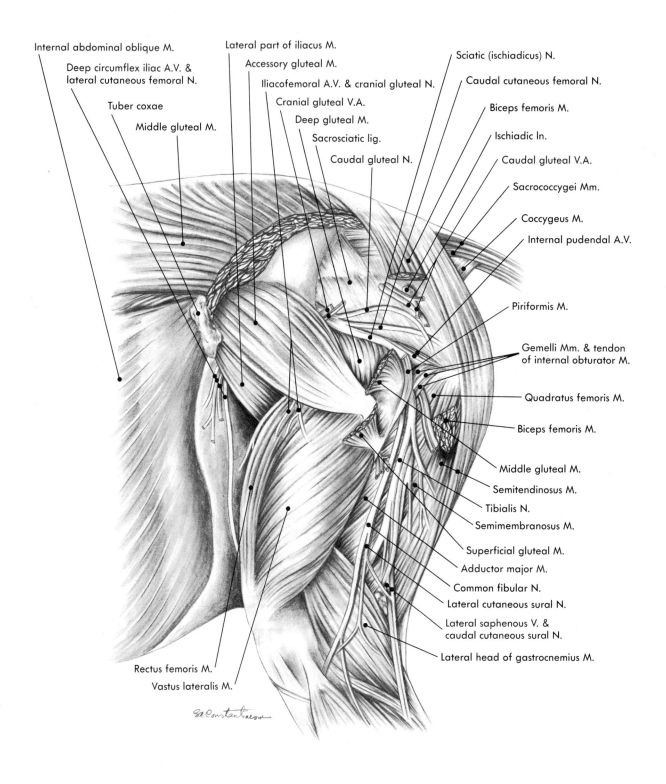

Fig. 4-12

Deep structures of lateral pelvis and thigh (left limb).

Examine the peritoneal reflections and the pelvic diaphragm. Identify and dissect the retractor penis M. or retractor clitoridis M., partially overlapped by the levator ani M. Some longitudinal muscle fibers of the rectum exceed it, running toward the first few caudal vertebrae as the rectococcygeus M. Identify the three parts of the external anal sphincter (that is, the superficial, deep, and cutaneous parts).

Now carefully continue the dissection of the vessels and the nerves, whose names and territories supplied vary with the sex of the specimen.

In the male, *notice* the prostatic A., the middle rectal A. (Aa.)., the ventral perineal and caudal rectal A., the A. of the penis, the A. of the bulb of the penis, the deep A. of the penis, and the dorsal A. of the penis. The first divisions of the pudendal N. are the cutaneous branches; then the deep and the superficial perineal Nn., the preputial and scrotal branches, and the dorsal N. of the penis. The caudal rectal Nn. supply the rectum and the anus by muscular and cutaneous branches. Identify and dissect all these structures.

In the mare, expose the vaginal A., the uterine br. with the caudal vesical A. (cranially) and the middle rectal A. (caudally), and the vestibular br., of the internal pudendal A. The ventral perineal A. branches into the caudal rectal A. (as in the male) and the dorsal labial br. The middle clitoral A. branches off of the obturator A. From the superficial perineal N. comes the labial Nn. A mammary br. and the dorsal N. of the clitoris are also present. Identify and dissect all these structures.

Palpate the transverse processes of the first few caudal vertebrae. On both aspects of them, dissect the dorsal and the ventral caudal intertransversarii Mm. On the dorsal aspect, dissect both the medial and lateral dorsal sacrococcygei Mm., and on the ventral aspect, dissect the lateral and the medial ventral sacrococcygei Mm. Dorsal and ventral to the transverse processes lie the dorsolateral and the ventrolateral caudal Aa. The median caudal A. lies between the paired medial ventral sacrococcygei Mm.

The coccygeus M. has already been transected. Now transect the levator ani M., reflect the stumps, and expose the origin of the retractor penis or clitoridis M. (Fig. 4-14).

Next, inspect the pelvic outlet (the perineum) (Figs. 4-15 and 4-16).

Before removing the genital tract from the pelvic cavity, palpate and identify the peritoneal reflections in the pelvic cavity, introducing one hand first between the sacrum and the rectum, then between the rectum and the genital organs, then between them and the urinary bladder, and finally between the urinary bladder and the floor of the pelvic cavity. The peritoneal reflections (pouches) are the following: the sacrorectal (divided in two pararectal fossae), the rectogenital, the vesicogenital, and the pubovesical pouches. Caudal to these reflections, only connective tissue surrounds the terminal segments of the digestive and urogenital apparatus. The so-called pelvic diaphragm (the connective tissue just mentioned) is anchored on the ischial arch, the medial aspect of the sacrosciatic ligg., and the caudal vertebrae, giving the viscera a stable position. Within the pelvic diaphragm, two muscles and two fasciae are considered here: the coccygeus and levator ani Mm. and the external and internal fasciae of the pelvic diaphragm.

Transect all the attachments of the pelvic diaphragm and free the internal male genitalia with their arterial, venous, and nervous supply already dissected. To identify the vessels and nerves easily, maintain their attachments to the viscera for as long as possible. Transect the three ligaments of the urinary bladder and the genital fold; free the ureters, then the rectum; transect the ischiocavernosus Mm. together with the crura of the penis; and choose between the following technique variations: (1) transect the penis approximately 15 cm distal to the ischial arch or (2) free the penis from all its attachments up to the most caudal cross-section you have already performed.

Identify the two lobes of the prostate, which are united by the median isthmus overlapping the ejaculatory ducts. Caudal to the prostate is the pelvic part of the urethra covered by the urethralis M.; examine this. Close to the ischiocavernosus Mm. and located between the urethralis M. and the bulbospongiosus M. are the paired bulbourethral glands, partially covered by the bulboglandularis M. (Fig. 4-17).

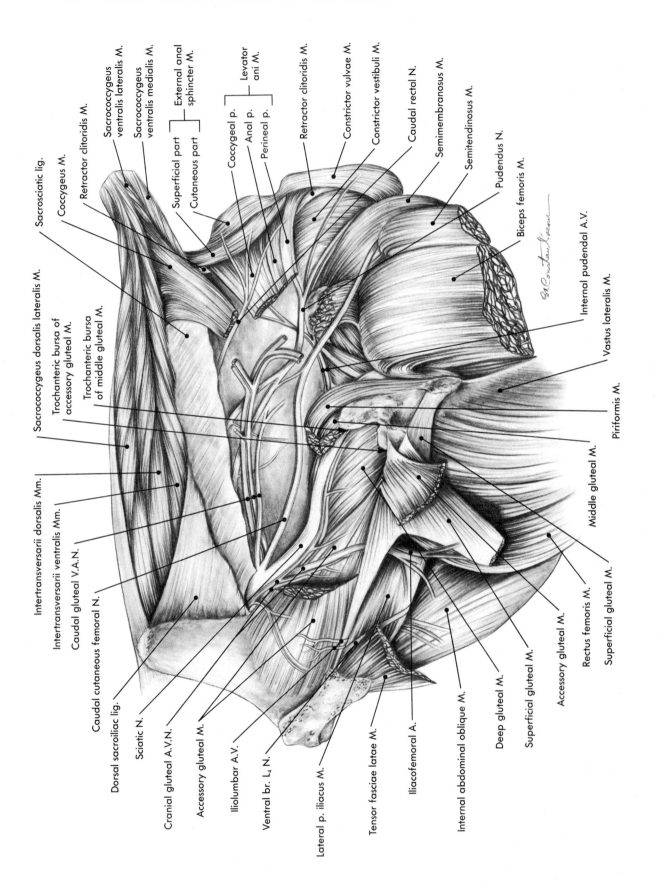

Fig. 4-13   Deepest structures of lateral pelvis, tail, anus, and vulva of the mare (left side).

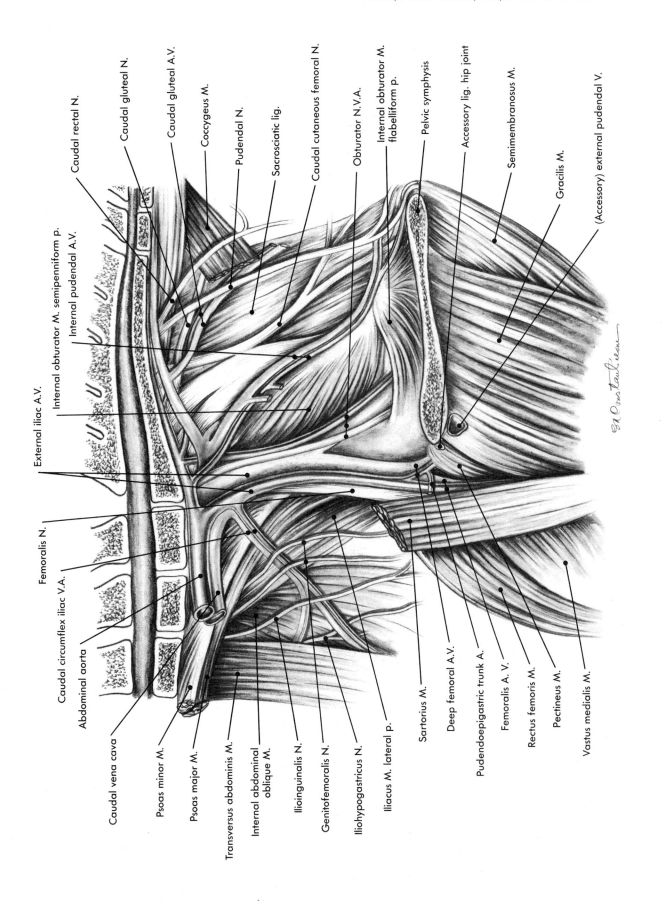

Caudal rectal N.

Caudal gluteal N.

Caudal gluteal A.V.

Coccygeus M.

Pudendal N.

Sacrosciatic lig.

Caudal cutaneous femoral N.

Obturator N.V.A.

Internal obturator M. flabelliform p.

Pelvic symphysis

Accessory lig. hip joint

Semimembranosus M.

Gracilis M.

(Accessory) external pudendal V.

Internal obturator M. semipenniform p.

Internal pudendal A.V.

External iliac A.V.

Femoralis N.

Caudal circumflex iliac V.A.

Abdominal aorta

Caudal vena cava

Psoas minor M.

Psoas major M.

Transversus abdominis M.

Internal abdominal oblique M.

Ilioinguinalis N.

Genitofemoralis N.

Iliohypogastricus N.

Iliacus M. lateral p.

Sartorius M.

Deep femoral A.V.

Pudendoepigastric trunk A.

Femoralis A. V.

Rectus femoris M.

Pectineus M.

Vastus medialis M.

Fig. 4-14

Right pelvis (medial aspect).

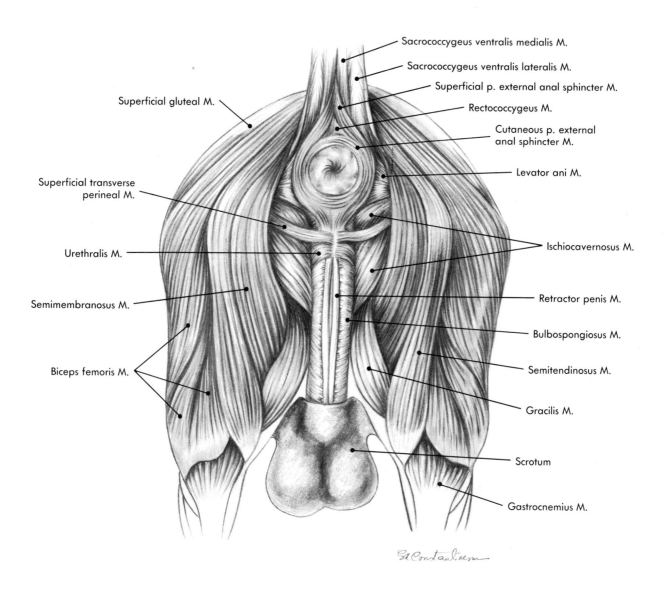

Superficial gluteal M.

Sacrococcygeus ventralis medialis M.

Sacrococcygeus ventralis lateralis M.

Superficial p. external anal sphincter M.

Rectococcygeus M.

Cutaneous p. external
anal sphincter M.

Levator ani M.

Superficial transverse
perineal M.

Urethralis M.

Semimembranosus M.

Biceps femoris M.

Ischiocavernosus M.

Retractor penis M.

Bulbospongiosus M.

Semitendinosus M.

Gracilis M.

Scrotum

Gastrocnemius M.

Fig. 4-15

Perineal region in the stallion.

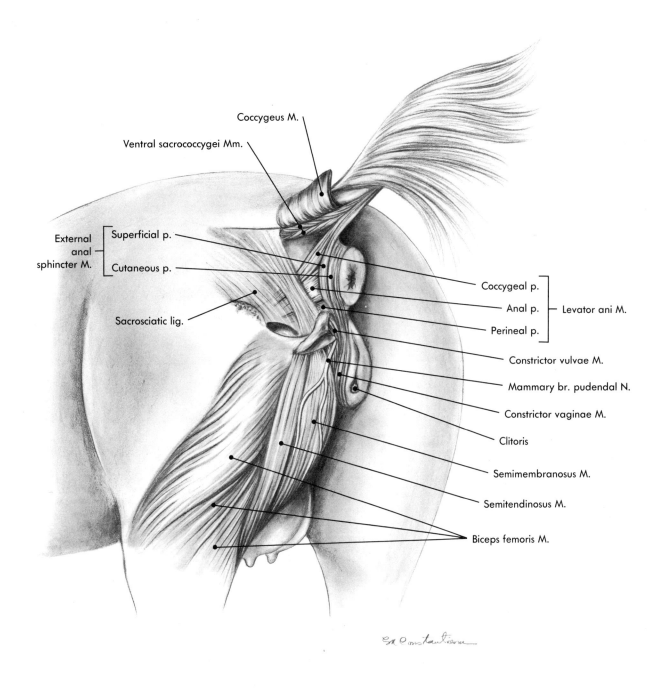

Coccygeus M.

Ventral sacrococcygei Mm.

External anal sphincter M.
- Superficial p.
- Cutaneous p.

Sacrosciatic lig.

Coccygeal p.
Anal p. — Levator ani M.
Perineal p.

Constrictor vulvae M.

Mammary br. pudendal N.

Constrictor vaginae M.

Clitoris

Semimembranosus M.

Semitendinosus M.

Biceps femoris M.

Perineal region in the mare.

Fig. 4-16

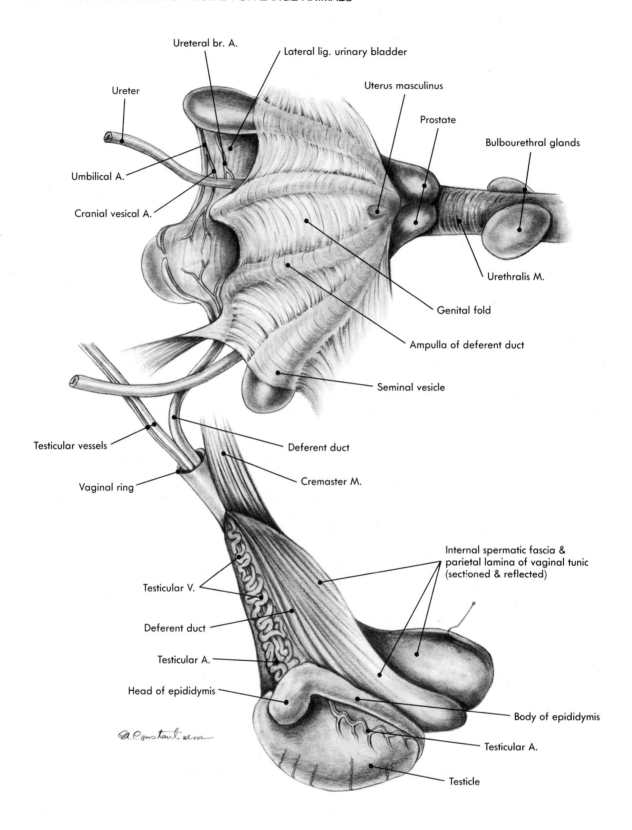

Ureteral br. A.

Lateral lig. urinary bladder

Ureter

Uterus masculinus

Prostate

Bulbourethral glands

Umbilical A.

Cranial vesical A.

Urethralis M.

Genital fold

Ampulla of deferent duct

Seminal vesicle

Testicular vessels

Deferent duct

Cremaster M.

Vaginal ring

Internal spermatic fascia &
parietal lamina of vaginal tunic
(sectioned & reflected)

Testicular V.

Deferent duct

Testicular A.

Head of epididymis

Body of epididymis

Testicular A.

Testicle

Fig. 4-17
Accessory genital glands (dorsal aspect); testicle and spermatic cord (left lateral aspect).

Make a longitudinal incision in the urinary bladder from the body through the neck of the organ, continuing to the pelvic part of the urethra. Examine the ureteral orifices and the longitudinal folds of the mucosa of the urinary bladder.

Notice that, between the two ureteral orifices and the neck of the urinary bladder toward the urethra, a small triangular area without mucosal folds is present. This is the trigonum vesicae.

On the dorsal wall of the urethra, identify the colliculus seminalis, a prominence onto which the two ejaculatory orifices open. *Notice* the numerous openings of the prostatic ductules on both sides of the colliculus. At the level of the bulbourethral glands, examine the openings of their ducts.

Continue the incision of the pars spongiosa of the urethra, and examine the corpus spongiosum penis, which starts with an enlargement named the bulb of the penis.

Review all the arteries, veins, and nerves supplying the structures already examined.

In a similar manner remove the female urogenital tract, including the vulva. Identify the pudendal A. and transect it close to its origin. Free the rectum from the subsacral area; free the urinary bladder and the genital tract; transect the pelvic diaphragm and the crura of the clitoris; free the vulva, the vestibulum, and the vagina; and remove all these structures through the pelvic inlet (Fig. 4-18, *D*).

Review the arteries, veins, and nerves related to the genital apparatus. Examine the relationship between the terminal portion of the urethra, the vagina (cranially), the vestibulum (caudally), and the external urethral orifice in between. Identify the constrictor vestibuli M., transect it, and expose the vestibular bulb (homologue of the corpus spongiosum penis).

Place the female genitalia in a natural position and make a longitudinal incision through the dorsal walls of the vestibulum, vagina, cervix, body of the uterus and one uterine horn (Fig. 4-18, *A*).

Examine the vulva with the two labia and the two commissures and the clitoris located on the floor of the vulva within the fossa clitoridis. Examine the glans clitoridis (homologue of the glans penis) and its relationship with the preputium clitoridis (homologue of the male prepuce) and the frenulum clitoridis (homologue of the male preputial frenulum) (Figs. 4-18, *B* and *C*). Examine the sinuses around the glans. Within the vestibulum, examine the small orifices of the vestibular glands. Cranial to the external urethral orifice, examine the transverse fold, which is comparable to the hymen. The vagina has no specific characteristics and ends near the vaginal segment of the cervix as the fornix vaginaè. Examine the external uterine opening (ostium) of the cervix, which is located in the middle of the prominent vaginal segment of the cervix. The opening is surrounded by folds of the mucosa.

Examine the cervix with its external and internal uterine openings (ostia), as well as the mucosa of the body and horns of the uterus.

Notice that the female horse is the only species in which the uterine body is not (partially) divided. Consequently, the most cranial extent of it can also be called the fundus uteri.

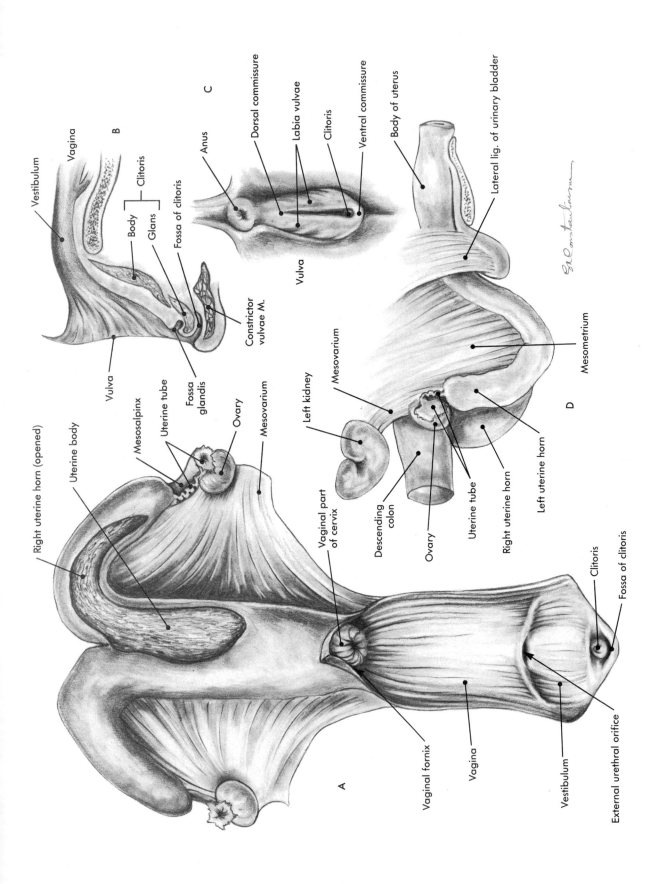

Fig. 4-18 **A,** Genital tract (dorsal aspect). **B,** Mediosagittal section through vulva and clitoris. **C,** Vulva. **D,** Left lateral view of ovary, salpinx, and uterus.

A procedure similar to that performed in horse is suggested. However, many characteristics are unique to the large ruminants. Examine the shape of the uterine horns (which is similar to the ram's horns). They are connected to each other by the intercornual ligg. (dorsal and ventral). The broad lig. is well developed, especially the proximal mesovarium, which is 25 to 30 cm long. In contrast to the mare, the round lig. of the uterus in the cow is located in the free border of the broad lig., extending up to the deep inguinal or vaginal ring (Fig. 4-19).

Large ruminants

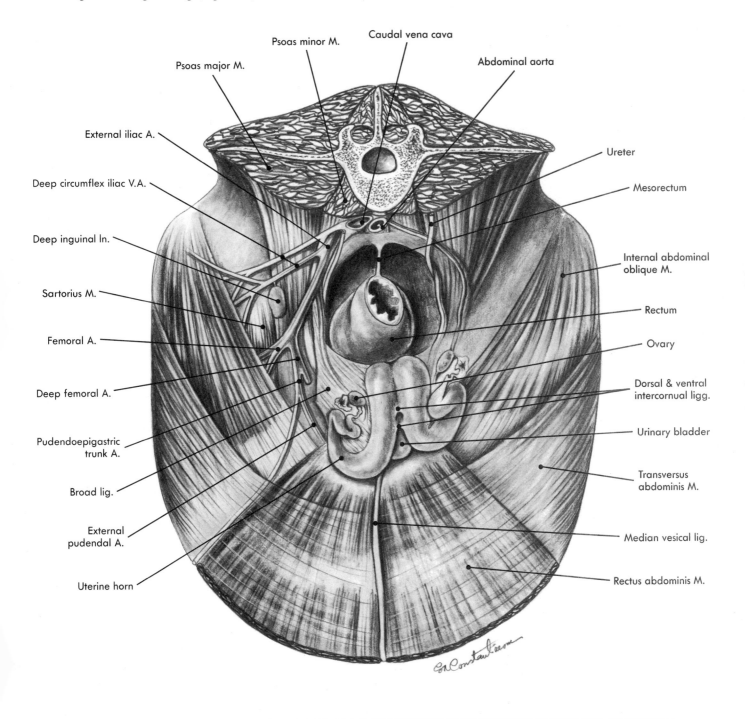

Pelvic inlet in the cow.

Fig. 4-19

Perform the dissection of the bull.

Make a longitudinal scrotodartoic incision; bluntly free the spermatic cord from the external spermatic fascia; transect the lig. of the tail of the epididymis; and examine the external cremaster M. and the position of the testicle. Palpate the ductus deferens and make a longitudinal incision in the internal spermatic fascia and the lamina parietalis of the vaginal tunic on the opposite side of the ductus deferens, penetrating the cavity of the vaginal tunic (Fig. 4-21). Examine the testicle, the epididymis, the testicular bursa, the ductus deferens, the vessels, and the serosal folds; compare them with those of the stallion (Fig. 4-22, *A* to *C*).

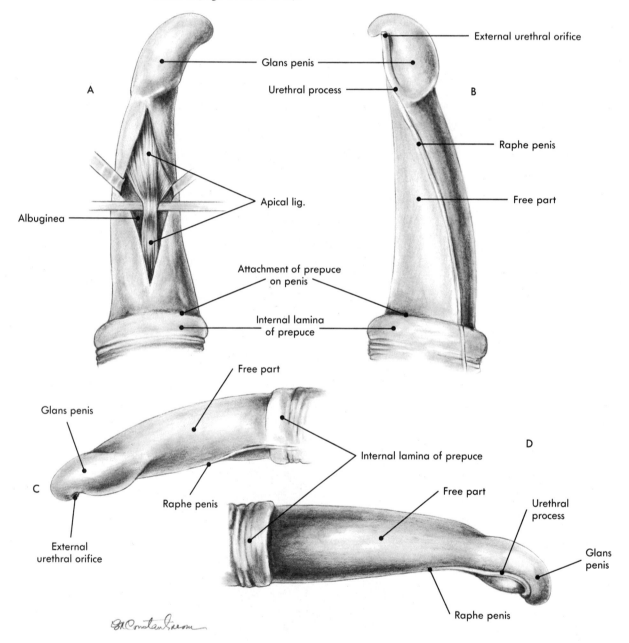

Fig. 4-20   **A,** Penis (dorsal aspect). **B,** Penis (ventral aspect). **C,** Penis (left lateral aspect). **D,** Penis (right lateral aspect).

Examine the exterior of the udder, the intermammary sulcus, and the teats.

In the dissection of the udder, first identify the two structures of origin, that is, the lateral and medial elastic laminae.

Examine the medial aspect of the gracilis and the adductor Mm. and the related symphyseal tendon, which extends from the prepubic tendon to the ischiadic symphysis. Ex-

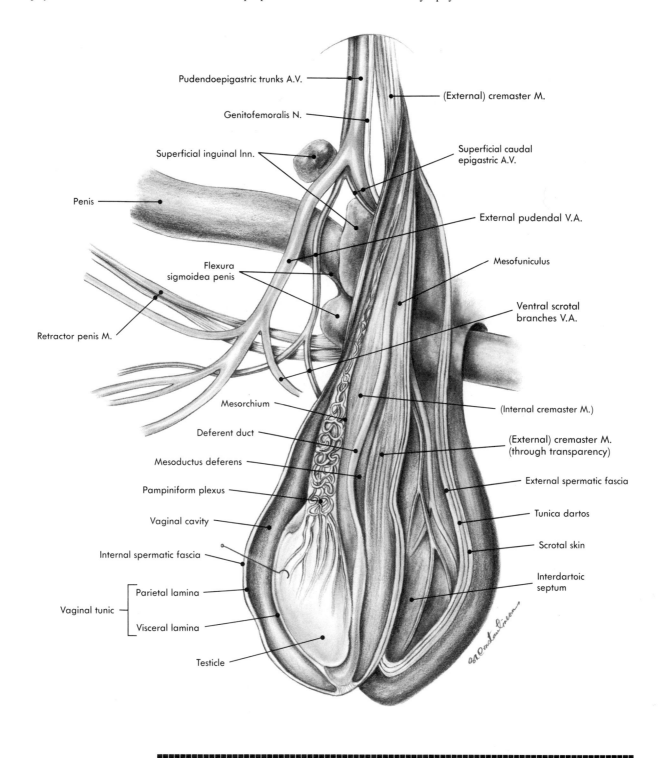

Right spermatic cord and adjacent structures.

Fig. 4-21

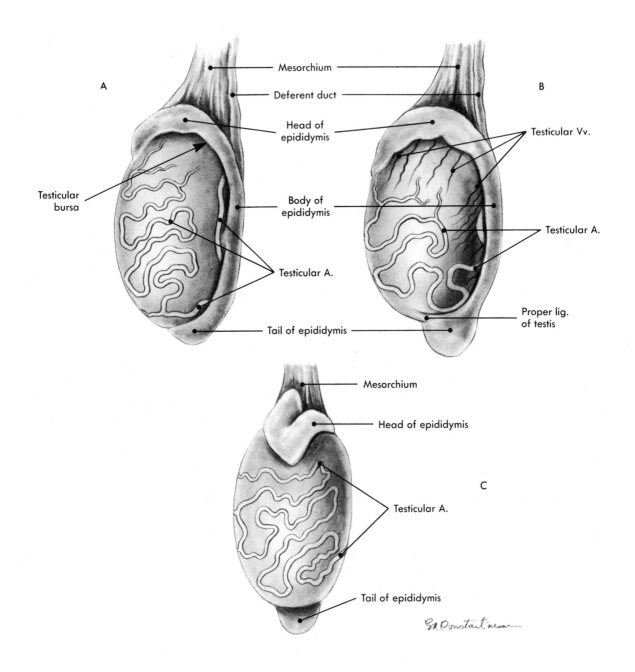

Fig. 4-22   **A,** Left testicle with testicular A. (caudolateral aspect). **B,** Left testicle with arterial and venous supply (caudolateral aspect). **C,** Left testicle with testicular A. (cranial view).

amine again the superficial inguinal ring. The lateral elastic laminae of the udder originate from the symphyseal tendon and the lateral border (crus) of the superficial inguinal ring. Follow the caudal extent of the abdominal tunic to its contact with the udder. Bluntly separate the two halves of the udder, exposing the medial laminae (which originate from the abdominal tunic) up to the fusion with fibers of the symphyseal tendon. Both the abdominal tunic and the symphyseal tendon are considered the origin of the medial elastic laminae of the udder (Fig. 4-23, *A*).

Remember! The suspensory apparatus of the udder is made up of the lateral and medial laminae.

Carefully remove one half of the udder from its base and from the symmetrical half, exposing the structures passing through the superficial inguinal ring (the external pudendal A.V. and the genitofemoral N.) and the associated structures (the superficial inguinal or mammary lnn. with their own vessels). Identify the two mammary Aa. and note their branching pattern. Cranial to the udder, as an anastomosis through inosculation (an anastomosis of 180 degrees between two already existing arteries or veins in which the direction of the blood flow is opposite), identify and dissect the caudal and cranial superficial epigastric Vv. in continuation with each other.

Notice that the caudal superficial epigastric V. is a branch of the external pudendal V., whereas the cranial V. is a branch of the superficial thoracic V., together forming the so-called milk vein. Sometimes the cranial superficial epigastric V. drains into the internal thoracic V., perforating the abdominal wall through the so-called milk well.

Although the *Nomina Anatomica Veterinaria* (N.A.V.) uses the synonym term *subcutaneous abdominal V.* for only the cranial superficial epigastric V. (the synonym for the caudal superficial epigastric V. is the *cranial mammary V.*), some authors have used the term *subcutaneous abdominal V.* to describe the cranial and caudal superficial epigastric Vv. together.

Notice that all the medial superficial branches of both the cranial and caudal superficial epigastric Vv. anastomose with each other on the cranial aspect of the udder, whereas the medial superficial branches of the perineal Vv. anastomose on the caudal aspect of the udder. The combination of these veins results in the venous mammary ring.

Reexamine the ventral cutaneous branches of the iliohypogastric, the ilioinguinal, and the genitofemoral Nn. supplying the udder.

Make an incision in a teat in its longitudinal axis, and examine the wall, which is composed of three layers: a skin, musculovascular, and mucosal layers. At the base of the teat, examine the venous ring. Inspect the papillary duct (the teat canal) and the papillary part of the lactiferous sinus (the teat sinus) (Fig. 4-23, *B*). Continue the incision within the mammary gland and expose the glandular part of the lactiferous sinus and the glandular tissue. The two parts of the lactiferous sinus may be partially separated by mucosal folds; examine them. Inspect the lactiferous ducts, and compare them to those of the previously dissected mare (Fig. 4-23, *C*).

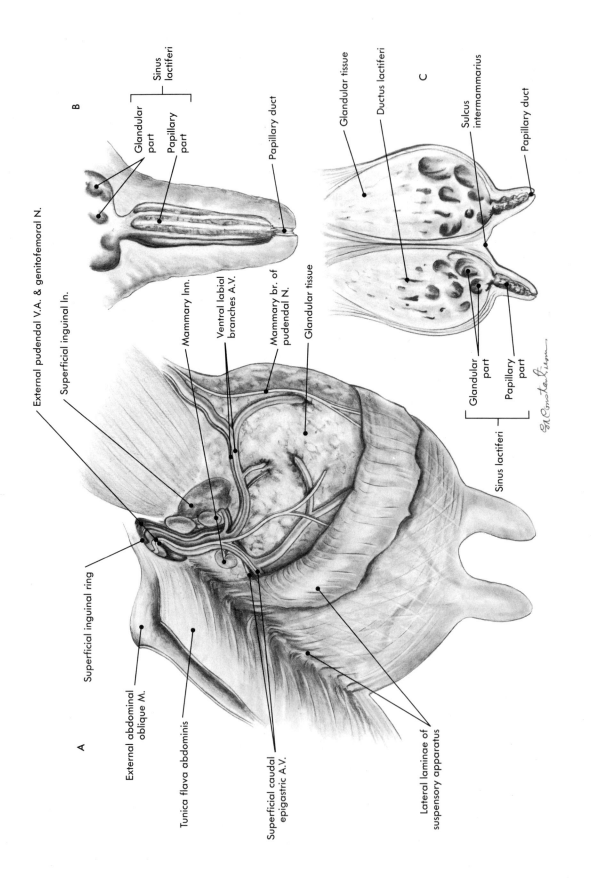

A, Mammary gland in the cow. **B,** Mammary papilla (teat). **C,** Transverse section to udder in the cow.

Fig. 4-23

For the dissection of the pelvic walls, the identification of the same landmark structures as in the horse is suggested.

Notice that in the ruminants the greater trochanter is not divided into cranial and caudal parts, and the third trochanter is not present.

Notice that the tuber ischiadicum of the ox has three processes: dorsal, lateral, and ventral. The proximal cutaneous br. of the pudendal N. branches out cranial to the dorsal process, whereas the distal br. becomes subcutaneous on the medial aspect of the tuber ischiadicum (between it and the ischiorectal fossa, midway between the dorsal and the ventral processes of the tuber ischiadicum) (Fig. 4-24).

Follow the same instructions suggested for the previous species and dissect the pelvic walls of the ox and cow.

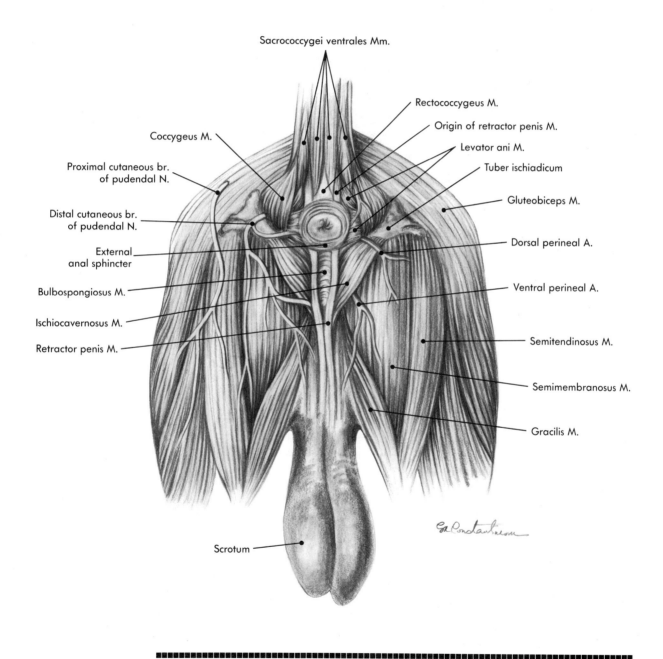

Perineal region in the bull.

Fig. 4-24

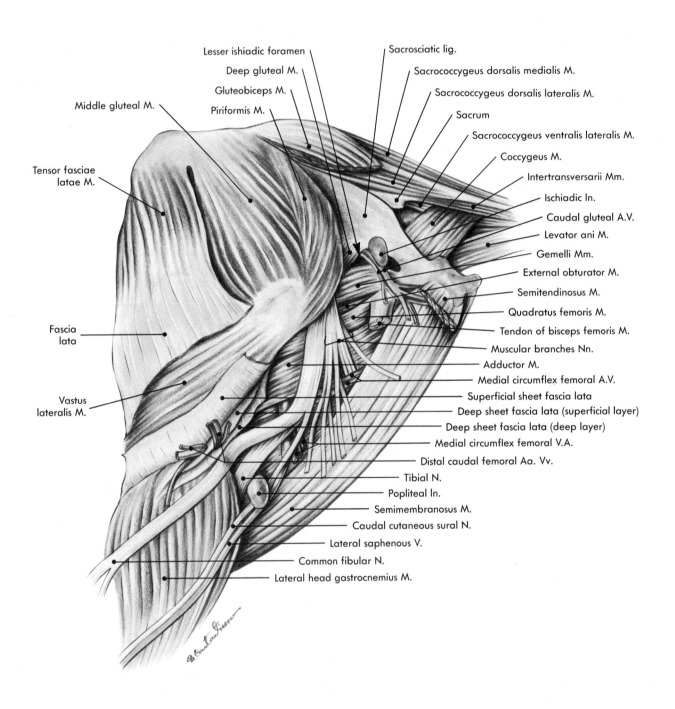

Fig. 4-25

Deep lateral structures of croup and thigh (left pelvic limb).

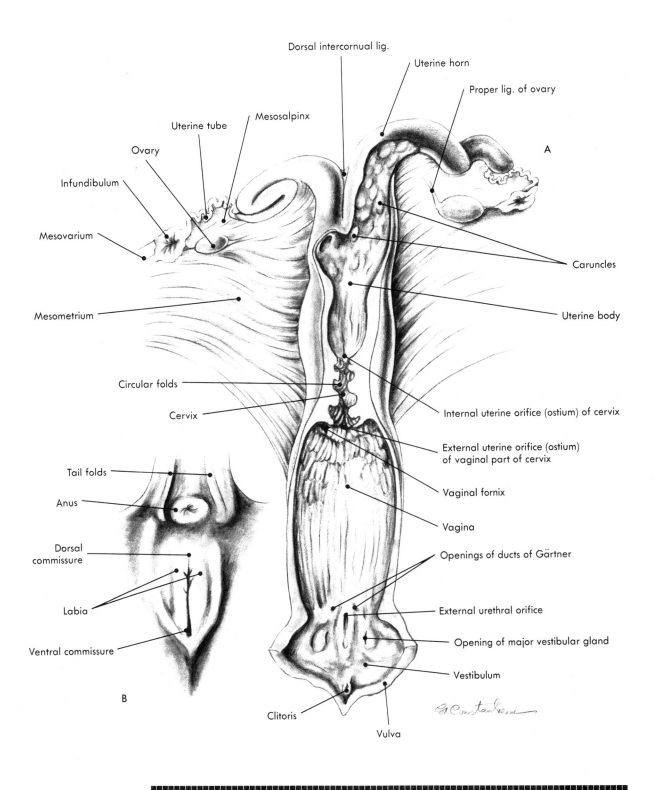

Dorsal intercornual lig.

Uterine horn

Proper lig. of ovary

Mesosalpinx

Uterine tube

Ovary

Infundibulum

Mesovarium

Mesometrium

Circular folds

Cervix

Tail folds

Anus

Dorsal
commissure

Labia

Ventral commissure

B

Clitoris

A

Caruncles

Uterine body

Internal uterine orifice (ostium) of cervix

External uterine orifice (ostium)
of vaginal part of cervix

Vaginal fornix

Vagina

Openings of ducts of Gärtner

External urethral orifice

Opening of major vestibular gland

Vestibulum

Vulva

A, Genital tract (dorsal aspect). **B,** Vulva.   **Fig. 4-26**

Sheep

▪▪▪▪▪▪▪▪▪▪

(Figs. 4-27 to 4-29)

Use the large ruminants as a model for dissecting the pelvic inlet and the pelvis in sheep. Review the section on the horse to apply that knowledge and those techniques to the small ruminants.

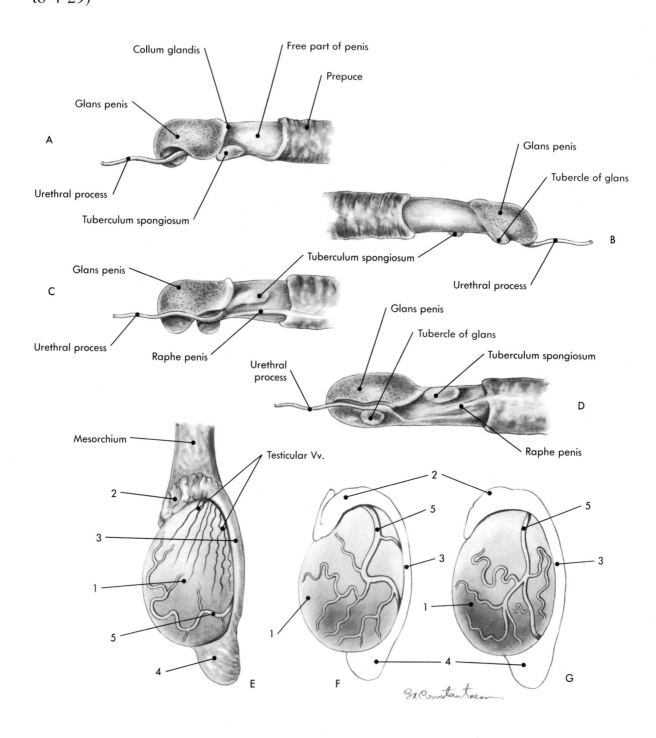

Fig. 4-27   **A,** Penis (left lateral aspect). **B,** Penis (right lateral aspect). **C,** Penis (left lateroventral aspect). **D,** Penis (ventral aspect). **E,** Testicle (caudolateral aspect). **F** and **G,** Variations of testicular A. *1,* Testicle; *2,* head of epididymis; *3,* body of epididymis; *4,* tail of epididymis; *5,* testicular A.

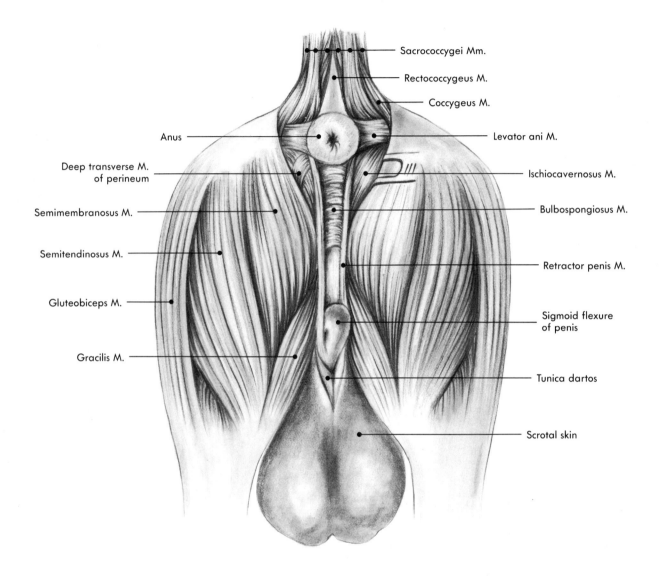

Sacrococcygei Mm.

Rectococcygeus M.

Coccygeus M.

Anus

Levator ani M.

Deep transverse M.
of perineum

Ischiocavernosus M.

Semimembranosus M.

Bulbospongiosus M.

Semitendinosus M.

Retractor penis M.

Gluteobiceps M.

Sigmoid flexure
of penis

Gracilis M.

Tunica dartos

Scrotal skin

Perineal region in the ram.

Fig. 4-28

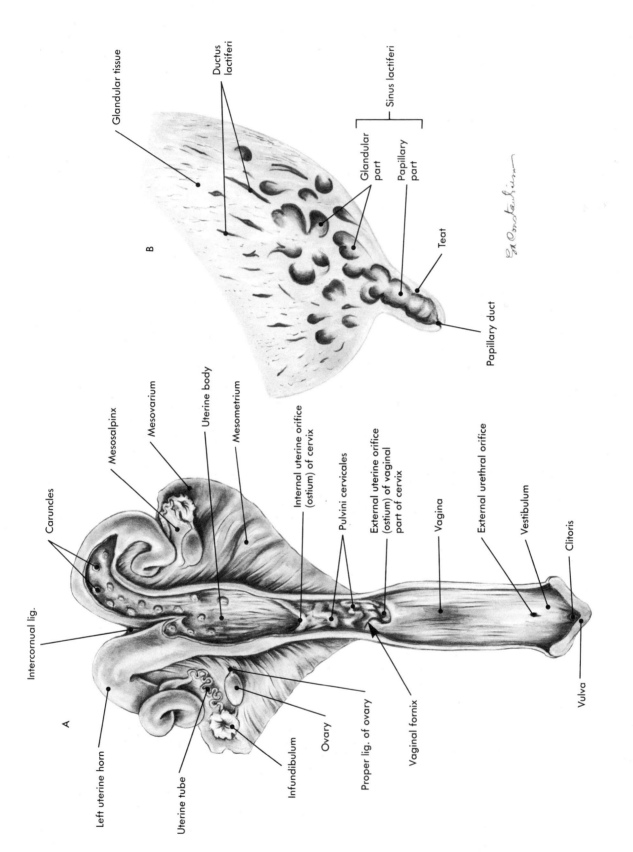

Fig. 4-29 **A**, Genital tract in ewe (dorsal view). **B**, Sagittal section through half of mammary gland.

Very few differences are encountered in the goat in comparison with the sheep. The pelvic inlet, pelvic walls, and perineum are similar.

In the female goat, the vestibular glands are not present. The ducts of Gärtner are similar to those of the cow, but they are not always present. The internal orifice of the cervix is similar to the ewe, that is, it is not distinct. The surface of the caruncles of the endometrium is planar. The vulva and clitoris are similar to those of the ewe.

The same paired inguinal mammary glands as in the ewe are observed in the female goat. No inguinal sinus is found in the goat. Compare the udder and especially the teats of the female goat to those in the ewe; they are much larger in the goat.

In the male goat, only the shape and the branching of the testicular A. are unique. The urethral process, glans, and free part of the penis are similar to the ram, but no tubercles or recess are observed. The superficial transverse perineal M. is represented by a vestigeal fibrous structure just ventral to the tuber ischiadicum.

Goat
■■■■■■■■
(Figs 4-30 to 4-33)

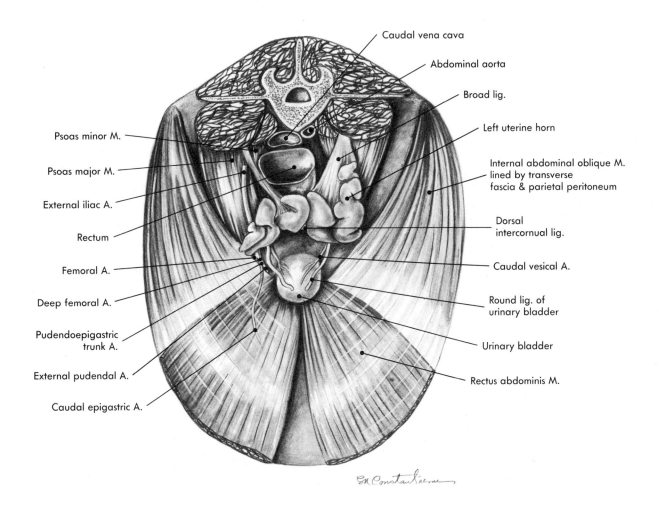

Pelvic inlet in the female goat.

Fig. 4-30

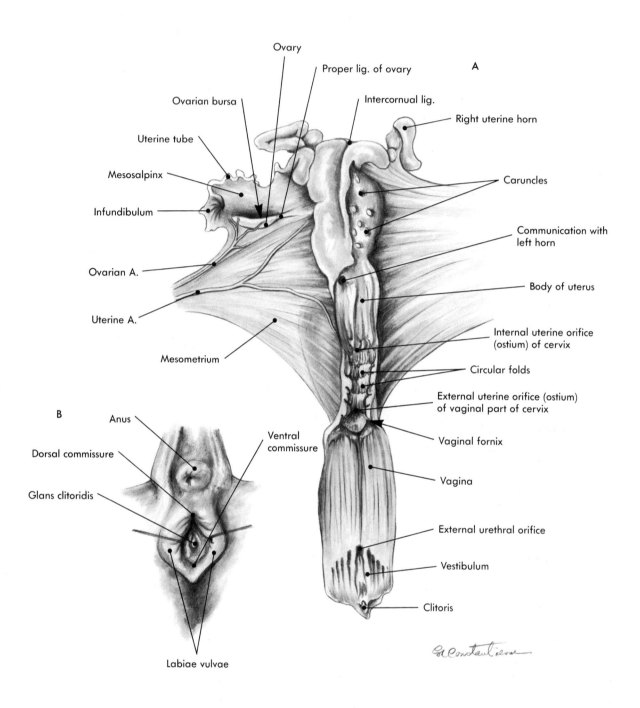

Ovary

Proper lig. of ovary

A

Intercornual lig.

Ovarian bursa

Right uterine horn

Uterine tube

Mesosalpinx

Caruncles

Infundibulum

Communication with left horn

Ovarian A.

Body of uterus

Uterine A.

Internal uterine orifice (ostium) of cervix

Circular folds

Mesometrium

External uterine orifice (ostium) of vaginal part of cervix

Vaginal fornix

B

Anus

Ventral commissure

Dorsal commissure

Vagina

Glans clitoridis

External urethral orifice

Vestibulum

Clitoris

Labiae vulvae

Fig. 4-31   **A,** Genital tract of the female goat (dorsal aspect). **B,** Vulva in the goat.

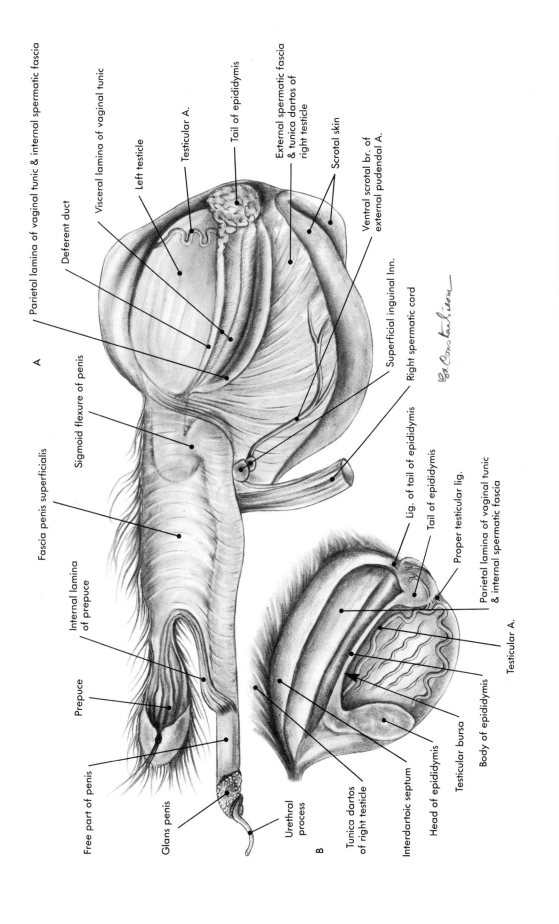

A, Penis and left testicle (lateral side). **B,** Left testicle (lateral aspect).

Fig. 4-32

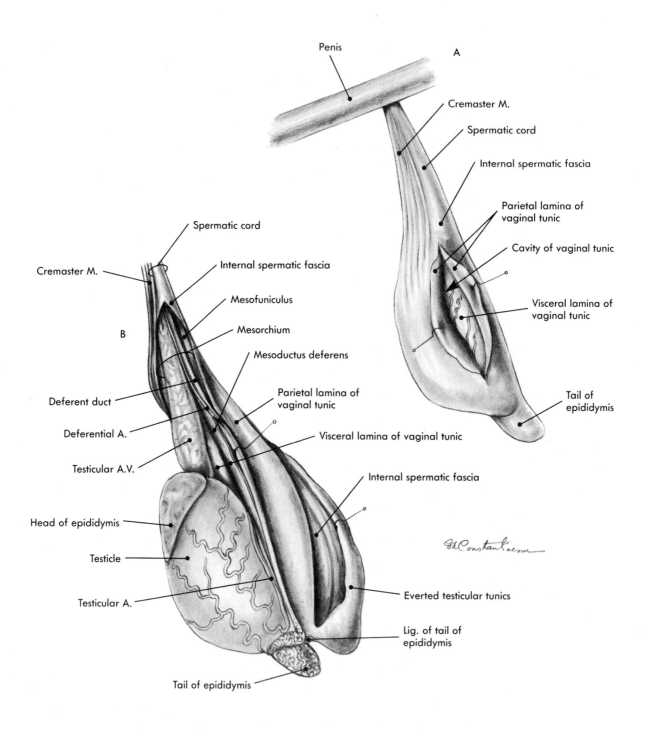

Penis

A

Cremaster M.

Spermatic cord

Internal spermatic fascia

Parietal lamina of vaginal tunic

Cavity of vaginal tunic

Visceral lamina of vaginal tunic

Tail of epididymis

Spermatic cord

Cremaster M.

Internal spermatic fascia

Mesofuniculus

B

Mesorchium

Mesoductus deferens

Deferent duct

Parietal lamina of vaginal tunic

Deferential A.

Visceral lamina of vaginal tunic

Testicular A.V.

Internal spermatic fascia

Head of epididymis

Testicle

Testicular A.

Everted testicular tunics

Lig. of tail of epididymis

Tail of epididymis

Fig. 4-33   **A,** Right testicle and spermatic cord (medial aspect). **B,** Right testicle and spermatic and testicular tunics.

Follow the directions suggested for the previous species and perform the dissection of the pig.

Pig
▪▪▪▪▪
(Figs. 4-34
to 4-38)

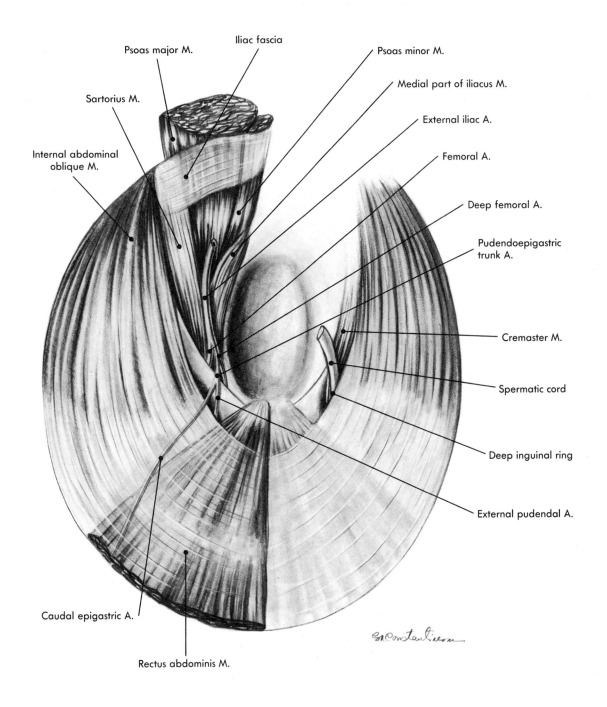

Psoas major M.

Iliac fascia

Psoas minor M.

Sartorius M.

Medial part of iliacus M.

External iliac A.

Internal abdominal oblique M.

Femoral A.

Deep femoral A.

Pudendoepigastric trunk A.

Cremaster M.

Spermatic cord

Deep inguinal ring

External pudendal A.

Caudal epigastric A.

Rectus abdominis M.

Pelvic inlet in the boar.

Fig. 4-34

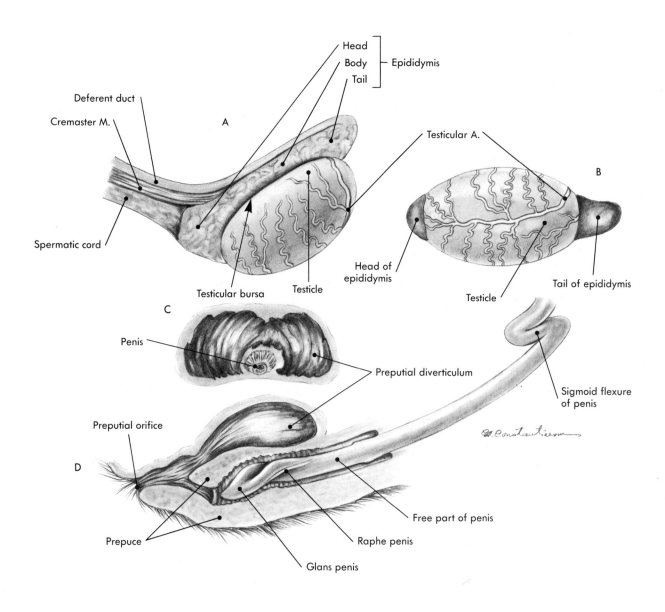

Fig. 4-35   **A,** Left testicle (lateral aspect). **B,** Testicle (ventrocaudal aspect). **C,** Transverse section through preputial diverticulum. **D,** Penis and prepuce (left lateral view).

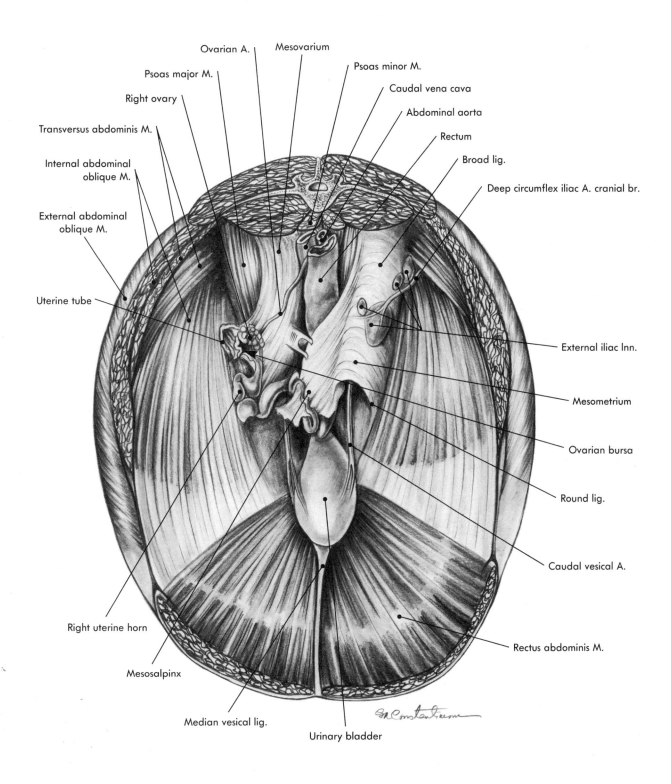

Ovarian A.

Mesovarium

Psoas minor M.

Caudal vena cava

Abdominal aorta

Rectum

Broad lig.

Deep circumflex iliac A. cranial br.

Psoas major M.

Right ovary

Transversus abdominis M.

Internal abdominal
oblique M.

External abdominal
oblique M.

Uterine tube

External iliac lnn.

Mesometrium

Ovarian bursa

Round lig.

Caudal vesical A.

Right uterine horn

Rectus abdominis M.

Mesosalpinx

Median vesical lig.

Urinary bladder

Pelvic inlet in the sow.

Fig. 4-36

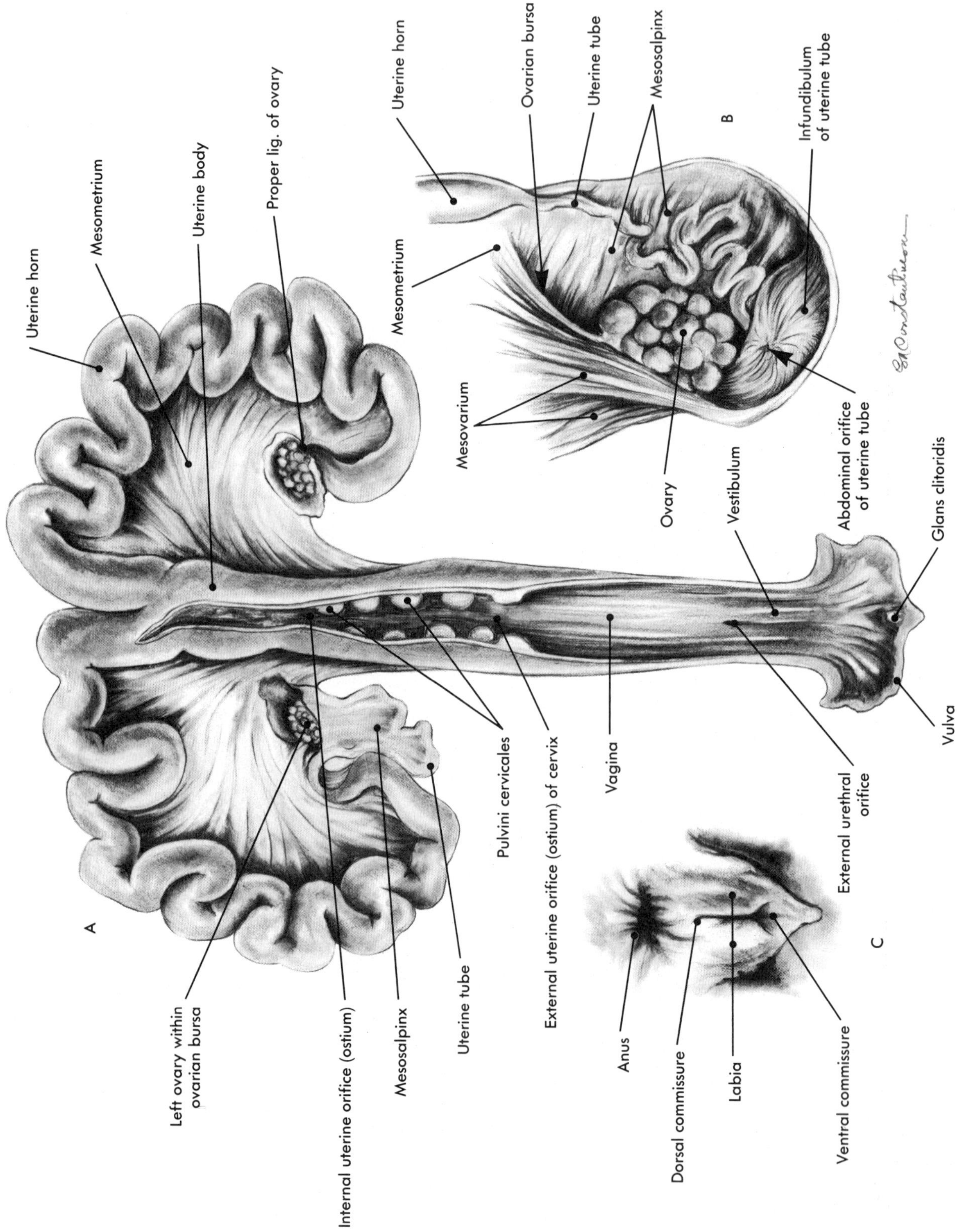

Uterine horn

Mesometrium

Uterine body

Proper lig. of ovary

Uterine horn

Mesometrium

Ovarian bursa

Uterine tube

Mesosalpinx

B

Infundibulum
of uterine tube

Mesovarium

Ovary

Vestibulum

Abdominal orifice
of uterine tube

Glans clitoridis

Vulva

A

Left ovary within
ovarian bursa

Internal uterine orifice (ostium)

Mesosalpinx

Uterine tube

Pulvini cervicales

External uterine orifice (ostium) of cervix

Vagina

External urethral
orifice

Anus

Dorsal commissure

Labia

Ventral commissure

C

A, Genital tract of sow (dorsal aspect). **B,** Right ovary with ovarian bursa. **C,** Vulva.

Fig. 4-37

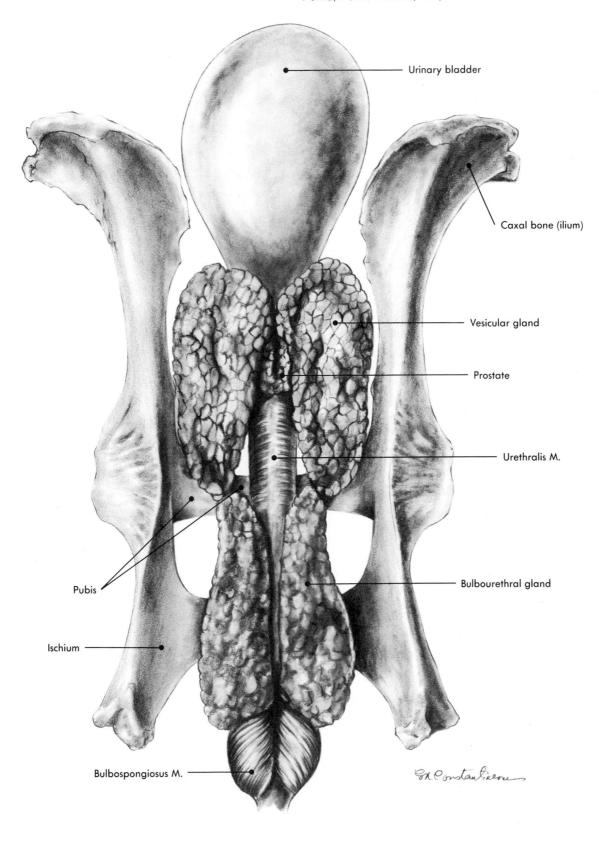

Urinary bladder

Caxal bone (ilium)

Vesicular gland

Prostate

Urethralis M.

Bulbourethral gland

Pubis

Ischium

Bulbospongiosus M.

Accessory genital glands of a mature boar (dorsal view).

Fig. 4-38

Chapter 5

Pelvic limb

██████████████████████████

Horse
██████████

Identify the following structures, which serve as landmarks, before starting the dissection (check them against a skeleton and either a living or an embalmed specimen): the patella; the crus; the lateral condyle of the tibia; the tibial crest (proximocranial aspect of the crus [located ventral to patella]); the common calcanean tendon (caudally); the body (shaft) of the tibia (on the medial aspect of the crus); the lateral and medial grooves, separating the common calcanean tendon from the rest of the crus; the two lateral grooves, separating the lateral digital extensor M. from the long digital extensor M. (cranially) and from the lateral digital flexor M. (caudally); the cranial border of tibia, separating it from the long digital extensor M.; the caudal border of the tibia, separating it from the lateral digital flexor and the medial digital flexor Mm.; the hock (the tibiotarsometatarsal joint); the calcaneus and the tuber calcanei (caudally); the distal end of the lateral ridge of the trochlea of the talus; the metatarsus III; the two metatarsal bones—II (medially) and IV (laterally), also called the splint bones; the interosseus medius, the tendons of the deep and superficial digital flexor Mm., and the grooves separating them; the ergot; the fetlock (the metatarsosesamophalangeal joint); the pastern (the area of the 1st phalanx); the coronet (the narrow area along the dorsal border of the hoof); and the hoof.

The exenteration (or the evisceration or removal) of the urogenital tract and rectum with the anus from the pelvic cavity has already been completed.

With a knife, make an incision along the linea alba up to the pubic symphysis. With a saw, separate the two pelvic limbs through the lumbosacral and coccygeal midline and the pelvic symphysis. Examine the vertebral canal and the spinal cord with the meninges.

Notice that in the horse, the conus medullaris ends at the level of S_2 and continues with the filum terminale surrounded by the last spinal nerves.

Remember! The combination of the filum terminale and the last sacral and caudal spinal nerves is called the cauda equina.

Examine the two origins of the pectineus M., which are separated by the accessory lig. of the hip joint and the fibrous ring surrounding the (accessory) external pudendal V., which perforates the origin of the gracilis M., as well as the vascular lacuna and the femoral triangle.

Remember! The femoral triangle represents the opening of the vascular lacuna on the medial aspect of the pelvic limb and is bordered by the sartorius M. (cranially), the gracilis and pectineus Mm. (caudally), the vastus medialis and medial part of the iliacus Mm. (deeply), and the lamina femoralis (superficially). The following structures are found within the femoral triangle: the femoral A.V., the saphenus N., and the deep inguinal lnn.

Transect the insertion of the external abdominal oblique aponeurosis from the prepubic tendon; remove the lamina femoralis, and widely expose the femoral triangle with all the structures inside.

Transect the gracilis M. through its middle and reflect the stumps. Examine the obturator A.N. supplying it. Deep to the gracilis M., the adductor minor (adductor brevis), adductor major (adductor magnus), and the semimembranosus Mm. are now exposed. Carefully separate the two adductor Mm. from each other, first noting the difference in color (the adductor minor M. is lighter than the adductor major M.) and the thin layer of connective tissue that separates them (Fig. 5-1).

Continue to dissect the femoral vessels, which obliquely cross the axis of the femur bone and pass between the two insertions of the adductor major M. through the femoral canal (canalis femoralis). Transect the long part of the adductor major M., and follow the femoral vessels through the femoral canal up to the level of the gastrocnemius M.

Notice that, ventral to the distal caudal femoral A. (only in cases where there is more than one caudal femoral A.), the femoral A. changes its name to the popliteal A.

Identify the medial circumflex femoral A., which is a branch of the deep femoral A.

The transection of the semimembranosus and semitendinosus Mm. is optional; however, it widely exposes the vessels and nerves located parallel with the caudal aspect of the femur. Also, transect the adductor major M. and the medial head of the gastrocnemius M. to expose the femoral and the popliteal Aa. Transect the iliopsoas M. from the lesser trochanter, the sartorius M. from the fascia iliaca, and the pectineus and adductor minor Mm. from their origin to better expose the deepest structures of the medial aspect.

Finish the dissection of the medial aspect of the thigh by dissecting the vastus medialis and rectus femoris Mm., which are located cranial to the sartorius M. Examine the femoral N. and vessels, which penetrate between the vastus medialis and rectus femoris Mm. (Fig. 5-2).

Turn the limb so that the lateral side is up. Carefully reflect the ventral stumps of the biceps femoris and semitendinosus Mm. as far as possible without breaking any vessels or nerves.

After rounding the hip joint, the sciatic N. gives off muscular branches, and then it splits into the common fibular (peroneal) and the tibial Nn. The common fibular N. runs toward the lateral aspect of the crus, overlapping the lateral head of the gastrocnemius M., whereas the tibial N. penetrates between the two heads of the gastrocnemius M. Identify the lateral cutaneous sural N., which is a branch of the common fibular N., and the caudal cutaneous sural N., which is a branch of the tibial N. Follow the latter up to the proximal extent of the crus, where it is accompanied by the lateral saphenous V. Identify the popliteal lnn. at the origin of the popliteal A.V. and the penetration of the tibial N. between the two heads of the gastrocnemius M.

Palpate the origin of the rectus femoris M. and the hip joint. At the lateral edge of the origin of the rectus femoris M. and close to the joint capsule, expose the articularis coxae M. To explore the lateral circumflex femoral vessels and widely expose the articularis coxae M., transect the quadriceps femoris M. in its proximal third, reflect the stumps, and separate those four muscles from one another.

For a better understanding of the dissection of the crus, first follow the superficial lamina of the fascia lata in the crus.

Remember! Dorsal to the patella, this lamina splits into a deep and a superficial layer.

Notice that the deep layer is inserted on the patella, whereas the superficial layer continues toward the hock on the cranial aspect of the crus, where it is called the superficial crural fascia and where it overlaps the proper (middle) crural fascia. The superficial crural fascia degenerates into loose connective tissue on the dorsal aspect of the metatarsal area.

The entire region is surrounded and protected by the proper (middle) crural fascia.

Notice that the aponeuroses of the gracilis, sartorius, biceps femoris, and semitendinosus Mm. are fused with one another and with the proper crural fascia; together, they

Text continued on p. 204.

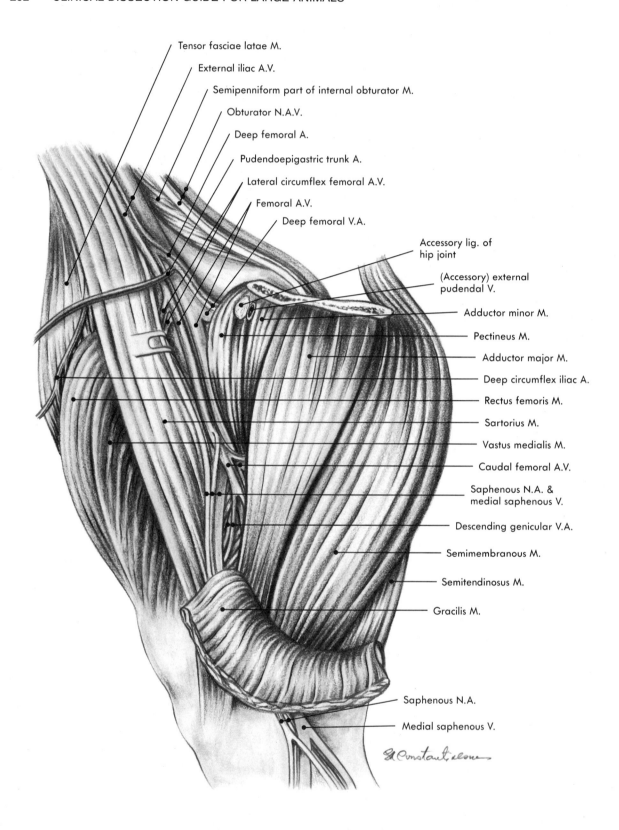

Tensor fasciae latae M.

External iliac A.V.

Semipenniform part of internal obturator M.

Obturator N.A.V.

Deep femoral A.

Pudendoepigastric trunk A.

Lateral circumflex femoral A.V.

Femoral A.V.

Deep femoral V.A.

Accessory lig. of hip joint

(Accessory) external pudendal V.

Adductor minor M.

Pectineus M.

Adductor major M.

Deep circumflex iliac A.

Rectus femoris M.

Sartorius M.

Vastus medialis M.

Caudal femoral A.V.

Saphenous N.A. & medial saphenous V.

Descending genicular V.A.

Semimembranous M.

Semitendinosus M.

Gracilis M.

Saphenous N.A.

Medial saphenous V.

Fig. 5-1   Medial aspect of right thigh (superficial).

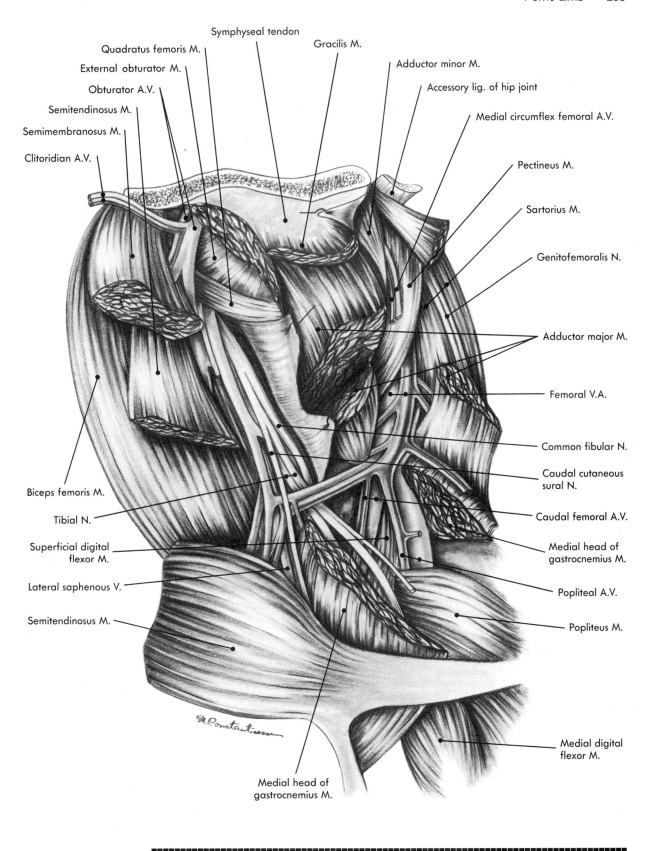

Symphyseal tendon

Quadratus femoris M.

Gracilis M.

External obturator M.

Adductor minor M.

Obturator A.V.

Accessory lig. of hip joint

Semitendinosus M.

Medial circumflex femoral A.V.

Semimembranosus M.

Pectineus M.

Clitoridian A.V.

Sartorius M.

Genitofemoralis N.

Adductor major M.

Femoral V.A.

Common fibular N.

Biceps femoris M.

Caudal cutaneous sural N.

Tibial N.

Caudal femoral A.V.

Superficial digital flexor M.

Medial head of gastrocnemius M.

Lateral saphenous V.

Popliteal A.V.

Semitendinosus M.

Popliteus M.

Medial digital flexor M.

Medial head of gastrocnemius M.

Medial aspect of left thigh (deep).

Fig. 5-2

are considered the tensors of this fascia. The proper crural fascia combines with the tendons of the biceps femoris and semitendinosus Mm. in the intermediate tendon to become part of the common calcanean tendon. In addition, the three extensor retinacula of the dorsal aspect of the hock are considered part of the same fascia. The proper crural fascia sends septa among the muscles and fuses with the periosteum of the medial aspect of the tibia. Deep to the proper crural fascia, the deep crural fascia is present and is a continuation of the deep lamina of the fascia lata. The deep crural fascia surrounds the deep caudal muscles of the crus.

Remember! The deep caudal muscles of the crus are represented by the popliteus M. and the three heads of the deep digital flexor M., which are the lateral head (lateral digital flexor M.), the medial head (medial digital flexor M.), and the caudal head (caudal tibial M.). The superficial caudal muscles of the crus are represented by the two heads—lateral and medial—of the gastrocnemius M., the soleus M. (together called the triceps surae M.), and finally the superficial digital flexor M. The four tendons combine together to make the calcanean tendon, which is part of the common calcanean tendon when joined with the intermediate tendon.

On the lateral aspect of the crus, transect the aponeurosis of the biceps femoris M. transversely at the level of the stifle joint, separating it from the proper crural fascia. The common fibular N. is exposed immediately under the fibrous structures. It crosses the lateral condyle of the tibia obliquely in a dorsoventral and caudocranial direction, over the lateral head of the gastrocnemius M. and the soleus M. (Fig. 5-3).

Continue the dissection of the saphenous A., medial saphenous V., and saphenous N. on the medial aspect of the crus from the thigh toward the hock.

Notice that they cross the attachment of the semitendinosus M. on the tibial crest and that all the vessels and nerves might have two or more main branches. Usually the medial saphenous V. and the saphenous A. each have a cranial and a caudal branch, as well as multiple communicating branches.

To free the muscles from their individual fascial covers, make an incision parallel to the long axis of each muscle at the level of their tendons. Introduce the blade of the scalpel through the incision line under the fascia in a parallel position to the muscle and continue the incision to the end of the muscle. The muscle will be entirely exposed after removing the fascia that surrounds it. During this procedure, protect and then dissect the vessels and nerves of the area.

Caution! The caudal tibial A.V. lie between the medial and lateral heads of the deep digital flexor M. and are overlapped by the medial head.

Carefully dissect the area at the medial aspect of the crus, distally, between the tibia and the calcaneus. The caudal branches of both the saphenous A. and medial saphenous V. anastomose with the corresponding caudal tibial vessels and, occasionally, with the distal branches of the caudal femoral A.V. in a double S-shaped A.V., respectively. From these anastomotic vessels branch the medial and the lateral plantar Aa. & Vv.

The tibial N. runs between the deep and the superficial groups of muscles, and is more easily exposed on the medial aspect of the crus. Dorsal to the hock it gives off the lateral and the medial plantar Nn. (Fig. 5-4).

Turn the pelvic limb so that the dorsal aspect of the hock is up. Carefully dissect the tendon of the long digital extensor M., which passes through the three extensor retinacula, surrounded by its own tendinous sheath. Examine the sheath and its relationships with the adjacent structures. Identify and dissect the two tendons of the cranial tibial M., which emerge from under the tendon of the fibularis tertius M. The latter protects the passage of the tendons of the cranial tibial M. by a fibrous ring. Identify and dissect the two tendons of the fibularis tertius M.

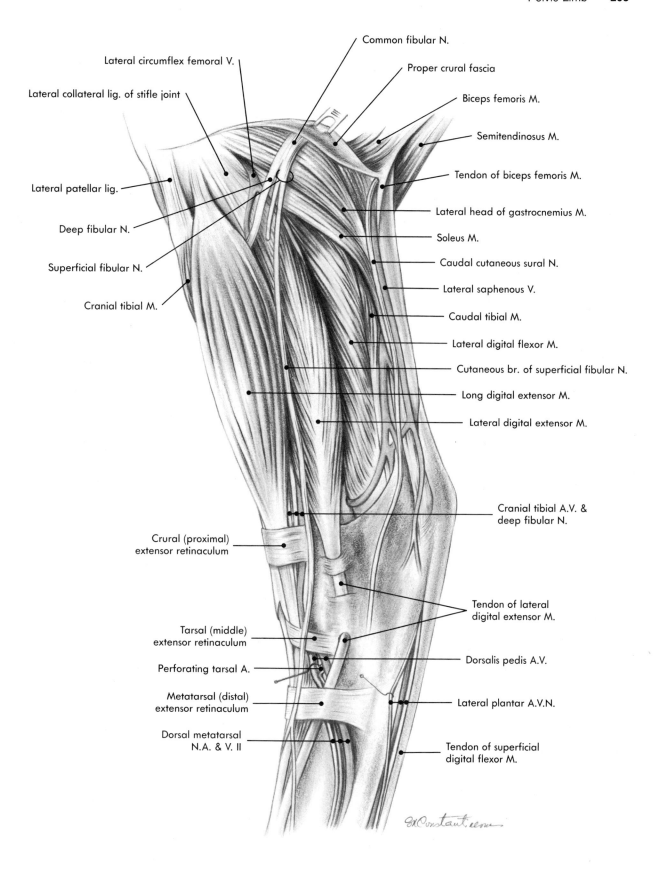

Common fibular N.

Proper crural fascia

Biceps femoris M.

Semitendinosus M.

Lateral circumflex femoral V.

Lateral collateral lig. of stifle joint

Tendon of biceps femoris M.

Lateral patellar lig.

Lateral head of gastrocnemius M.

Soleus M.

Deep fibular N.

Caudal cutaneous sural N.

Superficial fibular N.

Lateral saphenous V.

Cranial tibial M.

Caudal tibial M.

Lateral digital flexor M.

Cutaneous br. of superficial fibular N.

Long digital extensor M.

Lateral digital extensor M.

Cranial tibial A.V. & deep fibular N.

Crural (proximal) extensor retinaculum

Tendon of lateral digital extensor M.

Tarsal (middle) extensor retinaculum

Dorsalis pedis A.V.

Perforating tarsal A.

Lateral plantar A.V.N.

Metatarsal (distal) extensor retinaculum

Dorsal metatarsal N.A. & V. II

Tendon of superficial digital flexor M.

Left crus (lateral aspect, superficial).

Fig. 5-3

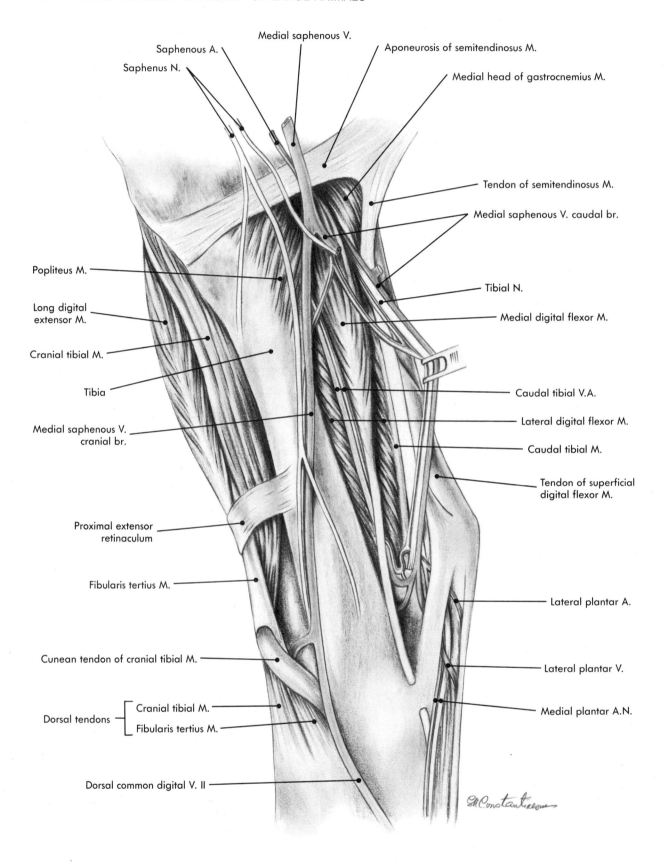

Saphenous A.

Saphenus N.

Medial saphenous V.

Aponeurosis of semitendinosus M.

Medial head of gastrocnemius M.

Tendon of semitendinosus M.

Medial saphenous V. caudal br.

Tibial N.

Medial digital flexor M.

Popliteus M.

Long digital extensor M.

Cranial tibial M.

Tibia

Medial saphenous V. cranial br.

Caudal tibial V.A.

Lateral digital flexor M.

Caudal tibial M.

Tendon of superficial digital flexor M.

Proximal extensor retinaculum

Fibularis tertius M.

Lateral plantar A.

Cunean tendon of cranial tibial M.

Lateral plantar V.

Medial plantar A.N.

Dorsal tendons { Cranial tibial M. / Fibularis tertius M.

Dorsal common digital V. II

Fig. 5-4

Right crus (medial aspect).

Notice that both the cranial tibial and fibularis tertius Mm. have one dorsal tendon. In addition, the cranial tibial M. has one more tendon (the cunean tendon), located medially, and the fibularis tertius M. one more tendon, located laterally.

Identify the cranial tibial A.V. and the deep fibular N., running between the cranial aspect of the tibia and the deep aspect of the cranial tibial M.

Examine the lateral aspect of the hock and the strong fascia, which binds the tendon of the lateral digital extensor M.

Caution! Do not open the canal that the tendon of the lateral digital extensor M. passes through. Leave the fascia untouched at this level between the crural and the tarsal extensor retinacula.

Make a vertical incision through the extensor digitalis brevis M. (between the tarsal and the metatarsal extensor retinacula) to expose the distal segments of the cranial tibial A.V. (the A.V. dorsalis pedis), and dissect the perforating tarsal A. and the dorsal metatarsal A. III, the perforating tarsal V., and the dorsal metatarsal V. II (Fig. 5-5).

Return to the stifle joint, dissect the three patellar ligg. and observe that there is no joint capsule between the patella and the tibia but only a large amount of fat (corpus adiposum infrapatellare). Instead, the femoropatellar joint capsule is present; its tensor is the tendon of the quadriceps femoris M. On both sides of the joint capsule, dissect the lateral and medial femoropatellar ligg. Examine the attachment of the biceps femoris tendon on the lateral femoropatellar lig. and on the lateral patellar lig. Also examine the attachment of the aponeuroses of the gracilis and sartorius Mm. on the medial patellar lig.

Expose the common origin of the long digital extensor and fibularis tertius (on the distal end of the femur) and the origin of the popliteus M. close and caudal to the previous two muscles. Examine the route of the popliteal tendon between the lateral meniscus and the lateral collateral lig. of the femorotibial joint.

If the medial head of the gastrocnemius M. has not yet been transected, do so now. The origin of the superficial digital flexor M. and part of the popliteus M. will be exposed. The popliteal A.V. travel between the popliteus M. and the femorotibial joint capsule (present only on the caudal aspect of the stifle). The popliteus M. widens toward the medial aspect of the crus.

Notice that the popliteal line (oblique in a dorsoventral and lateromedial direction) separates the insertions of the lateral digital flexor and popliteus Mm.

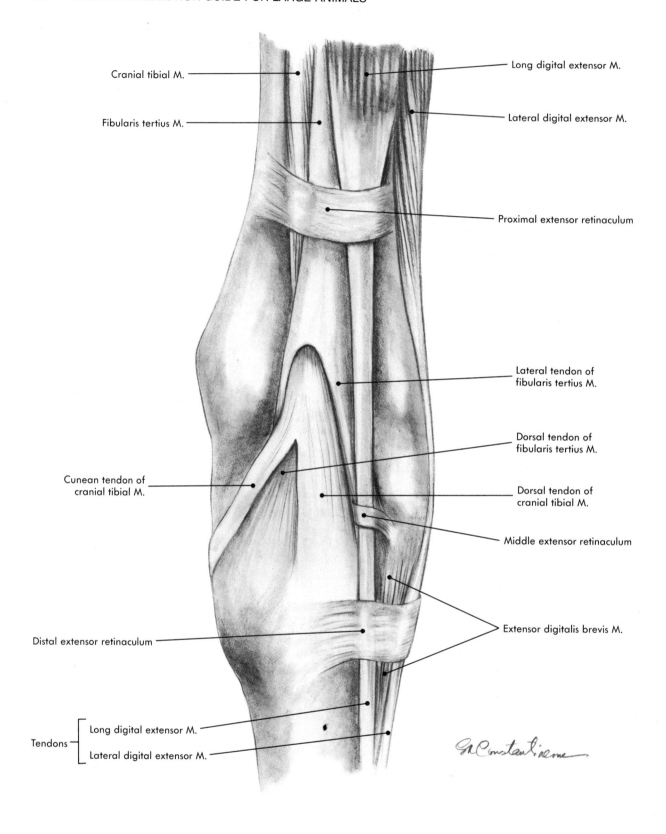

Cranial tibial M.

Fibularis tertius M.

Long digital extensor M.

Lateral digital extensor M.

Proximal extensor retinaculum

Lateral tendon of fibularis tertius M.

Dorsal tendon of fibularis tertius M.

Cunean tendon of cranial tibial M.

Dorsal tendon of cranial tibial M.

Middle extensor retinaculum

Distal extensor retinaculum

Extensor digitalis brevis M.

Tendons
Long digital extensor M.
Lateral digital extensor M.

Fig. 5-5

Tarsus (dorsal aspect of left hindlimb).

Transect the popliteus M. with a vertical incision, reflect the stumps, and carefully remove the femorotibial joint capsule, saving the vessels passing over it. Expose the two menisci, the meniscofemoral and the caudal lig. of the lateral meniscus and the caudal and cranial cruciate ligg. Explore the two synovial membranes between the femoral condyles and the two menisci (Figs. 5-6 to 5-9).

Turn the pelvic limb again so that the medial aspect is up. The medial tarsal fascia (at the level between the medial collateral lig. and the caudal border of the calcaneus) is called the flexor retinaculum. This structure protects two out of three tendons of the deep digital flexor M., that is, the lateral digital flexor and caudal tibial Mm., which are surrounded by a common tendinous sheath. The tendon of the medial digital flexor M. runs separately between the medial collateral lig. and the medial tarsal fascia, surrounded by its own tendinous sheath, and fuses with the other two tendons distal to the hock.

Return to the common calcanean tendon, and examine the relationship and different positions of the superficial digital flexor tendon in the middle of the crus and proximal to the calcaneus. The tendon of the superficial digital flexor M. surrounds the tendon of the triceps surae M. in a mediolateral direction, becoming superficial. Its attachment on the tuber calcanei is strong and protected by a tendinous bursa. The tendon continues its route as the most superficial tendon on the plantar aspect of the metatarsal area.

Examine the plantar fascia (in the metatarsal area) and its strong attachments to the splint bones. Remove it carefully to expose the tendons of the superficial and deep digital flexor Mm. and the interosseus medius M.

Notice that the plantar tarsometatarsal lig. continues as the accessory (check) lig. of the deep digital flexor M., which joins the tendon in the middle third of the metatarsus.

Examine the long plantar lig. on the plantar surface of the calcaneus.

The two plantar Nn. run parallel with the tendons of the digital flexor Mm. A communicating branch is observed between them in the middle of the metatarsus over the superficial digital flexor tendon (in an oblique position, dorsoventrally and mediolaterally). The plantar Nn. run together with the lateral and medial plantar Aa. & Vv. In addition, the dorsal metatarsal A. III and the dorsal common digital V. II are the largest vessels within the area and give off the digital Aa. & Vv.

Transect the superficial and deep digital flexor tendons; reflect the stumps; and examine and dissect the arterial and venous arches and branches, which anastomose and supply the related structures.

Do not dissect the fetlock and the digit. They will be dissected with the thoracic limb.

Turn the limb so that the dorsal aspect of the autopodium is up, and carefully dissect the dorsal metatarsal and dorsal digital Nn., which are terminal branches of the deep fibular N.

Caution! The tendon of the lateral digital extensor M. joins the tendon of the long digital extensor M. in the proximal third of the dorsal metatarsal area. Do not separate them.

At the most distal end of the splint bones, the plantar metatarsal Nn. become apparent and spread their fibers toward the dorsal aspect of the fetlock. Dissect them. These two nerves are the terminal branches of the deep branch of the lateral plantar N. after it supplies the interosseus medius M.

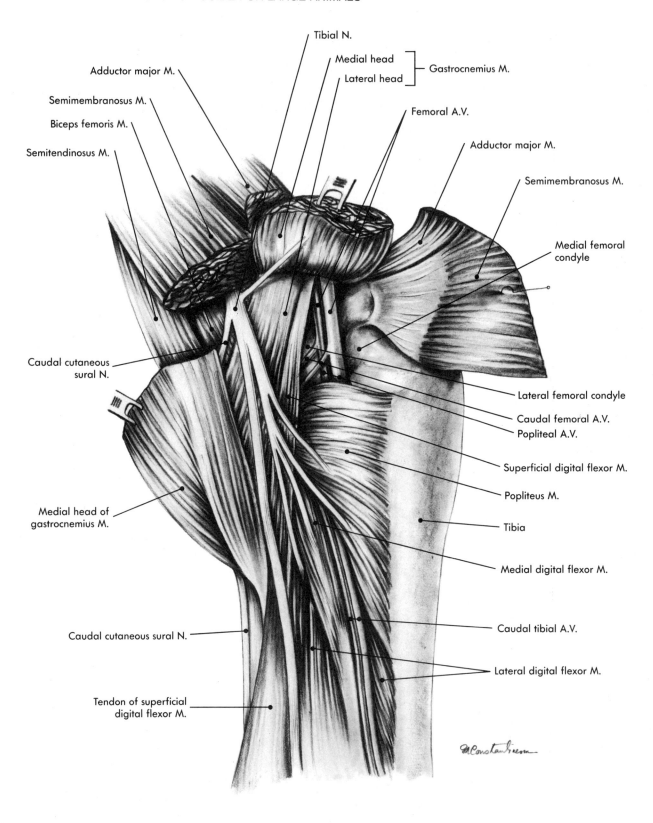

Tibial N.

Medial head
Lateral head $\big\}$ Gastrocnemius M.

Adductor major M.

Semimembranosus M.

Biceps femoris M.

Femoral A.V.

Semitendinosus M.

Adductor major M.

Semimembranosus M.

Medial femoral condyle

Caudal cutaneous sural N.

Lateral femoral condyle

Caudal femoral A.V.

Popliteal A.V.

Superficial digital flexor M.

Popliteus M.

Medial head of gastrocnemius M.

Tibia

Medial digital flexor M.

Caudal cutaneous sural N.

Caudal tibial A.V.

Lateral digital flexor M.

Tendon of superficial digital flexor M.

Fig. 5-6   Stifle and proximal crus of left pelvic limb (deep mediocaudal aspect).

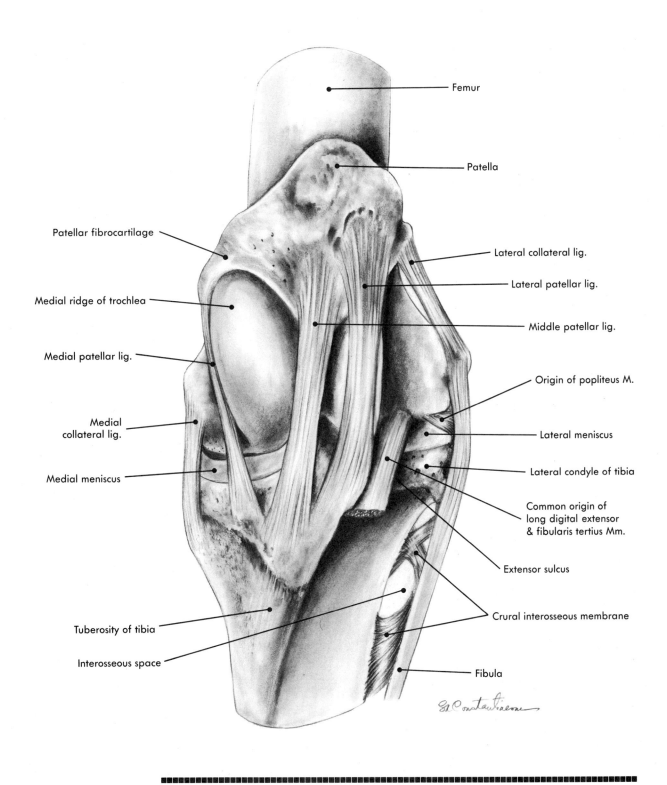

Femur

Patella

Patellar fibrocartilage

Lateral collateral lig.

Lateral patellar lig.

Medial ridge of trochlea

Middle patellar lig.

Medial patellar lig.

Origin of popliteus M.

Medial
collateral lig.

Lateral meniscus

Medial meniscus

Lateral condyle of tibia

Common origin of
long digital extensor
& fibularis tertius Mm.

Extensor sulcus

Tuberosity of tibia

Crural interosseous membrane

Interosseous space

Fibula

Stifle joint of left pelvic limb (cranial aspect).

Fig. 5-7

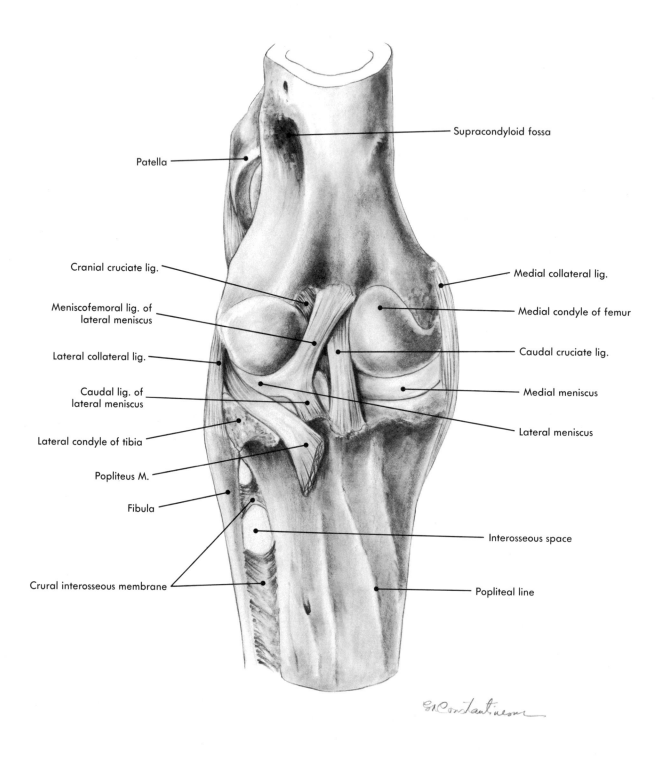

Patella

Cranial cruciate lig.

Meniscofemoral lig. of
lateral meniscus

Lateral collateral lig.

Caudal lig. of
lateral meniscus

Lateral condyle of tibia

Popliteus M.

Fibula

Crural interosseous membrane

Supracondyloid fossa

Medial collateral lig.

Medial condyle of femur

Caudal cruciate lig.

Medial meniscus

Lateral meniscus

Interosseous space

Popliteal line

Fig. 5-8   Stifle joint of left pelvic limb (caudal aspect).

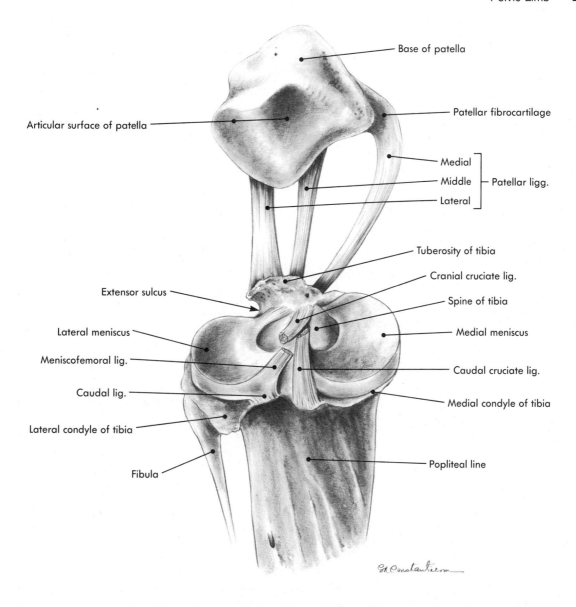

Base of patella

Patellar fibrocartilage

Articular surface of patella

Medial ⎤
Middle ⎬ Patellar ligg.
Lateral ⎦

Tuberosity of tibia

Cranial cruciate lig.

Extensor sulcus

Spine of tibia

Lateral meniscus

Medial meniscus

Meniscofemoral lig.

Caudal cruciate lig.

Caudal lig.

Medial condyle of tibia

Lateral condyle of tibia

Fibula

Popliteal line

Patella and proximal tibia of left pelvic limb (caudodorsal aspect).   **Fig. 5-9**

Return to the hock. After removing the dorsal joint capsule, explore each joint, identifying the bones and ligaments and attempting to understand the springlike mechanism of the talocrural joint.

Review the structures of the stifle joint, and focus your attention on the following: the complementary fibrocartilage of the patella; the characteristics of the femoral trochlea; the insertion of the quadriceps femoris M. on the patella; and the patellar ligg. Attempt to understand the locking mechanism of the stifle joint.

Disarticulate the hip joint and examine all the components that are specific for horse, such as the accessory lig. and the articularis coxae M. Also examine the lig. of the femoral head, the transverse lig. of the acetabulum, the acetabulum, and the femoral head with its fovea capitis.

Save the fetlock and digit to dissect these structures with those of the thoracic limb (Fig. 5-10).

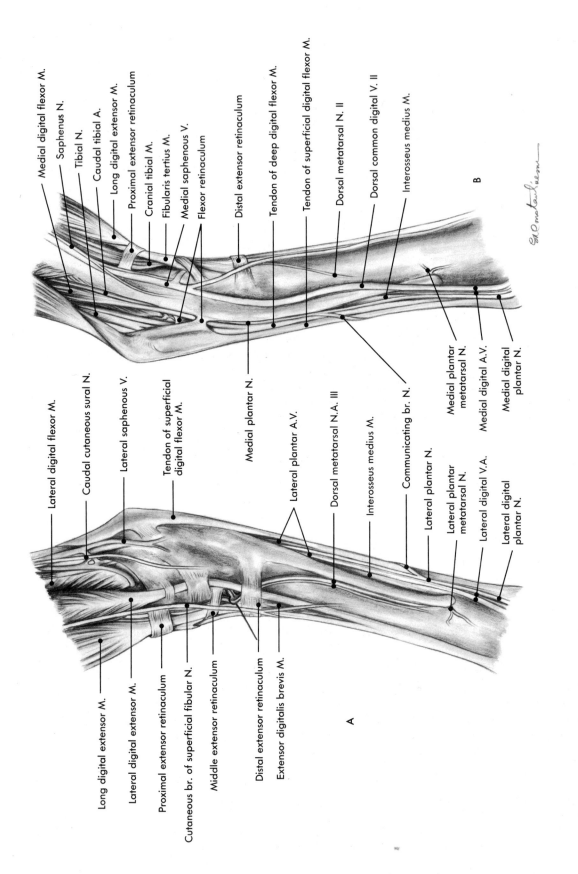

Medial digital flexor M.
Saphenus N.
Tibial N.
Caudal tibial A.
Long digital extensor M.
Proximal extensor retinaculum
Cranial tibial M.
Fibularis tertius M.
Medial saphenous V.
Flexor retinaculum
Distal extensor retinaculum
Tendon of deep digital flexor M.
Tendon of superficial digital flexor M.
Dorsal metatarsal N. II
Dorsal common digital V. II
Interosseus medius M.

B

Lateral digital flexor M.
Caudal cutaneous sural N.
Lateral saphenous V.
Tendon of superficial digital flexor M.
Medial plantar N.
Lateral plantar A.V.
Dorsal metatarsal N.A. III
Interosseus medius M.
Communicating br. N.
Lateral plantar N.
Lateral plantar metatarsal N.
Lateral digital V.A.
Lateral digital plantar N.

Medial plantar metatarsal N.
Medial digital A.V.
Medial digital plantar N.

Long digital extensor M.
Lateral digital extensor M.
Proximal extensor retinaculum
Cutaneous br. of superficial fibular N.
Middle extensor retinaculum
Distal extensor retinaculum
Extensor digitalis brevis M.

A

Fig. 5-10   **A**, Tarsometatarsal area (left lateral aspect). **B**, Tarsometatarsal area (left medial aspect).

Similar to the procedure for the horse, first identify the following structures of the large ruminants, which are used as landmarks: the patella; the crus; the lateral condyle of the tibia; the tibial crest; the common calcanean tendon; the body of the tibia; the hock; the calcaneus and the tuber calcanei; the malleolar bone; the metatarsal bone (III & IV fused); the dewclaws; the interosseus medius and the tendons of the deep and superficial digital flexor Mm. (separated by grooves); and the paired fetlocks, pasterns, coronets, and hooves.

Separate the two pelvic limbs, using the same technique as in the horse, and examine the vertebral canal and spinal cord. There are noticeable differences between the horse and the (large) ruminants in this area. Follow the same technique as in the horse.

Dissect the medial aspect of the thigh as shown in Fig. 5-11. Turn the limb so that the lateral side is up. Reflect the biceps femoris and semitendonosus Mm. ventrally and dissect the deep muscles, vessels, and nerves, which are similar to those of the horse.

To examine the hip joint, transect the deep gluteal and gemelli Mm. from the lateral and caudal aspects of the femur, respectively, and reflect them carefully off the joint capsule. Transect the quadratus femoris M. and the extrapelvic part of the external obturator M. from the caudal aspect of the femur, and reflect them. Transect the origin of the rectus femoris M., and reflect the muscle ventrally. The entire joint capsule and the iliofemoral lig. are now exposed. Make an incision in the joint capsule, and explore the acetabular cavity. Identify the lig. of the head of the femur, the transverse lig., and the pubofemoral lig. Examine the characteristics of the acetabulum and the proximal extremity of the femur.

Outline the two (not three, as in the horse) extensor retinacula, and remove the crural fasciae from the lateral aspect of the crus, saving the vessels and nerves. In comparison to the horse, one additional structure found in the large ruminants is the fibularis longus M. Another is the branching of the lateral saphenous V. in cranial and caudal branches. The cranial br. gives off the dorsal common digital Vv. II to IV, whereas the caudal br. anastomoses with the medial saphenous V.

Notice that the long digital extensor M. has two portions that continue as two distinct tendons.

Transect the lateral head of the gastrocnemius M. and the superficial digital flexor M. from their origins on the caudal aspect of the femur, and reflect the muscles caudally to expose the tibial N. (Fig. 5-12).

Transect the fibularis tertius M. and the lateral part of the extensor pedis longus M. in the middle, and reflect the two stumps cranially. Then transect the fibularis longus M., and reflect its two stumps. The motor branches of the deep fibular N., the cranial tibial A.Vv. (paralleling the medial part of the extensor pedis longus M.), and the cutaneous branch of the superficial fibular N. are now exposed.

On the caudal aspect of the stifle joint, follow the course of the femoral A. after it gives off the distal caudal femoral A. that penetrates between the two heads of the gastrocnemius M. This is the popliteal A. Follow its course farther up to where it branches into the two tibial Aa. (cranial and caudal). To better expose the cranial tibial A., transect the popliteus M.

Notice that the cranial tibial A. of the ox is at least twice as large as the caudal tibial A.

On the medial aspect of the crus remove the crural fasciae and expose the muscles, vessels, and nerves. The medial saphenous V. normally has only one branch (the caudal one), which anastomoses with the lateral saphenous V. (Fig. 5-13).

Turn the limb so that the dorsal aspect of the autopodium is up, and carefully dissect the structures.

Continue the dissection of the tarsal, metatarsal, and fetlock areas. The digital areas are similar to those of the thoracic limb and are described in the following chapter.

In the tarsometatarsal area, the branching of the vessels and nerves is a characteristic of the ruminants, with no significant differences between the large and small ruminants (Fig. 5-14).

Large ruminants
■■■■■■■■■■■■■■■■■■
(Figs. 5-11 to 5-14)

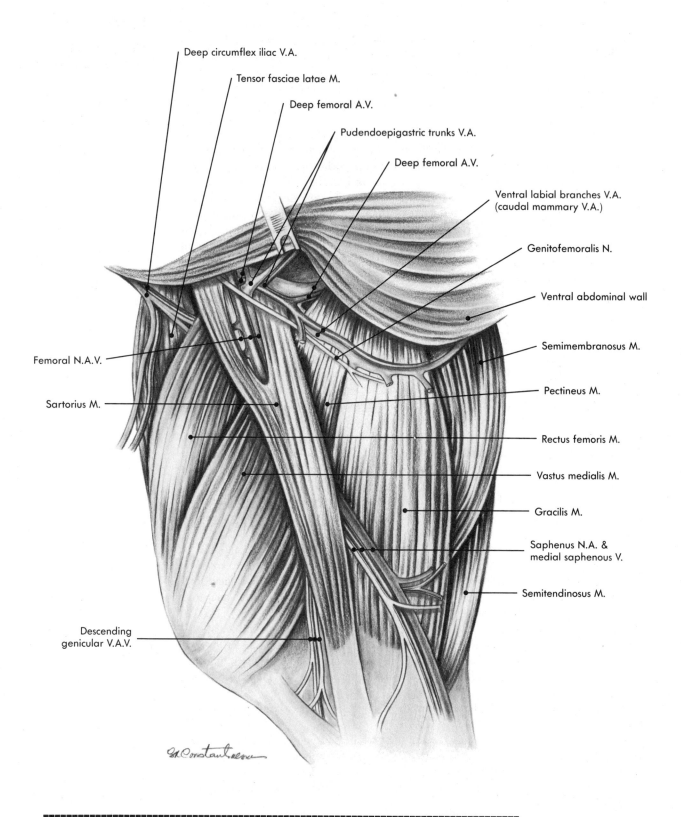

Fig. 5-11

Right thigh (superficial, medial aspect).

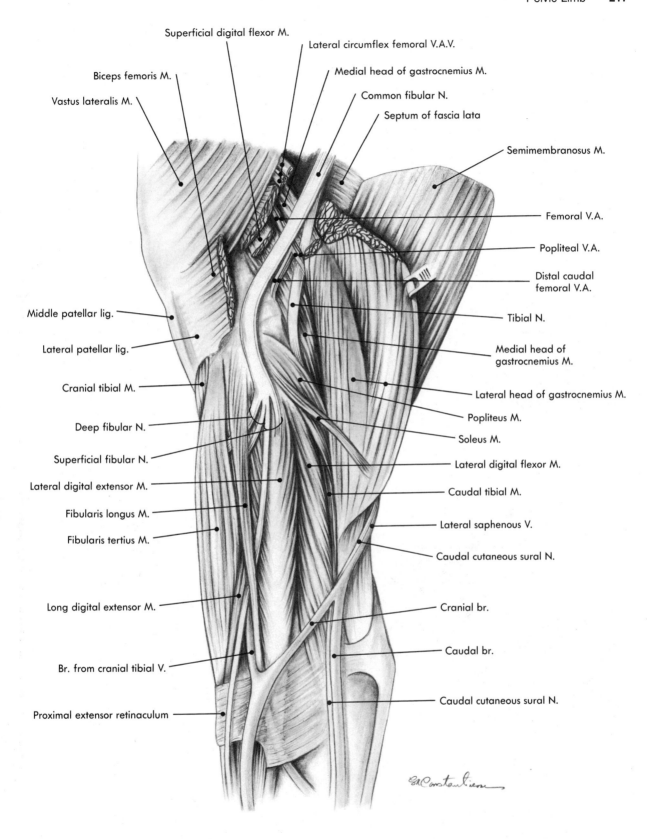

Superficial digital flexor M.

Biceps femoris M.

Vastus lateralis M.

Lateral circumflex femoral V.A.V.

Medial head of gastrocnemius M.

Common fibular N.

Septum of fascia lata

Semimembranosus M.

Femoral V.A.

Popliteal V.A.

Distal caudal femoral V.A.

Tibial N.

Middle patellar lig.

Lateral patellar lig.

Medial head of gastrocnemius M.

Cranial tibial M.

Lateral head of gastrocnemius M.

Deep fibular N.

Popliteus M.

Superficial fibular N.

Soleus M.

Lateral digital extensor M.

Lateral digital flexor M.

Fibularis longus M.

Caudal tibial M.

Fibularis tertius M.

Lateral saphenous V.

Caudal cutaneous sural N.

Long digital extensor M.

Cranial br.

Caudal br.

Br. from cranial tibial V.

Caudal cutaneous sural N.

Proximal extensor retinaculum

Stifle and crus (deep, left lateral aspect).

Fig. 5-12

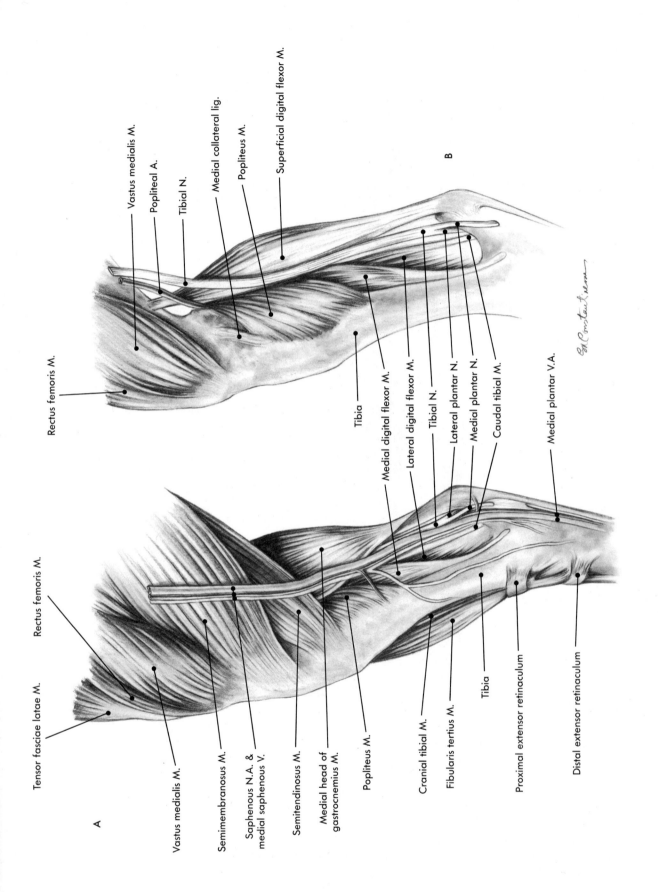

Vastus medialis M.

Popliteal A.

Tibial N.

Medial collateral lig.

Popliteus M.

Superficial digital flexor M.

Rectus femoris M.

B

Tibia

Medial digital flexor M.

Lateral digital flexor M.

Tibial N.

Lateral plantar N.

Medial plantar N.

Caudal tibial M.

Medial plantar V.A.

Tensor fasciae latae M.

Rectus femoris M.

Vastus medialis M.

Semimembranosus M.

Saphenous N.A. & medial saphenous V.

Semitendinosus M.

Medial head of gastrocnemius M.

Popliteus M.

Cranial tibial M.

Fibularis tertius M.

Tibia

Proximal extensor retinaculum

Distal extensor retinaculum

A

Fig. 5-13   **A,** Right crus (superficial, medial aspect). **B,** Right crus (deep, medial aspect).

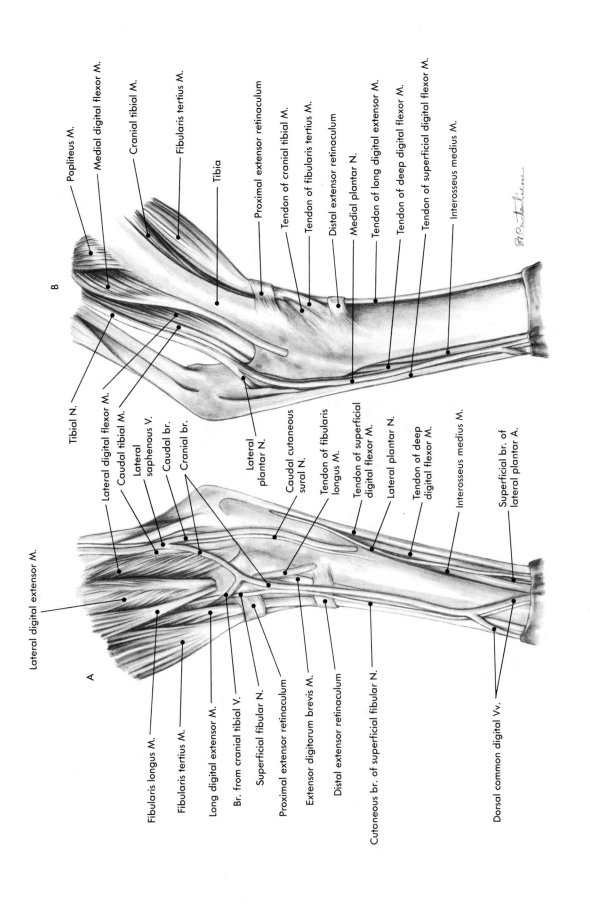

B

Popliteus M.

Medial digital flexor M.

Cranial tibial M.

Fibularis tertius M.

Tibia

Proximal extensor retinaculum

Tendon of cranial tibial M.

Tendon of fibularis tertius M.

Distal extensor retinaculum

Medial plantar N.

Tendon of long digital extensor M.

Tendon of deep digital flexor M.

Tendon of superficial digital flexor M.

Interosseus medius M.

Tibial N.

Lateral digital flexor M.

Caudal tibial M.

Lateral saphenous V.

Caudal br.

Cranial br.

Lateral plantar N.

Caudal cutaneous sural N.

Tendon of fibularis longus M.

Tendon of superficial digital flexor M.

Lateral plantar N.

Tendon of deep digital flexor M.

Interosseus medius M.

Superficial br. of lateral plantar A.

A

Lateral digital extensor M.

Fibularis longus M.

Fibularis tertius M.

Long digital extensor M.

Br. from cranial tibial V.

Superficial fibular N.

Proximal extensor retinaculum

Extensor digitorum brevis M.

Distal extensor retinaculum

Cutaneous br. of superficial fibular N.

Dorsal common digital Vv.

Fig. 5-14

A, Tarsometatarsal area (left lateral aspect). **B,** Tarsometatarsal area (left medial aspect).

Sheep

(Figs. 5-15 and 5-16)

Follow the directions for dissection given in the section on the large ruminants.

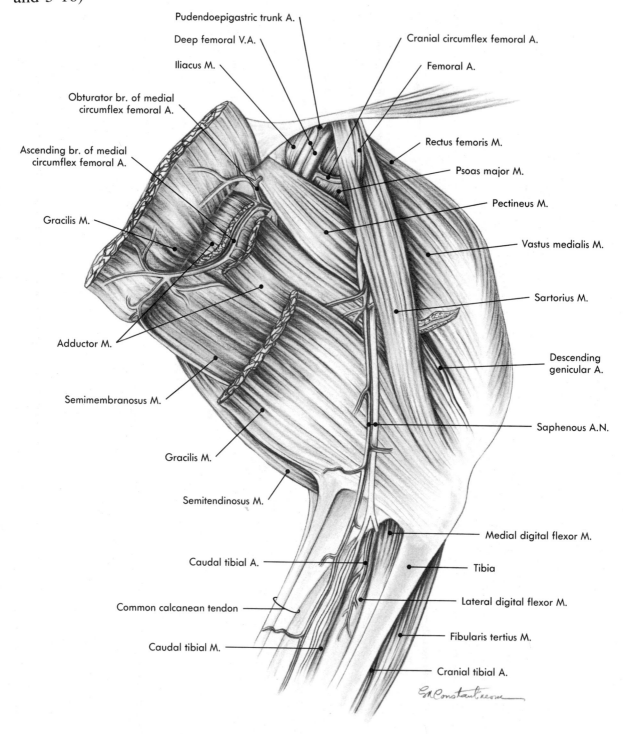

Pudendoepigastric trunk A.

Deep femoral V.A.

Iliacus M.

Obturator br. of medial circumflex femoral A.

Ascending br. of medial circumflex femoral A.

Gracilis M.

Adductor M.

Semimembranosus M.

Gracilis M.

Semitendinosus M.

Caudal tibial A.

Common calcanean tendon

Caudal tibial M.

Cranial circumflex femoral A.

Femoral A.

Rectus femoris M.

Psoas major M.

Pectineus M.

Vastus medialis M.

Sartorius M.

Descending genicular A.

Saphenous A.N.

Medial digital flexor M.

Tibia

Lateral digital flexor M.

Fibularis tertius M.

Cranial tibial A.

Fig. 5-15

Thigh and proximal crus (left medial aspect).

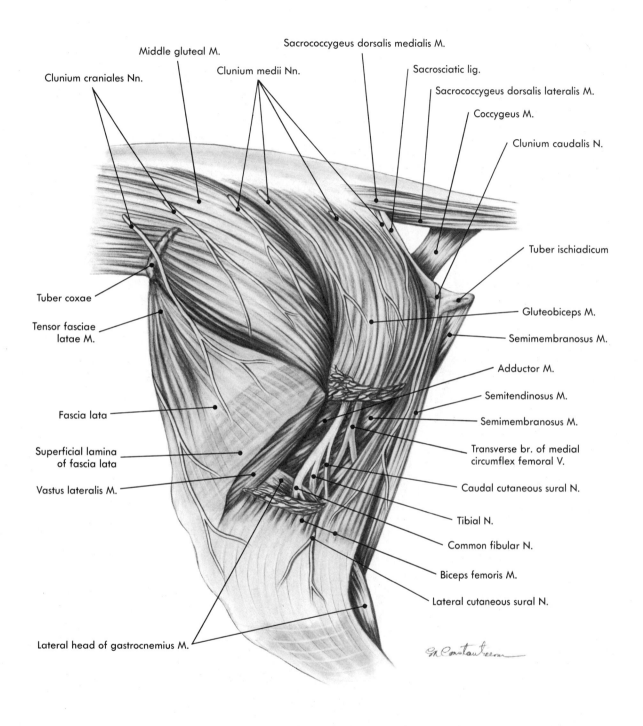

Clunium craniales Nn.

Middle gluteal M.

Clunium medii Nn.

Sacrococcygeus dorsalis medialis M.

Sacrosciatic lig.

Sacrococcygeus dorsalis lateralis M.

Coccygeus M.

Clunium caudalis N.

Tuber ischiadicum

Tuber coxae

Tensor fasciae latae M.

Gluteobiceps M.

Semimembranosus M.

Adductor M.

Semitendinosus M.

Semimembranosus M.

Fascia lata

Transverse br. of medial circumflex femoral V.

Superficial lamina of fascia lata

Caudal cutaneous sural N.

Vastus lateralis M.

Tibial N.

Common fibular N.

Biceps femoris M.

Lateral cutaneous sural N.

Lateral head of gastrocnemius M.

Croup and thigh (left lateral aspect).

Fig. 5-16

Goat
■■■■■■■
(Figs. 5-17
to 5-19)

Follow the same steps as in the large ruminants and the sheep.

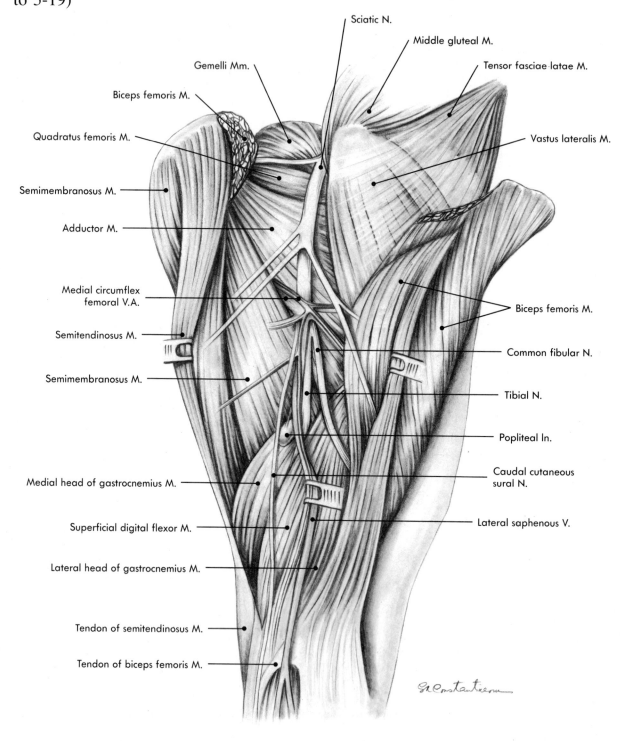

Fig. 5-17

Thigh (deep, right lateral aspect).

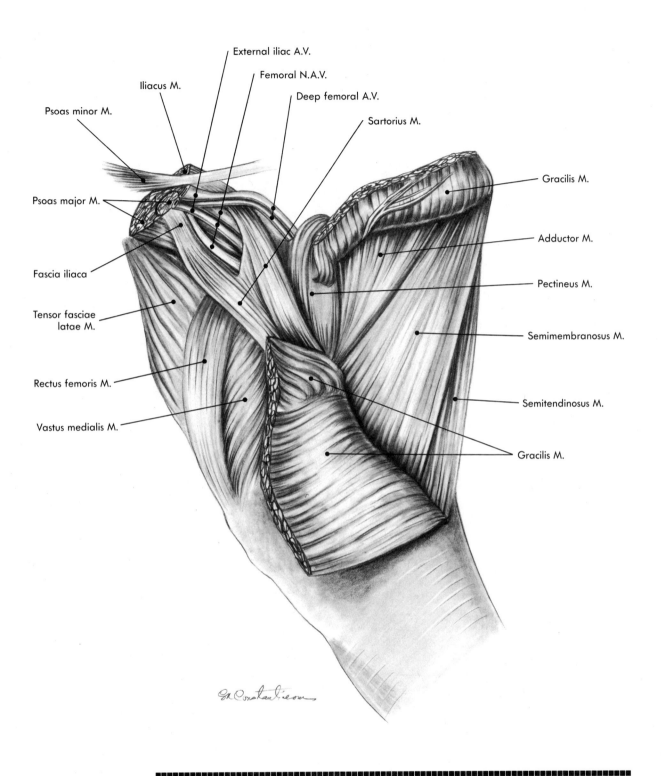

Thigh (deep, right medial aspect).

Fig. 5-18

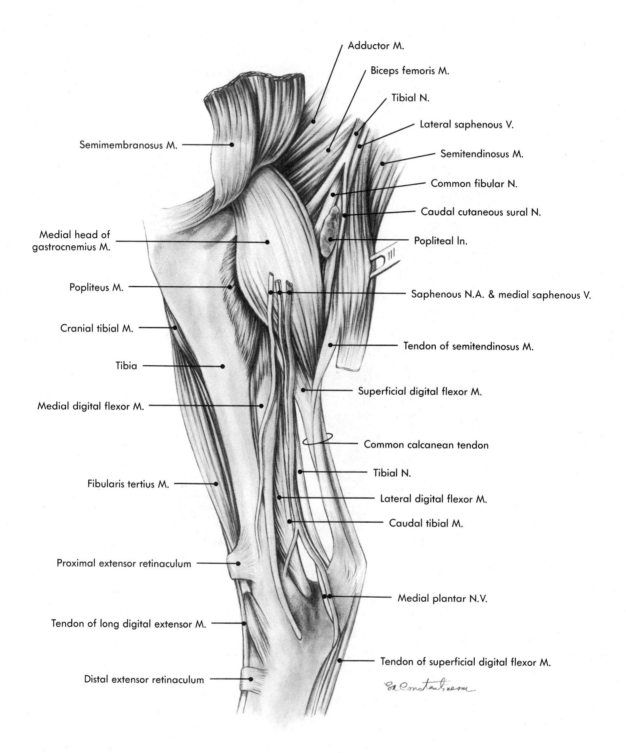

Adductor M.

Biceps femoris M.

Tibial N.

Lateral saphenous V.

Semimembranosus M.

Semitendinosus M.

Common fibular N.

Caudal cutaneous sural N.

Medial head of gastrocnemius M.

Popliteal ln.

Popliteus M.

Saphenous N.A. & medial saphenous V.

Cranial tibial M.

Tendon of semitendinosus M.

Tibia

Medial digital flexor M.

Superficial digital flexor M.

Common calcanean tendon

Tibial N.

Fibularis tertius M.

Lateral digital flexor M.

Caudal tibial M.

Proximal extensor retinaculum

Medial plantar N.V.

Tendon of long digital extensor M.

Tendon of superficial digital flexor M.

Distal extensor retinaculum

Fig. 5-19

Crus (right medial aspect).

Dissect the pelvic limb of the pig as suggested for the previous species.

Pig
(Figs. 5-20 to 5-24)

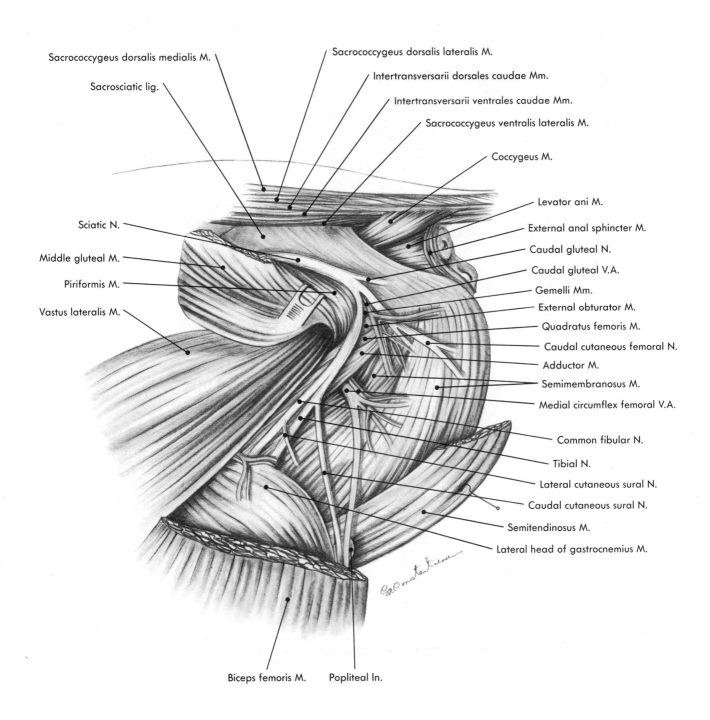

Sacrococcygeus dorsalis medialis M.

Sacrococcygeus dorsalis lateralis M.

Intertransversarii dorsales caudae Mm.

Intertransversarii ventrales caudae Mm.

Sacrosciatic lig.

Sacrococcygeus ventralis lateralis M.

Coccygeus M.

Levator ani M.

Sciatic N.

External anal sphincter M.

Caudal gluteal N.

Middle gluteal M.

Caudal gluteal V.A.

Piriformis M.

Gemelli Mm.

Vastus lateralis M.

External obturator M.

Quadratus femoris M.

Caudal cutaneous femoral N.

Adductor M.

Semimembranosus M.

Medial circumflex femoral V.A.

Common fibular N.

Tibial N.

Lateral cutaneous sural N.

Caudal cutaneous sural N.

Semitendinosus M.

Lateral head of gastrocnemius M.

Biceps femoris M.   Popliteal ln.

Croup and thigh (deep, left lateral aspect).

Fig. 5-20

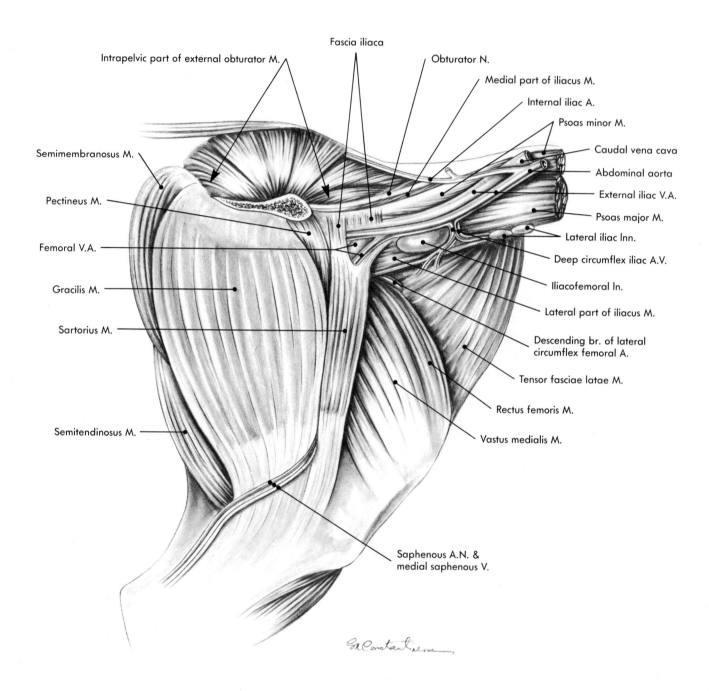

Fascia iliaca

Intrapelvic part of external obturator M.

Obturator N.

Medial part of iliacus M.

Internal iliac A.

Psoas minor M.

Semimembranosus M.

Caudal vena cava

Abdominal aorta

Pectineus M.

External iliac V.A.

Psoas major M.

Femoral V.A.

Lateral iliac lnn.

Deep circumflex iliac A.V.

Gracilis M.

Iliacofemoral ln.

Sartorius M.

Lateral part of iliacus M.

Descending br. of lateral circumflex femoral A.

Tensor fasciae latae M.

Rectus femoris M.

Semitendinosus M.

Vastus medialis M.

Saphenous A.N. & medial saphenous V.

Fig. 5-21

Pelvis and thigh (left medial aspect).

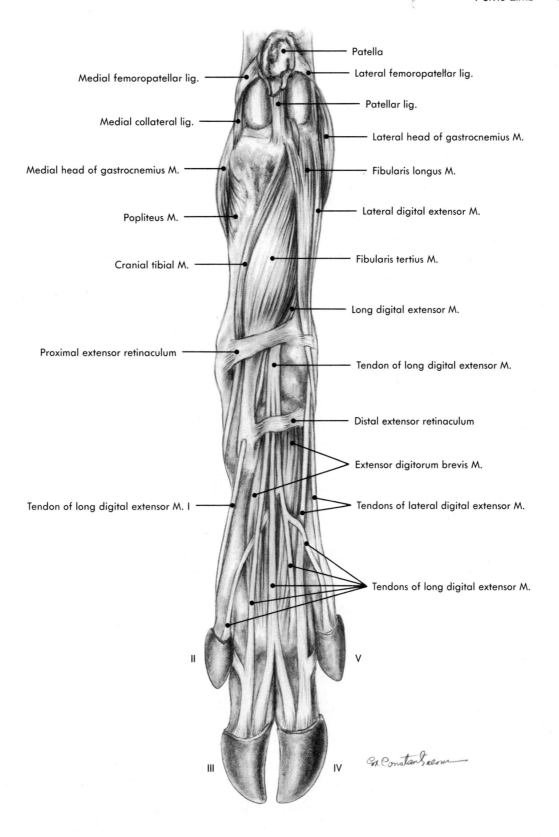

Patella

Lateral femoropatellar lig.

Medial femoropatellar lig.

Patellar lig.

Medial collateral lig.

Lateral head of gastrocnemius M.

Medial head of gastrocnemius M.

Fibularis longus M.

Popliteus M.

Lateral digital extensor M.

Cranial tibial M.

Fibularis tertius M.

Long digital extensor M.

Proximal extensor retinaculum

Tendon of long digital extensor M.

Distal extensor retinaculum

Extensor digitorum brevis M.

Tendon of long digital extensor M. I

Tendons of lateral digital extensor M.

Tendons of long digital extensor M.

II

V

III

IV

Crus and autopodium, left pelvic limb (dorsal aspect).

Fig. 5-22

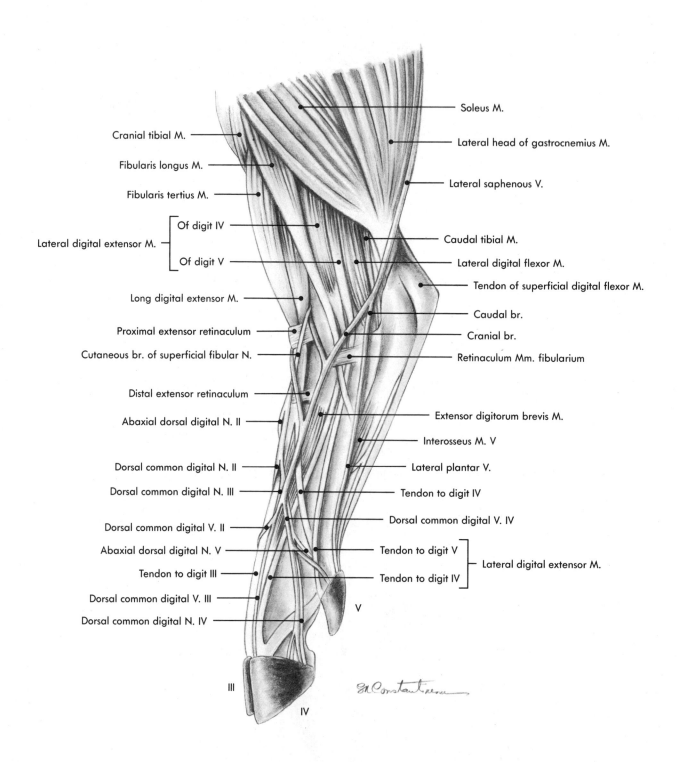

Cranial tibial M.

Fibularis longus M.

Fibularis tertius M.

Lateral digital extensor M. — { Of digit IV / Of digit V }

Long digital extensor M.

Proximal extensor retinaculum

Cutaneous br. of superficial fibular N.

Distal extensor retinaculum

Abaxial dorsal digital N. II

Dorsal common digital N. II

Dorsal common digital N. III

Dorsal common digital V. II

Abaxial dorsal digital N. V

Tendon to digit III

Dorsal common digital V. III

Dorsal common digital N. IV

Soleus M.

Lateral head of gastrocnemius M.

Lateral saphenous V.

Caudal tibial M.

Lateral digital flexor M.

Tendon of superficial digital flexor M.

Caudal br.

Cranial br.

Retinaculum Mm. fibularium

Extensor digitorum brevis M.

Interosseus M. V

Lateral plantar V.

Tendon to digit IV

Dorsal common digital V. IV

Tendon to digit V

Tendon to digit IV

Lateral digital extensor M.

III

IV

V

Fig. 5-23   Crus and autopodium (left lateral aspect).

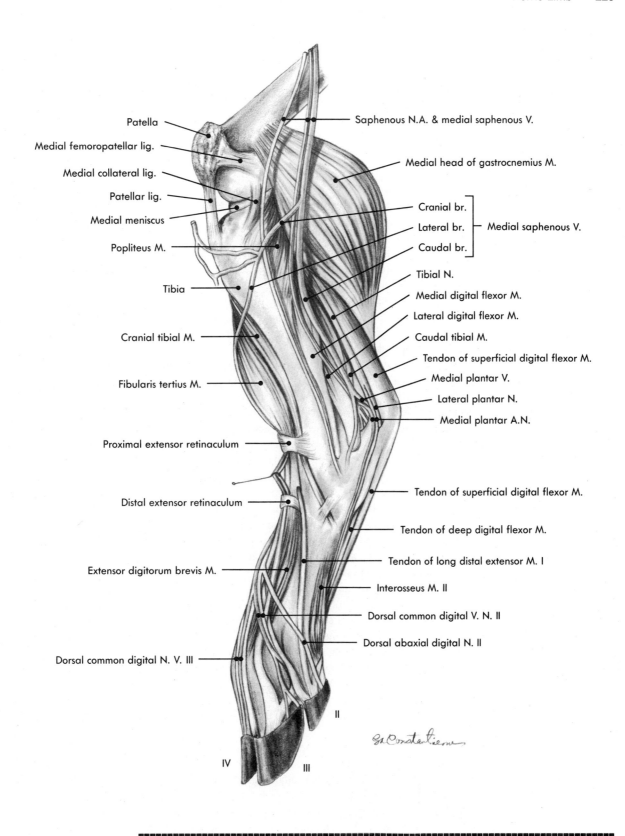

Patella

Medial femoropatellar lig.

Medial collateral lig.

Patellar lig.

Medial meniscus

Popliteus M.

Tibia

Cranial tibial M.

Fibularis tertius M.

Proximal extensor retinaculum

Distal extensor retinaculum

Extensor digitorum brevis M.

Dorsal common digital N. V. III

Saphenous N.A. & medial saphenous V.

Medial head of gastrocnemius M.

Cranial br.
Lateral br.   ⎤ Medial saphenous V.
Caudal br.   ⎦

Tibial N.

Medial digital flexor M.

Lateral digital flexor M.

Caudal tibial M.

Tendon of superficial digital flexor M.

Medial plantar V.

Lateral plantar N.

Medial plantar A.N.

Tendon of superficial digital flexor M.

Tendon of deep digital flexor M.

Tendon of long distal extensor M. I

Interosseus M. II

Dorsal common digital V. N. II

Dorsal abaxial digital N. II

II

IV

III

Crus and autopodium (right medial aspect).

Fig. 5-24

Chapter 6

Thoracic limb

Horse

As with the dissection of the pelvic limb, some important landmarks should first be identified. Check them with a skeleton.

Turn the thoracic limb so that the lateral aspect is up, and identify by palpation the following landmarks: the caudal angle of the scapula, the tuber of the scapular spine, the supraglenoid tubercle, the greater tubercle of the humerus, the deltoid tuberosity, the humeral crest, the olecranon, the elbow, the accessory carpal bone, and the carpal joint. The landmarks located within the metapodium and the acropodium are similar to those found in the pelvic limb.

In the antebrachium, identify the two longitudinal grooves: the craniolateral groove, which separates the extensor carpi radialis M. from the common digital extensor M., and the caudolateral groove, which separates the common digital extensor M. from the extensor carpi ulnaris M.

Notice that the lateral digital extensor M. lies in the caudolateral groove.

Palpate the accessoriometacarpal lig. (the distal lig. of the accessory carpal bone), the third and fourth metacarpal bones, the interosseus medius M., and the tendons of the deep and superficial digital flexor Mm.

Turn the limb so that the medial aspect is up, and identify by palpation the following structures: the coracoid process of the scapula, the lesser tubercle of the humerus, the teres major tuberosity, the elbow, the radial and olecranon fossae, the medial epicondyle, the radial tuberosity, the interosseous space between the radius and ulna, the body (shaft) of the radius, and the carpal joint; the medial aspect of the metapodium and acropodium is similar to that of the pelvic limb and contralateral limb.

In the antebrachial area, palpate the medial aspect of the radius; the extensor carpi radialis M. extends up to the cranial border of the bone, whereas the flexor carpi radialis M. corresponds to the area between the caudal border of the radius and the vertical line passing through the chestnut.

Skin the limb up to the fetlock and proceed with the removal of the hoof. While one student holds the limb firmly in contact with the table and a second holds the fetlock and the pastern, a third student sections the hoof through the quarters with a saw. The line of section is parallel to the long axis of the digit. Use of a wide bladed saw is suggested. In addition the thumb of your free hand should make contact with the blade to avoid an accident. Section the hoof as deep as the third phalanx (the sound is different when the saw is penetrating the bone). With large pliers, first remove the heel and then the rest of the wall, the frog, and the sole. Skin the digit. The hoof might also be removed by boiling the digit for at least 1 hour. With the limb tied in a vince, the hoof can then be pulled out easily.

Begin the dissection of the medial aspect of the thoracic limb by examining the pectoral muscles you have already transected while removing the limb. Dissect them carefully to separate them from each other and to save the vessels and nerves. Identify the lateral pectoral groove between the medial border of the brachiocephalicus M. and the lateral border of the descending pectoral M. Identify the remnants of the rhomboideus cervicis and thoracis Mm. and of the serratus ventralis cervicis and thoracis Mm. at the junction where the limb was totally removed from the body wall (Fig. 6-1).

Notice the median pectoral groove, which separates the two symmetrical descending pectoral Mm.

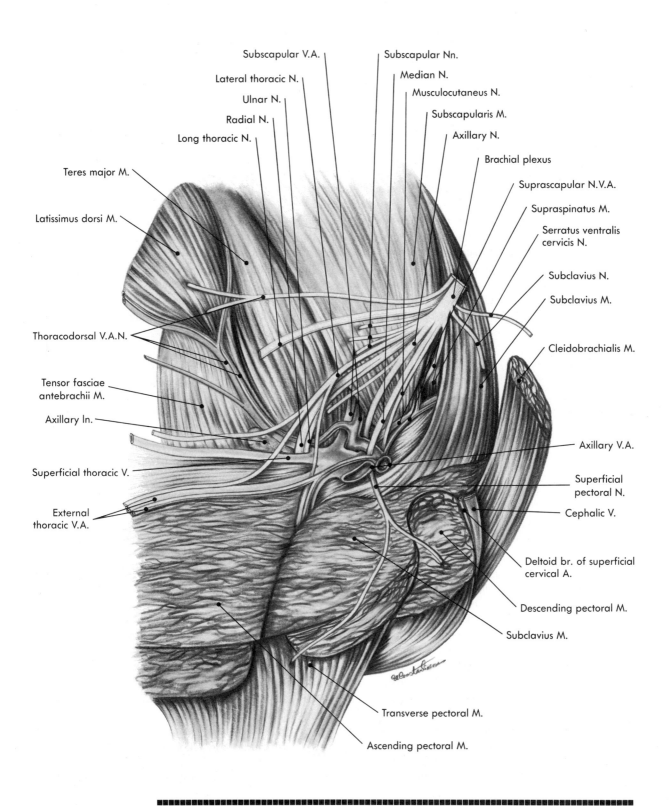

Subscapular V.A.

Subscapular Nn.

Lateral thoracic N.

Median N.

Ulnar N.

Musculocutaneus N.

Radial N.

Subscapularis M.

Long thoracic N.

Axillary N.

Teres major M.

Brachial plexus

Latissimus dorsi M.

Suprascapular N.V.A.

Supraspinatus M.

Serratus ventralis
cervicis N.

Subclavius N.

Subclavius M.

Thoracodorsal V.A.N.

Cleidobrachialis M.

Tensor fasciae
antebrachii M.

Axillary ln.

Axillary V.A.

Superficial thoracic V.

Superficial
pectoral N.

External
thoracic V.A.

Cephalic V.

Deltoid br. of superficial
cervical A.

Descending pectoral M.

Subclavius M.

Transverse pectoral M.

Ascending pectoral M.

Brachial plexus and related structures.

Fig. 6-1

Deep to the lateral pectoral groove, identify the subclavius M. The brachial segment of the cephalic V. and the deltoid br. of the superficial cervical A. lie in this groove.

Separate the descending pectoral M. from the transverse pectoral M. The descending pectoral M. is attached to the humeral crest, whereas the transverse pectoral M. exceeds the elbow and its fibers fuse with the superficial antebrachial fascia (located only on the medial aspect of the forearm). The transverse pectoral M. is the tensor of this fascia. Identify and dissect the external thoracic A.V., which are occasionally accompanied by a branch of the lateral thoracic N. (see Fig. 6-1).

Remove the ascending pectoral M., saving its attachments on both the greater and lesser tubercles of the humerus. Reflect the descending and transverse pectoral Mm. ventrally to expose the medial aspect of the arm. The medial brachial fascia covers and protects all the structures, sending two separate sheets around the coracobrachialis M. and around the vessels and nerves. Make an incision in the transverse pectoral M. and the superficial antebrachial fascia in the longitudinal axis of the forearm, and reflect the stumps of the fascia, protecting the antebrachial segment of the cephalic V. and the musculocutaneous N. The proper antebrachial fascia should be exposed.

Examine the medial aspect of the shoulder joint and the loose fascia that surrounds the brachial plexus and the associated vessels. Identify the omohyoideus M., which attaches to the fascia. Examine the two sheets of the fascia around the suprascapularis and the subclavius Mm. Make an incision in each sheet in the longitudinal axis of the corresponding muscles and expose them.

Remember! The superficial pectoral M. consists of the descending and transverse pectoral Mm., whereas the deep pectoral M. is considered the ascending pectoral M.; in addition, the subclavius M. is also a deep structure.

Identify the axillary A.V. If the pectoral Mm. are not reflected or yet removed, pull out the brachial plexus and identify as many of the nerves and muscles they supply as possible.

Remember! The brachial plexus provides the muscles that connect the limb to the body wall with specific nerves. The cranial pectoral Nn. and caudal pectoral Nn. supply the muscles located cranial and caudal to the long axis of the limb, respectively. These nerves and muscles include the following: the N. for the serratus ventralis cervicis and rhomboideus cervicis Mm. (the dorsal scapular N.); the N. for the subclavius M. (the subclavius N.) and the N. for the superficial pectoral Mm. (the cranial pectoral Nn.); the thoracodorsal N., which supplies the latissimus dorsi M.; the long thoracic N., which supplies the serratus ventralis thoracis M.; the lateral thoracic N., which supplies the cutaneus trunci M.; and the caudal pectoral Nn., which supply the ascending pectoral M.

Identify and dissect these structures.

The other nerves in this area supply the intrinsic muscles of the thoracic limb, such as (in craniocaudal order): the suprascapular N., the subscapular N., and the axillary N. (dorsally), and the musculocutaneous, median, ulnar, and radial Nn. (ventrally).

Follow the course of the suprascapular N., which penetrates the space between the supraspinatus and subscapularis Mm. and runs beside the suprascapular A.V. Examine the subscapular N., which divides into many branches before entering the subscapular M. The axillary N. penetrates between the subscapularis and the teres major Mm. at the level of the caudal aspect of the shoulder joint. Identify the subscapular A.V. and their branches, the caudal circumflex humeral A.V., which run parallel to the axillary N. Identify the thoracodorsal A.V., which travel parallel to the corresponding nerve, and the superficial thoracic V. (or spur vein). Identify the axillary lnn. Palpate the caudal aspect of the shoulder joint and expose the articularis humeri M. Identify the origin of the brachialis M.

Notice that the thoracodorsal A.V. are branches of the axillary vessels, whereas (in the horse) the superficial thoracic V. is only a branch of the thoracodorsal V. and has no arterial satellite.

The dorsal nerves are now exposed and dissected, together with the vessels and muscles. Next, dissect the last four nerves with the brachial A.V. (Vv.) and the muscles of the medial aspect of the arm. To dissect the nerves and vessels, it is necessary to reflect and partially or totally remove the pectoral Mm. In addition, make an incision in the medial brachial fascia in the long axis of the coracobrachialis M. (the tendon of which originates from the coracoid process of the scapula), and expose the vessels and nerves.

Several landmarks are important here: the musculocutaneus and median Nn. travel in front of the brachial A., whereas the ulnar and radial Nn. are located caudal to the artery. The musculocutaneus N. crosses the lateral aspect of the axillary A., whereas the median N. crosses its medial aspect; they communicate with each other, making a loop that resembles a hinge (or sling) around the artery and is called the "axillary loop" (ansa axillaris) (Fig. 6-2).

Notice that the nerve supplying the superficial pectoral M. has its origin in the axillary loop.

Carefully dissect the proximal muscular branch of the musculocutaneus N., which penetrates between the two parts of the coracobrachialis M., together with the cranial circumflex humeral A.V. This nerve supplies the coracobrachialis and biceps brachii Mm.

Following the cranial aspect of the brachial A. distally, the bicipital A., which supplies the biceps brachii M., and the transerve cubital A. are shown. From the caudal border of the brachial A., two arteries originate: the deep brachial and collateral ulnar Aa. The deep brachial A. penetrates the space between the humerus, the medial and long heads of the triceps brachii M., and the common insertion of the latissimus dorsi, tensor fasciae antebrachii, and teres major Mm. on the teres major tuberosity. The radial N. penetrates the same area. Follow its course as far as possible into the spiral groove of the humerus (the groove of the brachial M.), where it is accompanied by the collateral radial A., which is a branch of the deep brachial A.

The collateral ulnar A. originates from the brachial A. at an acute angle and crosses the ventral part of the tensor fasciae antebrachii M. on its deep aspect and accompanies the homologous vein and the ulnar N. toward the medial aspect of the olecranon.

Remove the remainder of the brachial fascia from around the biceps brachii M. and from the tensor fasciae antebrachii M.

Notice that the medial brachial fascia surrounds the biceps brachii and brachialis Mm., separately.

Caution! In the horse, the tensor fasciae antebrachii M. is comprised of a muscular portion and an aponeurosis; the latter is intimately covered by the medial brachial fascia.

Because of its intimate relationship with the musculocutaneus N., use care when dissecting the median N. At approximately the same level of the bicipital A., the musculocutaneus N. separates from the median N. and distributes its distal muscular branch (to the brachialis M.) and the medial cutaneous antebrachial N. (see Fig. 6-2).

Pull out the latissimus dorsi M. and gently separate it from the tensor fasciae antebrachii M. by introducing your hand between both the muscular part and the aponeurosis of the tensor fasciae antebrachii M. in the angle between the latissimus dorsi M., the caudal border of the tensor fasciae antebrachii M., and the long head of the triceps brachii M. The aponeurosis of the tensor fasciae antebrachii M. fuses with that of the latissimus dorsi M.; together, they form a fibrous sulcus for guiding the tendon of the teres major M. Examine all these structures and their common insertion on the teres major tuberosity (Fig. 6-3).

Caution! Do not make an incision and do not transect the aponeurosis or the muscular portion of the tensor fasciae antebrachii M. Dissect and reflect only the muscle located toward the caudal border of the arm so that the space at the caudal aspect of the humerus is enlarged.

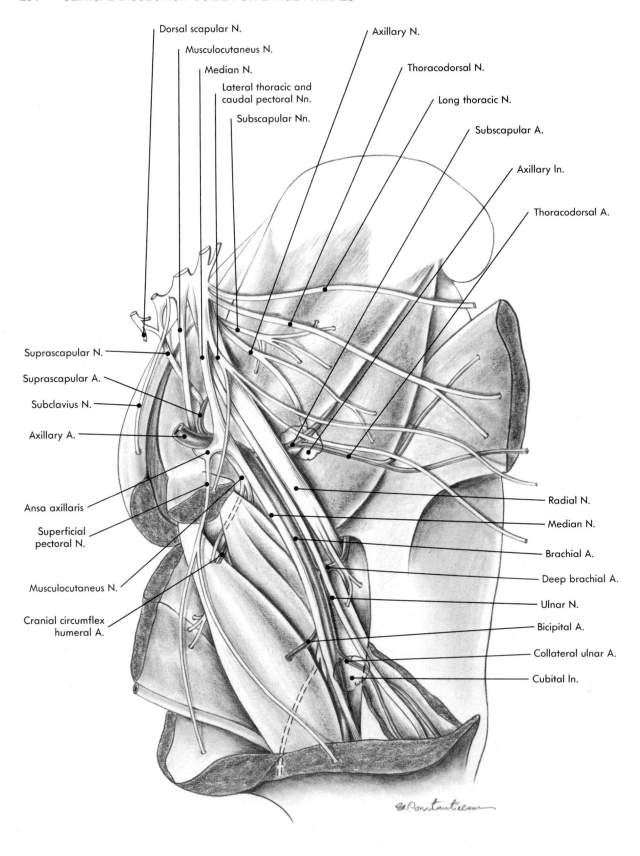

Dorsal scapular N.

Musculocutaneus N.

Median N.

Lateral thoracic and
caudal pectoral Nn.

Subscapular Nn.

Axillary N.

Thoracodorsal N.

Long thoracic N.

Subscapular A.

Axillary ln.

Thoracodorsal A.

Suprascapular N.

Suprascapular A.

Subclavius N.

Axillary A.

Ansa axillaris

Superficial
pectoral N.

Musculocutaneus N.

Cranial circumflex
humeral A.

Radial N.

Median N.

Brachial A.

Deep brachial A.

Ulnar N.

Bicipital A.

Collateral ulnar A.

Cubital ln.

Brachial plexus with muscles outlined.

Fig. 6-2

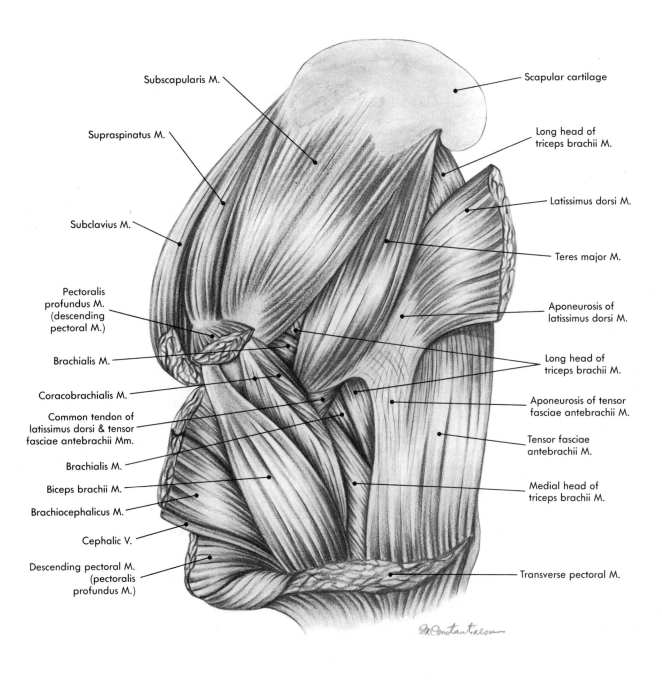

Subscapularis M.

Supraspinatus M.

Subclavius M.

Pectoralis
profundus M.
(descending
pectoral M.)

Brachialis M.

Coracobrachialis M.

Common tendon of
latissimus dorsi & tensor
fasciae antebrachii Mm.

Brachialis M.

Biceps brachii M.

Brachiocephalicus M.

Cephalic V.

Descending pectoral M.
(pectoralis
profundus M.)

Scapular cartilage

Long head of
triceps brachii M.

Latissimus dorsi M.

Teres major M.

Aponeurosis of
latissimus dorsi M.

Long head of
triceps brachii M.

Aponeurosis of tensor
fasciae antebrachii M.

Tensor fasciae
antebrachii M.

Medial head of
triceps brachii M.

Transverse pectoral M.

Muscles of shoulder and arm (medial aspect).

Fig. 6-3

Again turn the thoracic limb so that the lateral aspect is up; this enables the scapular and brachial areas to be dissected before the forearm.

Caution! When dissecting the muscles of the lateral scapular area, do not remove the lateral axillary fascia. For aesthetic purposes, leave it intact.

Carefully dissect the deltoid M. and reflect it caudally to expose the insertions of the infraspinatus and teres minor Mm.

Notice that the infraspinatus M. has one superficial and one deep portion. Transect the superficial and expose the deep portion. Reflect the superficial portion dorsally as far as possible to expose the suprascapular N.

Caudal to the teres minor M., the axillary N. and caudal circumflex humeral A.V. are visible. One of the branches of the axillary N. exceeds the deltoid M. caudal to the deltoid tuberosity and becomes the cranial cutaneous antebrachial N.

Dissect the long and lateral heads of the triceps brachii M. Transect the lateral head of the triceps brachii M. in its middle, and carefully reflect the stumps to expose the radial N., which is accompanied by the collateral radial A.V. At the ventral border of this muscle, the radial N. becomes the lateral cutaneous antebrachial N. At the deep aspect of the lateral head of the triceps brachii M., identify a portion of the long head of the triceps brachii M., the anconeus M., and a portion of the brachialis M. (between the deltoid tuberosity and anconeus M.) (Fig. 6-4).

Dissect all the cutaneous antebrachial nerves. Identify the two tendons of the extensor carpi ulnaris M. Just proximal to the accessory carpal bone, dissect the dorsal br. of the ulnar N. Identify and dissect the cephalic (antebrachial portion) and the accessory cephalic Vv. Saving as many cutaneous branches as possible, remove the proper antebrachial fascia in the following manner: identify the tendons of the extensor carpi radialis M.; common and lateral digital extensor Mm. and extensor carpi ulnaris M. (on the craniolateral aspect); and the tendons of the flexor carpi ulnaris and flexor carpi radialis Mm. (on the caudomedial aspect) just proximal to the carpus. Make incisions in the fascia on the long axis of each tendon. Introduce the blade of the scalpel between the fascia and the tendons and continue the incisions up to the elbow. Reflect the fascia from each muscle and remove it in a manner similar to that described in Chapter 5.

Concentrate on the elbow. Turn the limb so that the cranial aspect is up, and identify the lacertus fibrosus.

Caution! Leave the two connections of the lacertus fibrosus attached to the biceps brachii and to the extensor carpi radialis Mm. The fibrous structure is crossed or paralleled on the medial aspect by the medial cutaneous antebrachial N. and by the accessory cephalic and median cubital Vv.

Continue the dissection of the brachial A.V., crossing the medial aspect of the elbow obliquely. Proximal to the radial insertion of the biceps brachii M., identify the transverse cubital A.V., which surrounds the elbow craniolaterally. Dissect the median cubital V., which connects the cephalic and brachial Vv.

Caution! The median N. and the brachial A. switch positions from the brachial to the elbow regions, with the nerve becoming caudal to the artery.

The vessel and nerve course deep to the origin of the flexor carpi radialis M. Don't continue their dissection at this time.

Leave a small portion of the proper antebrachial fascia attached to the distal extension of the tensor fasciae antebrachii M. (Fig. 6-5). If you have not already done so, dissect and reflect the cranial border of this muscle caudally to expose the deep aspect. The ulnar N. and the collateral ulnar vessels (with the cubital lnn. at their origin) as well as the medial head of the triceps brachii M. and the anconeus M. are now widely exposed. Identify the muscular branches and the caudal cutaneous antebrachial N. as branches of the ulnar N.

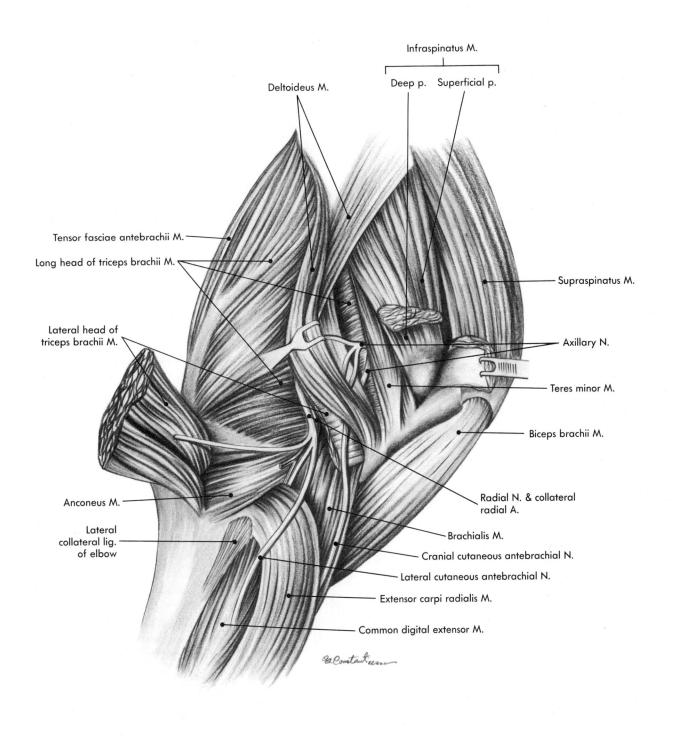

Muscles, vessels, and nerves of shoulder (lateral aspect, deep level).

Fig. 6-4

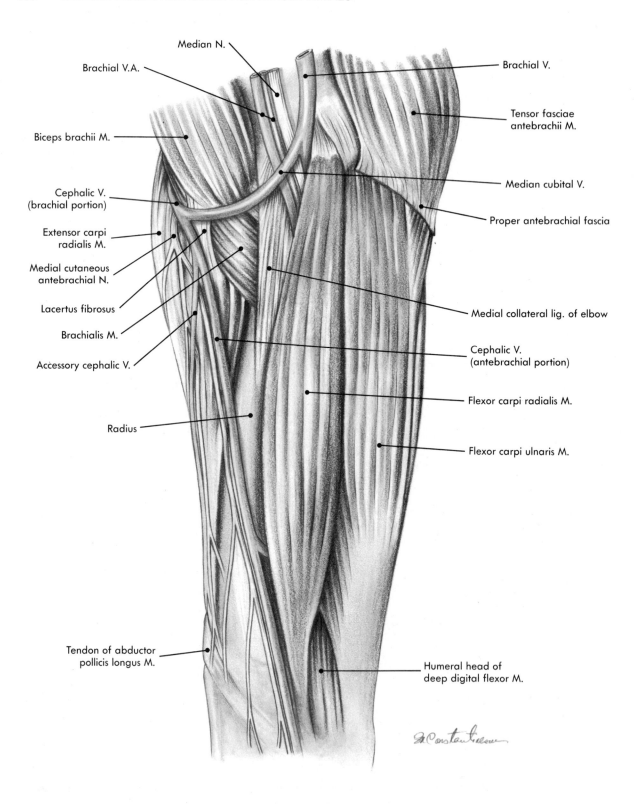

Median N.

Brachial V.A.

Brachial V.

Biceps brachii M.

Tensor fasciae antebrachii M.

Cephalic V. (brachial portion)

Median cubital V.

Extensor carpi radialis M.

Proper antebrachial fascia

Medial cutaneous antebrachial N.

Lacertus fibrosus

Medial collateral lig. of elbow

Brachialis M.

Cephalic V. (antebrachial portion)

Accessory cephalic V.

Flexor carpi radialis M.

Radius

Flexor carpi ulnaris M.

Tendon of abductor pollicis longus M.

Humeral head of deep digital flexor M.

Fig. 6-5

Forearm (medial aspect, superficial level).

Dissect the medial collateral lig. of the elbow and examine its location, twisted fibers, direction, and possible muscular fibers of the vestigial pronator teres M. The insertion of the brachialis M. slips between the medial collateral lig. and the radius. Dissect both the humeral and ulnar heads of the flexor carpi ulnaris M.; between them, fibers of the ulnar N., segments of the collateral ulnar vessels, and the ulnar head of the deep digital flexor M. are visible. Carefully dissect the caudal border of the forearm, especially the space between the flexor carpi ulnaris and extensor carpi ulnaris Mm., to expose the tendon of the ulnar head of the deep digital flexor M., the ulnar N., and the collateral ulnar vessels. Dissect both flexor carpi radialis and flexor carpi ulnaris Mm., and transect them at a point one-hand width proximal to the carpus. Carefully reflect the stumps for a better dissection of the "median" and "ulnar" structures and their connections.

Begin the dissection with the distal extent of the brachial A.V. at the deep aspect of the flexor carpi radialis M. Identify the common interosseous A.V. at the level of the interosseous space between the radius and ulna.

Remember! Distal to the common interosseous vessels, the names of the brachial A.V. change to the median A.V. (Fig. 6-6).

Identify the deep antebrachial A., which is the first branch of the median A. Dorsal to the carpus, the median A. branches into the proximal radial A., the radial A., and the palmar br. The latter anastomoses with the collateral ulnar A. It then perforates the flexor retinaculum on the medial aspect of the accessory carpal bone and runs together with the vein (palmar br.) and the lateral palmar N. (from the median N.) after it has received the palmar br. from the ulnar N. The radial A.V. perforate the flexor retinaculum laterally and close to the tendon of the flexor carpi radialis M. Dissect these vessels and nerves, and make two corresponding sections through the flexor retinaculum by following the directions of the vessels and nerve.

The median A.N. course within the carpal canal and are protected by the double tendinous sheath of the deep and superficial digital flexor tendons. To expose them, make an incision in the flexor retinaculum midway between the radial vessels and the accessory carpal bone.

Notice that the carpal canal has a dorsal wall (represented by the palmar carpal joint capsule, which is filled with cartilage), a lateral wall (represented by the accessory carpal bone and its ligaments), and a mediopalmar wall (represented by the flexor retinaculum) (Figs. 6-7 to 6-9).

Proximal to the carpus, locate the fibrous connection between the superficial digital flexor M. and the caudal aspect of the radius. This is the proximal check or accessory lig. or the check/accessory lig. of the superficial digital flexor M. Separate the superficial digital flexor M. from the humeral head of the deep digital flexor M. The latter makes a groove that is oriented medially and that receives the superficial digital flexor M. (see Fig. 6-6).

Caution! The humeral head of the deep digital flexor M. has three portions. Do not separate them from one another.

Identify the small radial head of the deep digital flexor M. in the distal half of the caudal aspect of the radius.

Turn the limb so that the lateral aspect is up, and dissect all the muscles, including the abductor pollicis longus M. as far as the proximal extent of the carpus. The tendons of these muscles course between the dorsal joint capsule of the carpus and the extensor retinaculum, which are structures that create common septa between the tendons (Fig. 6-10).

Notice that both the flexor and the extensor retinacula are considered parts of the carpal fascia.

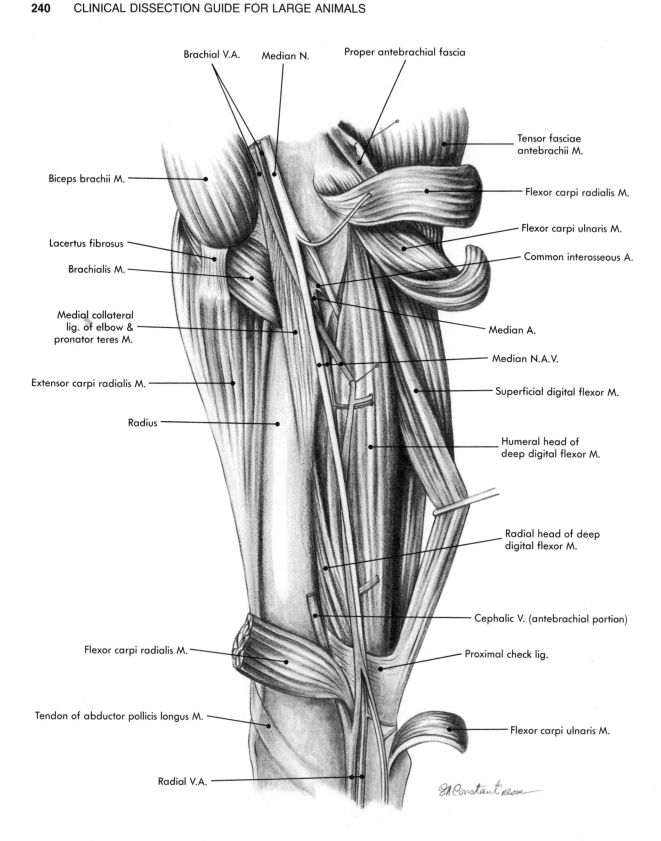

Brachial V.A.

Median N.

Proper antebrachial fascia

Tensor fasciae antebrachii M.

Biceps brachii M.

Flexor carpi radialis M.

Flexor carpi ulnaris M.

Lacertus fibrosus

Common interosseous A.

Brachialis M.

Medial collateral lig. of elbow & pronator teres M.

Median A.

Median N.A.V.

Extensor carpi radialis M.

Superficial digital flexor M.

Radius

Humeral head of deep digital flexor M.

Radial head of deep digital flexor M.

Cephalic V. (antebrachial portion)

Flexor carpi radialis M.

Proximal check lig.

Tendon of abductor pollicis longus M.

Flexor carpi ulnaris M.

Radial V.A.

Fig. 6-6

Forearm (medial aspect, deep level).

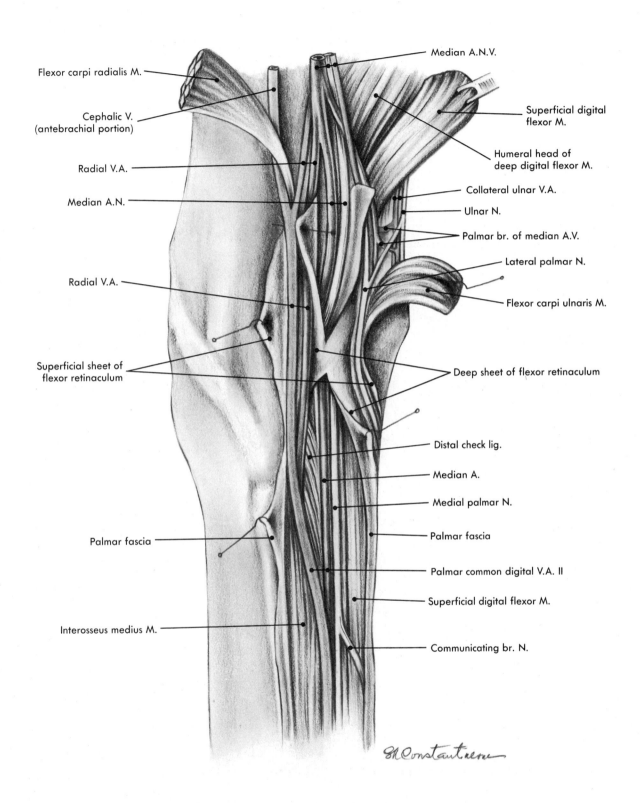

Flexor carpi radialis M.

Cephalic V.
(antebrachial portion)

Radial V.A.

Median A.N.

Radial V.A.

Superficial sheet of
flexor retinaculum

Palmar fascia

Interosseus medius M.

Median A.N.V.

Superficial digital
flexor M.

Humeral head of
deep digital flexor M.

Collateral ulnar V.A.

Ulnar N.

Palmar br. of median A.V.

Lateral palmar N.

Flexor carpi ulnaris M.

Deep sheet of flexor retinaculum

Distal check lig.

Median A.

Medial palmar N.

Palmar fascia

Palmar common digital V.A. II

Superficial digital flexor M.

Communicating br. N.

Carpus of right limb (mediopalmar aspect).

Fig. 6-7

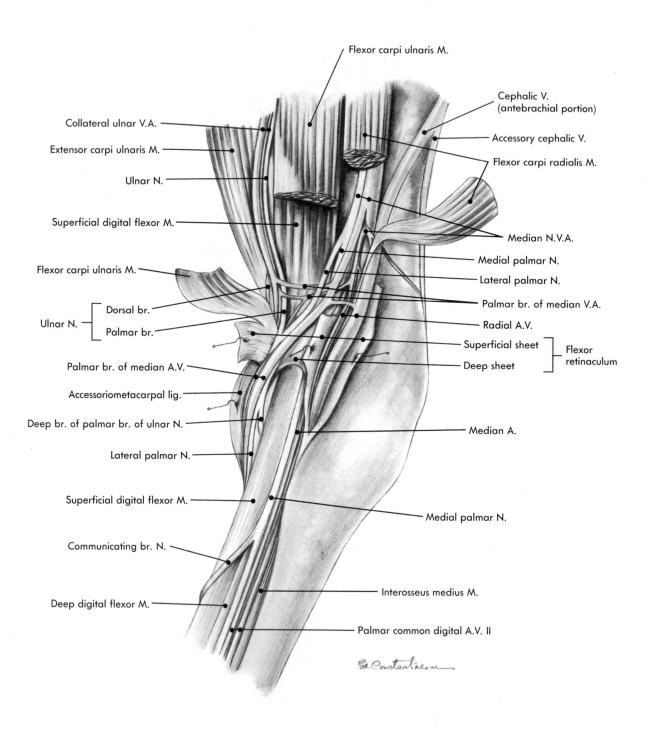

Flexor carpi ulnaris M.

Cephalic V. (antebrachial portion)

Collateral ulnar V.A.

Accessory cephalic V.

Extensor carpi ulnaris M.

Flexor carpi radialis M.

Ulnar N.

Superficial digital flexor M.

Median N.V.A.

Medial palmar N.

Lateral palmar N.

Flexor carpi ulnaris M.

Palmar br. of median V.A.

Ulnar N. — Dorsal br.

Radial A.V.

Palmar br.

Superficial sheet — Flexor retinaculum

Deep sheet

Palmar br. of median A.V.

Accessoriometacarpal lig.

Median A.

Deep br. of palmar br. of ulnar N.

Lateral palmar N.

Superficial digital flexor M.

Medial palmar N.

Communicating br. N.

Interosseus medius M.

Deep digital flexor M.

Palmar common digital A.V. II

Fig. 6-8

Carpus of left limb (mediopalmar aspect).

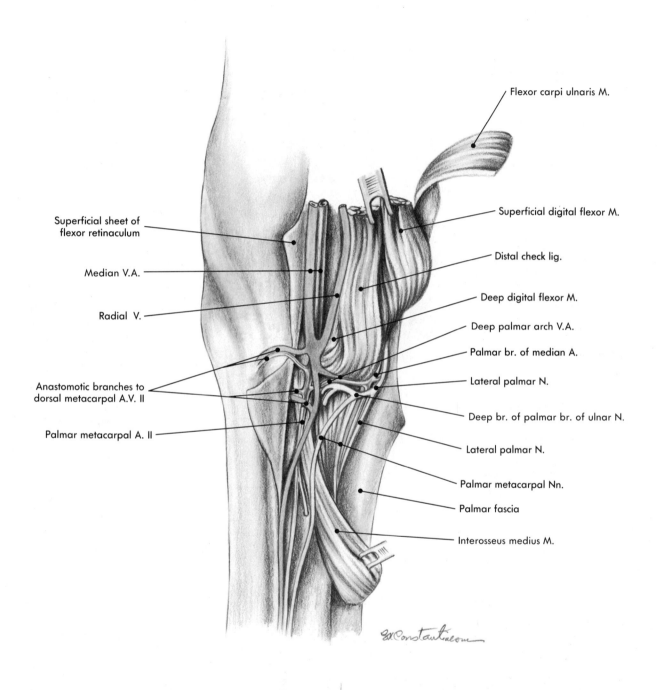

Flexor carpi ulnaris M.

Superficial digital flexor M.

Superficial sheet of flexor retinaculum

Distal check lig.

Median V.A.

Deep digital flexor M.

Radial V.

Deep palmar arch V.A.

Palmar br. of median A.

Lateral palmar N.

Anastomotic branches to dorsal metacarpal A.V. II

Deep br. of palmar br. of ulnar N.

Lateral palmar N.

Palmar metacarpal A. II

Palmar metacarpal Nn.

Palmar fascia

Interosseus medius M.

Carpometacarpal area of right limb (mediopalmar aspect, deep level).   **Fig. 6-9**

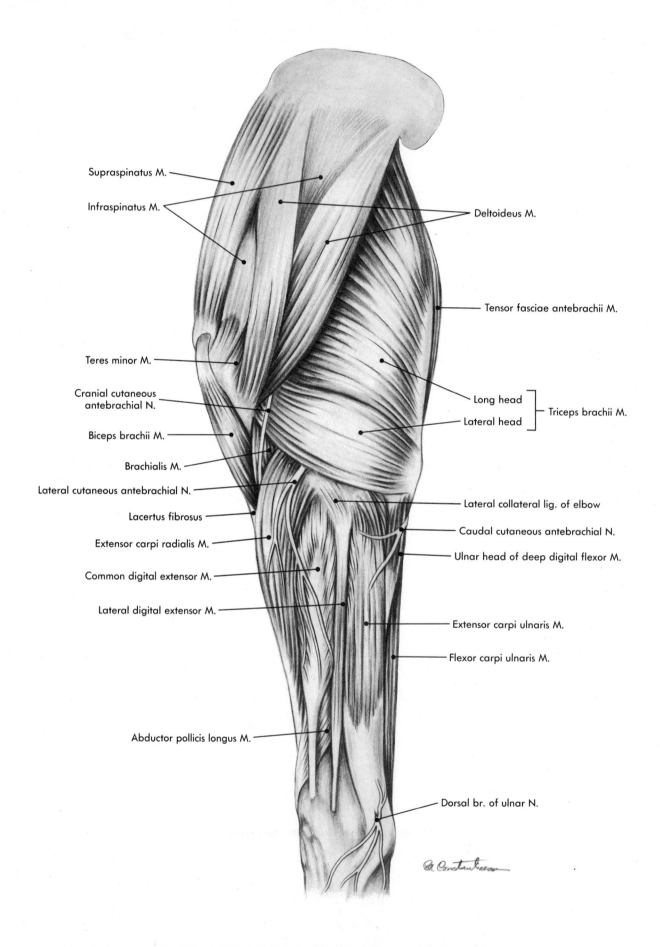

Supraspinatus M.

Infraspinatus M.

Deltoideus M.

Tensor fasciae antebrachii M.

Teres minor M.

Long head
Lateral head
Triceps brachii M.

Cranial cutaneous
antebrachial N.

Biceps brachii M.

Brachialis M.

Lateral cutaneous antebrachial N.

Lacertus fibrosus

Extensor carpi radialis M.

Common digital extensor M.

Lateral digital extensor M.

Lateral collateral lig. of elbow

Caudal cutaneous antebrachial N.

Ulnar head of deep digital flexor M.

Extensor carpi ulnaris M.

Flexor carpi ulnaris M.

Abductor pollicis longus M.

Dorsal br. of ulnar N.

Fig. 6-10

Left limb up to carpus (lateral aspect, superficial structures).

Transect the common digital extensor M. in the middle of the forearm, and reflect the stumps. The muscular branches of the radial N., which supply the common and lateral digital extensor Mm., the accessory M. of the common digital extensor M., the extensor carpi radialis M., the abductor pollicis longus M., and the extensor carpi ulnaris M., are now exposed. Identify the transverse cubital A. and/or the collateral radial A., which run between the extensor carpi radialis and the common digital extensor Mm., and the cranial interosseous A., which runs between the common and the lateral digital extensor Mm. Both parallel the two borders of the abductor pollicis longus M. (Fig. 6-11).

Begin the dissection of the metacarpus with the palmar aspect. Identify the palmar fascia as the continuation of the flexor retinaculum in the metacarpus. It is strongly attached to the splint bones (the metacarpal bones II [medial] and IV [lateral]). Transect the fascia from its attachments, and remove it carefully, saving the vessels and nerves paralleling or crossing the digital flexor tendons.

One by one, follow the structures already identified and exposed in the carpal area. Begin with the lateral palmar N., and pull it out to expose the origin of the deep branch of palmar br. of the ulnar N. on the medial aspect of the accessoriometacarpal lig. Continue to dissect both the lateral and medial palmar Nn. and their communicating br., which obliquely crosses the tendon of the superficial digital flexor M. in a proximodistal and mediolateral direction. The medial palmar N. is accompanied by the palmar common digital A.V. II, which are the largest vessels in the area. The lateral palmar N. is paralleled by the superficial branches of the palmar branches of the median A.V., which continue as the palmar common digital A.V. III.

Transect the superficial and deep digital flexor tendons in the middle of metacarpus, and reflect their proximal stumps. Identify the distal check or accessory lig. or the check/accessory lig. of the deep digital flexor M. Transect and reflect it dorsally. Between it and the interosseus medius M., deep palmar arterial and venous arches and other vessels are exposed, along with the origin of the lateral and medial palmar metacarpal Nn. Dissect and identify all the vessels and nerves of the palmar metacarpal area and the lateral and medial interosseous Mm. up to the fetlock (Fig. 6-12).

The following section is also applicable to the pelvic limb. The term *palmar* should be replaced by the term *plantar*. Identify the ligament connecting the ergot to the distal end of the splint bone and the ligament connecting the ergot to the cartilage of the hoof (the lig. of the ergot). Between the two ligaments of the ergot, the fascia of the digital cushion may be observed; it is the most superficial palmar digital structure beneath the skin.

Carefully dissect the branches and divisions of the digital V.A. and the palmar digital N. (Fig. 6-13).

The fasciae, tendons, ligaments, and bones of the digit are deep to the vessels and nerves. Transect the fascia of the digital cushion along the lig. of the ergot, and reflect it. Follow the palmar fascia, and examine its relationship with the tendon of the superficial digital flexor M.; they are fused, and consequently, the tendinous sheath surrounding the deep digital flexor tendon cannot also surround the superficial digital flexor tendon as does the double tendinous sheath in the carpal canal. The palmar fascia continues in the digital area as the palmar digital fascia, which is strongly attached by three pairs of insertions. The first pair is attached to the abaxial aspects of the proximal sesamoid bones (the palmar annular lig. or superficial transverse metacarpal lig.). The second and third pairs are attached to the proximal and distal extremities of the first phalanx (P. I), respectively (the so-called proximal digital annular lig.).

Notice that the palmar digital fascia is a continuous structure. Demonstrate this on your specimen (Fig. 6-14).

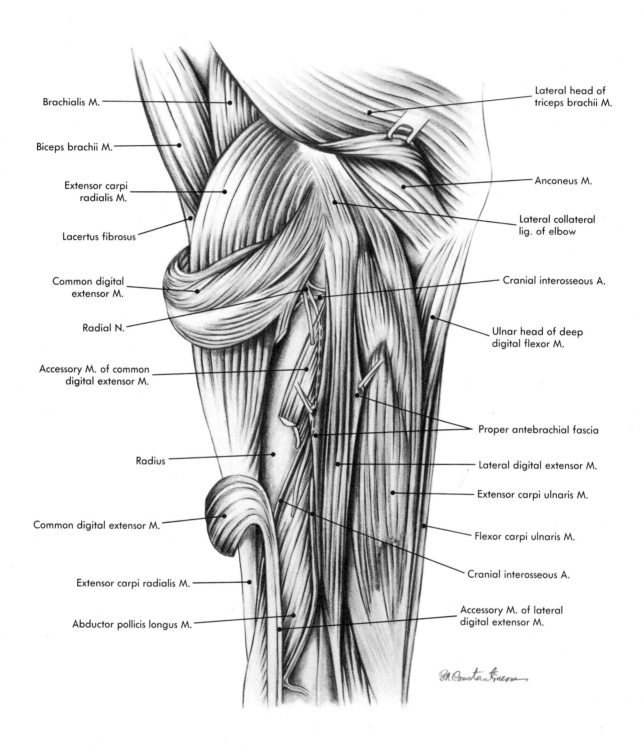

Brachialis M.

Biceps brachii M.

Extensor carpi radialis M.

Lacertus fibrosus

Common digital extensor M.

Radial N.

Accessory M. of common digital extensor M.

Radius

Common digital extensor M.

Extensor carpi radialis M.

Abductor pollicis longus M.

Lateral head of triceps brachii M.

Anconeus M.

Lateral collateral lig. of elbow

Cranial interosseous A.

Ulnar head of deep digital flexor M.

Proper antebrachial fascia

Lateral digital extensor M.

Extensor carpi ulnaris M.

Flexor carpi ulnaris M.

Cranial interosseous A.

Accessory M. of lateral digital extensor M.

Fig. 6-11   Forearm (lateral aspect, deep level).

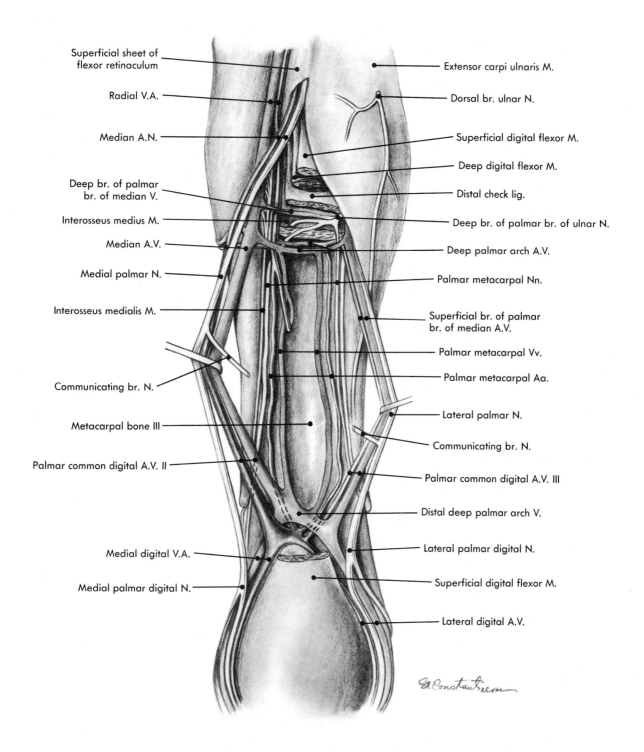

Superficial sheet of flexor retinaculum

Radial V.A.

Median A.N.

Deep br. of palmar br. of median V.

Interosseus medius M.

Median A.V.

Medial palmar N.

Interosseus medialis M.

Communicating br. N.

Metacarpal bone III

Palmar common digital A.V. II

Medial digital V.A.

Medial palmar digital N.

Extensor carpi ulnaris M.

Dorsal br. ulnar N.

Superficial digital flexor M.

Deep digital flexor M.

Distal check lig.

Deep br. of palmar br. of ulnar N.

Deep palmar arch A.V.

Palmar metacarpal Nn.

Superficial br. of palmar br. of median A.V.

Palmar metacarpal Vv.

Palmar metacarpal Aa.

Lateral palmar N.

Communicating br. N.

Palmar common digital A.V. III

Distal deep palmar arch V.

Lateral palmar digital N.

Superficial digital flexor M.

Lateral digital A.V.

Metacarpal area of right limb (palmar aspect, deep level).

Fig. 6-12

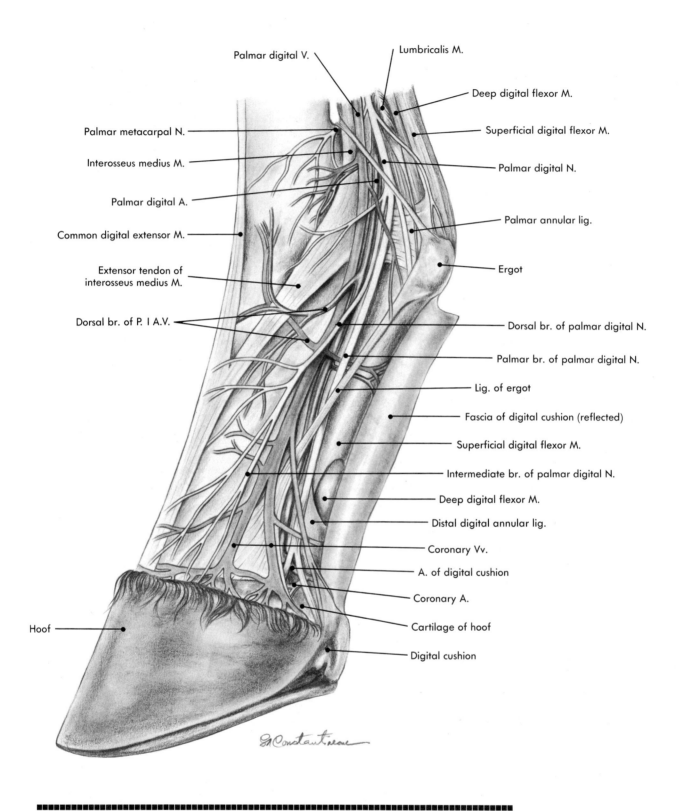

Palmar digital V.

Lumbricalis M.

Deep digital flexor M.

Palmar metacarpal N.

Superficial digital flexor M.

Interosseus medius M.

Palmar digital N.

Palmar digital A.

Palmar annular lig.

Common digital extensor M.

Ergot

Extensor tendon of
interosseus medius M.

Dorsal br. of palmar digital N.

Dorsal br. of P. I A.V.

Palmar br. of palmar digital N.

Lig. of ergot

Fascia of digital cushion (reflected)

Superficial digital flexor M.

Intermediate br. of palmar digital N.

Deep digital flexor M.

Distal digital annular lig.

Coronary Vv.

A. of digital cushion

Coronary A.

Cartilage of hoof

Hoof

Digital cushion

Fig. 6-13

Vessels and nerves of digit of right limb (medial aspect).

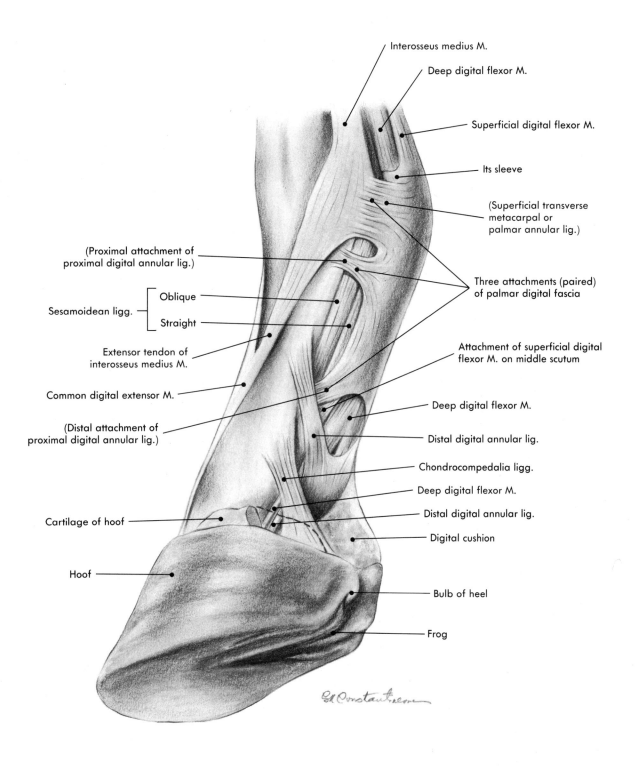

Interosseus medius M.

Deep digital flexor M.

Superficial digital flexor M.

Its sleeve

(Superficial transverse
metacarpal or
palmar annular lig.)

(Proximal attachment of
proximal digital annular lig.)

Three attachments (paired)
of palmar digital fascia

Sesamoidean ligg. — Oblique

Straight

Extensor tendon of
interosseus medius M.

Attachment of superficial digital
flexor M. on middle scutum

Common digital extensor M.

Deep digital flexor M.

(Distal attachment of
proximal digital annular lig.)

Distal digital annular lig.

Chondrocompedalia ligg.

Deep digital flexor M.

Cartilage of hoof

Distal digital annular lig.

Digital cushion

Hoof

Bulb of heel

Frog

Tendons of digit of right limb (mediopalmar aspect superficial level).

Fig. 6-14

The superficial digital flexor tendon forms a ring (or sleeve) around the deep digital flexor tendon in the fetlock area. At the level of the pastern joint (the proximal interphalangeal joint), it bifurcates to insert on the middle scutum. Between the two insertions of the superficial digital flexor tendon and the two insertions of the distal digital annular lig., the tendinous sheath of the deep digital flexor M. and the tendon are visible.

Notice that the two insertions of the distal digital annular lig. overlap the two insertions of the superficial digital flexor tendon and the distal insertions of the palmar digital fascia.

Follow the interosseus medius M., and examine its double insertion on the abaxial aspects of the proximal sesamoid bones. Note the two extensor branches (tendons), which obliquely cross the first phalanx (P. I) to meet the tendon of the common digital extensor M.

The following technique is suggested. Dissect the digital vessels, nerves, and fibrous structures on one side and then turn the limb so that the symmetrical side is up. If you need to practice, do the same dissection again and then remove the vessels and nerves along with the ergot and its ligaments. Continue the dissection of the fascia, tendons, and exposed ligaments, and identify them.

Next, dissect the deep structures that are proximal to the hoof. Transect the three attachments of the palmar digital fascia and the insertion of the superficial digital flexor tendon on the middle scutum; reflect them together. The ring of the superficial digital flexor tendon is now entirely exposed, as is the tendon of the deep digital flexor M. Examine the palmar ligaments of the proximal interphalangeal joint (three pairs of attachments on P. I). Identify the middle scutum, which is the fibrocartilaginous plate that covers the proximal palmar aspect of the second phalanx (P. II) and which provides a smooth gliding surface for the deep digital flexor tendon. The straight (superficial) sesamoidean lig. is attached to the middle scutum. Transect this ligament, and reflect the proximal stump to expose and dissect the oblique (middle), cruciate, and short sesamoidean ligg. Finally, examine the palmar lig. (formerly the intersesamoidean lig.) and the metacarpointersesamoidean lig. (the elastic structure connecting the proximal end of the previous ligament to the palmar aspect of the third metacarpal bone [Mc. III]). The palmar lig. is reinforced by the proximal scutum, which is the concave and smooth fibrocartilage that allows the safe gliding of the digital flexor tendons (Fig. 6-15).

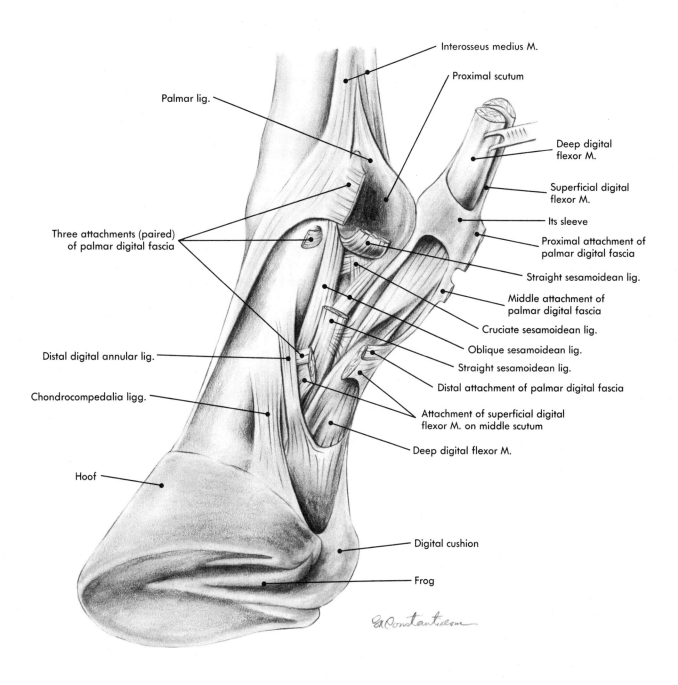

Tendons of digit of right limb (mediopalmar aspect deep level).   **Fig. 6-15**

The hoof has already been removed and saved for further examination. Identify the three main components of the hoof: the wall, frog, and sole (Fig. 6-16). Examine the external and internal surfaces (Fig. 6-17). These structures represent the continuation of the epidermis.

The wall is divided externally into a frontal region (the toe), two lateral and medial parts on the sides (the quarters), and the most palmar areas (the heels).

Notice that the toe is taller than the heels.

The heels bend (the angles of the heels) and continue forward as two symmetrical, thin, and pointed laminae called bars. Between the wall and the bars, the two angles of the sole are visible. The bars parallel the two crura of the frog and are separated from the frog by the two paracuneal grooves.

Examine the frog with its V-shaped crura that meet dorsally in the apex and are separated by the central (cuneal) groove. The palmar extremities of the crura are large and round and are called the bulbs of the heels.

The sole and frog are turned toward the ground, whereas the limb is in a resting position (see Fig. 6-17). The sole fills the space between the solear surface of the wall, the bars, and the frog. The bars separate the frog (except the apex) from the sole. At the junction of the sole with the wall, regardless of the color of the horn, a white zone is observed. The interdigitations between the nonpigmented horny (unsensitive) lamellae and the lamellae of the laminar corium (sensitive, living tissue) are referred to as the white zone, which has clinical importance. The living tissue is described later.

The internal aspect of the wall consists of the following three different structures (starting from the coronary border of the hoof): (1) the perioplic groove (1 mm wide), which receives and protects the perioplic corium; (2) the coronary groove (1 cm wide), which receives the coronary corium; and (3) the 550 to 600 horny lamellae (laminae), which interdigitate with the same number of lamellae (laminae) of the laminar corium. The horny lamellae also continue at the level of the bars. The perioplic and coronary grooves have very fine holes for the papillae of the perioplic and coronary corium.

The internal aspects of the frog and sole consist of very small holes that perforate the entire surface to receive the papillae of the solear corium. The frog has the following characteristics: one central spine (the cuneal spine), which corresponds to the central (cuneal) groove (from the external aspect); two lateral grooves, which correspond to the crura of the frog (from the external aspect); and two lateral ridges, which correspond to the paracuneal grooves (from the external aspect) (see Fig. 6-17).

Make a vertical section through the wall, and examine the three layers of the epidermis (the horny wall): the external layer (the stratum externum), which is represented by the periople (a very thin protective layer produced by the perioplic corium); the internal layer (stratum internum), which is represented by the horny lamellae (nonpigmented); and the middle layer (stratum medium), which is the thickest layer and is represented by pigmented horn (see Fig. 6-17).

The continuation of the dermis (sensitive tissue) within the hoof is comprised of five adjacent structures (Fig. 6-18).

Remember! The dermis, also known as corium, consists of a rich vascular connective tissue. The corium has the same relation to the hoof as the dermis has to the skin.

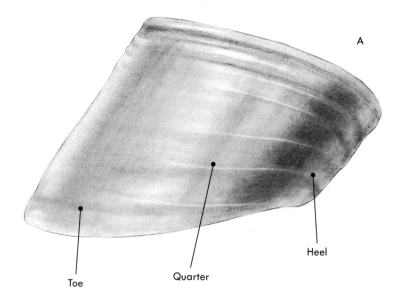

A

Heel

Toe

Quarter

B

Apex

Crura

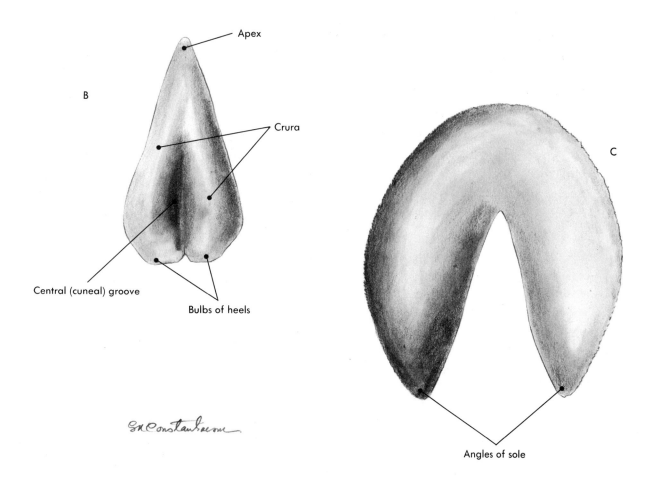

C

Central (cuneal) groove

Bulbs of heels

Angles of sole

A, Wall. **B,** Frog. **C,** Sole.   **Fig. 6-16**

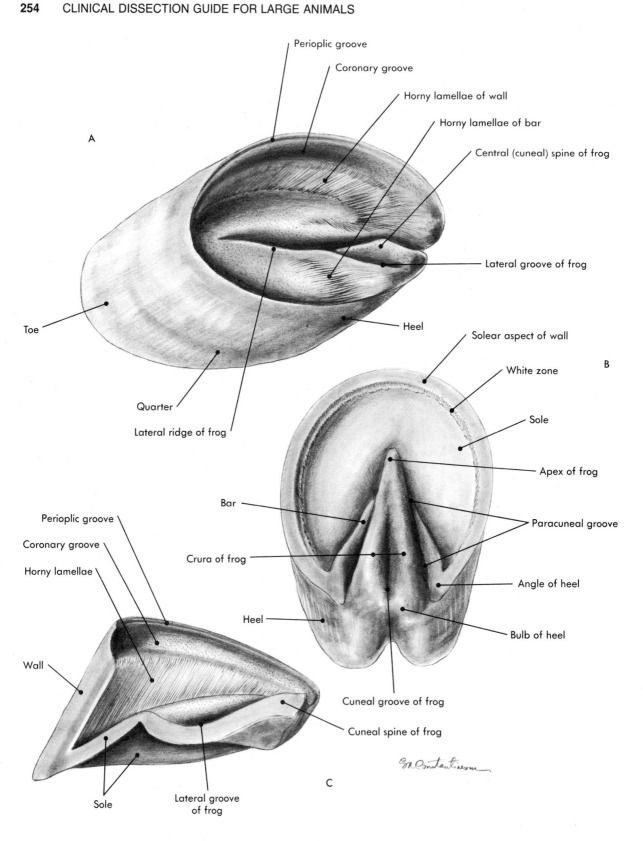

Fig. 6-17 **A,** Internal aspect of hoof. **B,** Solear aspect of hoof. **C,** Median section through hoof (internal aspect).

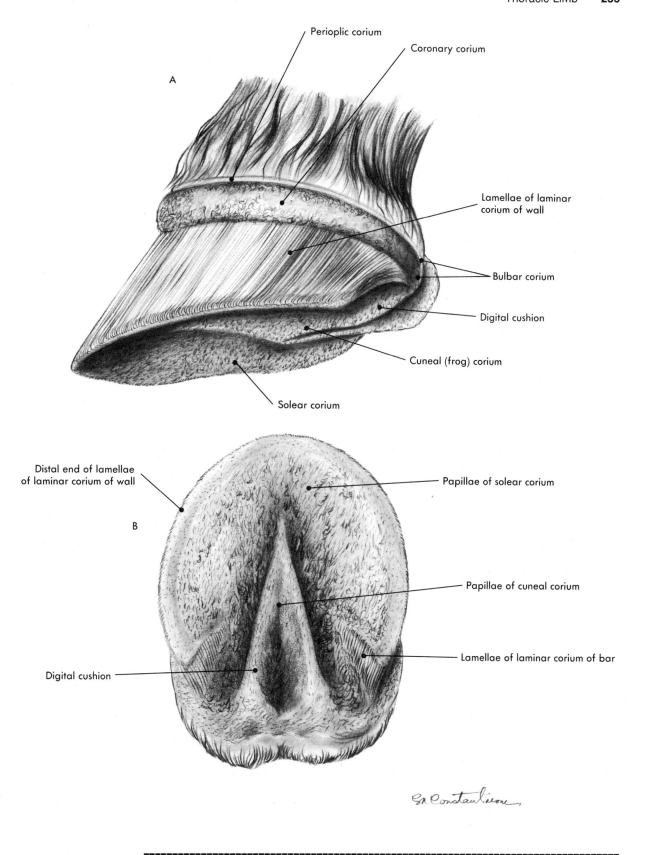

Perioplic corium

Coronary corium

A

Lamellae of laminar corium of wall

Bulbar corium

Digital cushion

Cuneal (frog) corium

Solear corium

Distal end of lamellae of laminar corium of wall

Papillae of solear corium

B

Papillae of cuneal corium

Lamellae of laminar corium of bar

Digital cushion

A, Corium of digit (lateroventral aspect). **B,** Corium of digit (ventral aspect).

Fig. 6-18

Three out of the five structures of the dermis are located in contact with the internal aspect of the wall; they are the perioplic corium, the coronary corium, and the laminar corium. At the solear aspect, the cuneal corium (the corium of the frog) and the solear corium (the corium of the sole) are the main structures; the laminar corium of the bars is also present.

The perioplic corium (1 mm wide) consists of many fine papillae. The perioplic corium continues as the bulbar corium toward the heels and is protected within the perioplic groove.

The coronary corium (1 cm wide) bears papillae that are similar to the perioplic corium, and it continues in a manner similar to bulbar corium. It is protected within the coronary groove and lies on a supportive subcutaneous tissue, called the coronary cushion.

The laminar corium, which is equal in number to the horny lamellae, covers the parietal aspect of the third phalanx (P. III) and the corresponding area of the bars (on the solear aspect) between the cuneal corium and solear corium. This structure is called the laminar corium of the bars (see Fig. 6-18).

The cuneal corium and solear corium bear fine papillae and join the laminar corium to form a continuous envelope for the deep structures.

The anatomical structures located between the dermis and the bones, the ligaments, and/or the tendons are the digital cushion and the two symmetrical cartilages of the hoof.

The digital cushion is a pyramidal fibroelastic structure that makes contact with the frog, the cartilages of the hoof, and the distal digital annular lig. The base of the pyramid, which is protected by the bulbs of the heels, is partly subcutaneous (Fig. 6-19, A).

The two fibrocartilages of the hoof are slightly rectangular, with the dorsal half attached at the palmar processes of P. III and the palmar half remaining free. This palmar half is attached to the corresponding side of the digital cushion. With the exception of the proximal (convex) borders, which exceed the coronary border of the hoof, the lateral aspects of the cartilages are covered by the dermis of the hoof (Fig. 6-19, B and C).

Examine all these structures.

Turn the limb so that the palmar side is up. Keeping the distal interphalangeal joint in extension, remove the digital cushion, make 1 or 2 cross-sections, and examine them. The palmar venous plexus is located under the cuneal corium and solear corium and also on the axial surfaces of the cartilages of the hoof. A similar venous network is located between the dorsal aspect of P. III and the corresponding dermis. Piercing the cartilages of the hoof, the palmar venous plexuses communicate with the coronary plexuses; the entire system is considered to be the origin of the coronary Vv.

Identify the chondrocompedalia ligg. (the ligaments that extend from the cartilage of the hoof and that are called chondro- plus the name of the bone to which they are attached). On the side without vessels and nerves, remove the cartilage in a mediolateral direction, keeping it in abduction (Fig. 6-19).

Caution! Save the collateral lig. of the distal interphalangeal joint and the collateral sesamoid (navicular) lig.

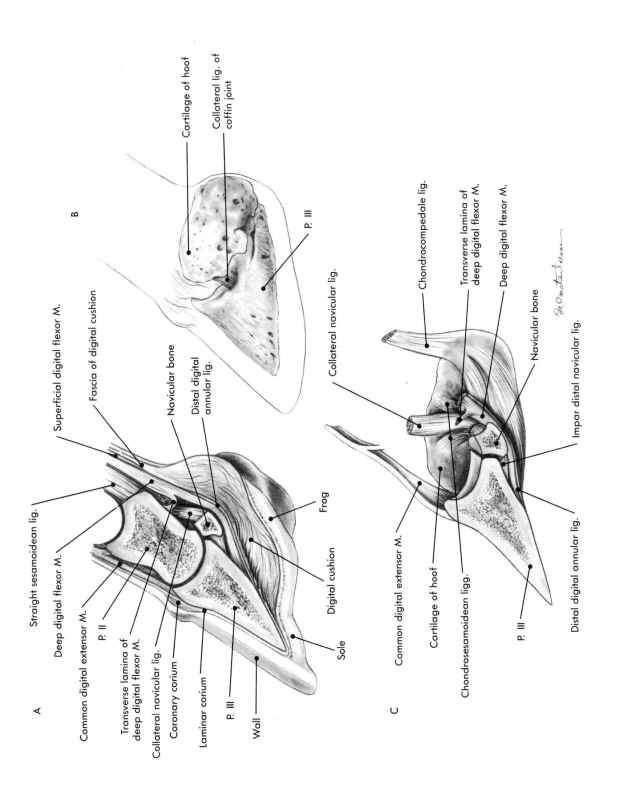

A, Median section through digit. **B,** Topography of cartilage of hoof (abaxial aspect). **C,** Chondrocompedalia ligg.

Fig. 6-19

Dissect the area between the distal digital annular lig. and the collateral sesamoid (navicular) lig., and examine the tendon of the deep digital flexor M., which attaches to the palmar aspect of P. II (the transverse lamina). This attachment separates the navicular bursa from the tendinous sheath of the digital flexor tendons proximal to the transverse lamina. Examine the fibers of the collateral sesamoid lig. at the point where they join the symmetrical fibers at the palmar aspect of the navicular bone. These fibers are fused with the transverse lamina or are independently attached on P. II distal to the transverse lamina. This fibrous curtain or hinge, as well as the navicular bone and the unpaired distal sesamoid (navicular) lig., separates the navicular bursa from the synovial membrane of the second interphalangeal joint.

Transect the attachment of the distal digital annular lig. from P. I. Transect the deep digital flexor tendon at the level where it becomes superficial between the two insertions of the superficial digital flexor tendon, and reflect it distally. The middle scutum and the transverse lamina are now fully exposed. Examine the complementary fibrocartilage of the proximal palmar border of P. II and the palmar ligaments.

Make a longitudinal incision through the deep digital flexor tendon and its transverse lamina, and reflect them to examine the deep structures. The hinge made by the fibers of the two collateral sesamoid ligg., the navicular bone, and the unpaired distal sesamoid lig. are now visible. The flexor surface of the navicular bone is covered by a fibrocartilage, which extends proximally and fuses its fibers with the hinge made by the collateral sesamoid ligg. to form the distal scutum. The distal scutum provides a smooth gliding surface for the tendon of the deep digital flexor m. Examine these structures (Fig. 6-20, *A*).

Turn the limb so that the side without vessels and nerves is up. Dissect and examine the collateral ligaments of *all* the digital joints. Check with a skeleton as many times as is necessary.

Turn the limb so that the opposite side is up, and continue the dissection of the digital A.V.N. in the area protected by the hoof (Fig. 6-20, *B*). The dorsal br. of the digital A. runs in the parietal groove of P. III. The continuation of the digital A. on the palmar aspect enters the solear foramen, anastomoses with its opposite artery, and gives off numerous fine arteries that perforate P. III and finally unite into the terminal arch surrounding the solear border of P. III (Fig. 6-20, *C*).

Transect the deep digital flexor tendon at its insertion on the semilunar line and the flexor surface of P. III; examine the solear foramen and the solear groove (see Fig. 6-20, *C*).

At the carpal joint, dissect and examine the dorsal and palmar joint capsules, the collateral ligg. and the four ligaments of the accessory carpal bone. Identify the carpal and metacarpal bones on a skeleton and on your specimen.

Dissect and examine the cranial fibrous joint capsule of the elbow joint, which has no corresponding structure on the caudal aspect. The only structure present both cranially and caudally is the humeroradial synovial membrane.

After the entire thoracic limb is dissected and if you do not plan to save it for other purposes, disarticulate it joint by joint to examine the articular surfaces of the bones.

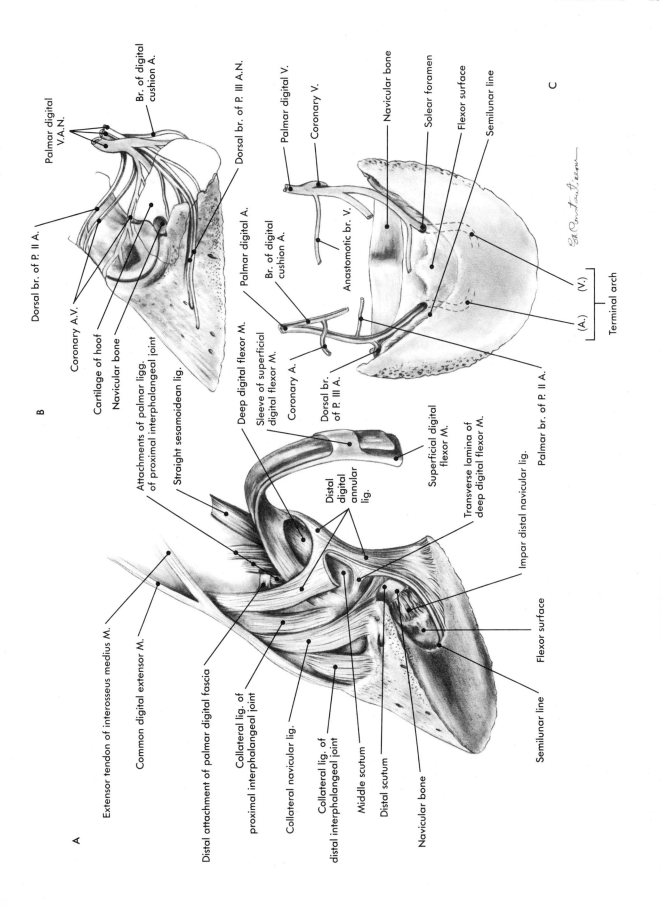

A, Lateroventral aspect of deep structures of the digit. **B,** Terminal branches of palmar digital A.V.N. (lateral/medial aspect). **C,** Terminal branches of palmar digital vessels (ventral view).

Fig. 6-20

A

Extensor tendon of interosseus medius M.

Common digital extensor M.

Distal attachment of palmar digital fascia

Collateral lig. of proximal interphalangeal joint

Collateral navicular lig.

Collateral lig. of distal interphalangeal joint

Middle scutum

Distal scutum

Navicular bone

Semilunar line

Deep digital flexor M.

Sleeve of superficial digital flexor M.

Distal digital annular lig.

Superficial digital flexor M.

Transverse lamina of deep digital flexor M.

Impar distal navicular lig.

Flexor surface

B

Palmar digital V.A.N.

Br. of digital cushion A.

Dorsal br. of P. II A.

Coronary A.V.

Cartilage of hoof

Navicular bone

Attachments of palmar ligg. of proximal interphalangeal joint

Straight sesamoidean lig.

Dorsal br. of P. III A.N.

Palmar digital A.

Br. of digital cushion A.

Coronary A.

Dorsal br. of P. III A.

Palmar br. of P. II A.

C

Palmar digital V.

Coronary V.

Anastomotic br. V.

Navicular bone

Solear foramen

Flexor surface

Semilunar line

Terminal arch

(V.)

(A.)

Large ruminants

Identify the following landmarks on your specimen. On the lateral aspect of the thoracic limb, identify the cranial and caudal angles of the scapula; the spine of the scapula; the acromion; the supraglenoid tubercle; the greater tubercle of the humerus; the deltoid tuberosity; the olecranon; the elbow; the carpus; the accessory carpal bone; the metacarpus; the dewclaw; the proximal sesamoid bones; the metacarposesamophalangeal joint; the proximal phalanx (P. I); the middle phalanx (P. II); and the hoof. On the lateral aspect of the forearm, identify the three (not two, as in the horse) vertical grooves separating (in a craniocaudal order) the extensor carpi radialis M. from the common digital extensor M., the common from the lateral digital extensor M., and the lateral digital extensor M. from the extensor carpi ulnaris M. On the lateral aspect of the metacarpus, identify the grooves separating the palmar aspect of the metacarpal bones from the interosseus medius M. and the interosseus medius M. from the digital flexor tendons.

On the medial aspect of the limb, identify the coracoid process of the scapula, the lesser tubercle and the medial epicondyle of the humerus, the radial tuberosity, the body of the radius, the carpus, and the same structures of the metapodium and the acropodium as on the lateral aspect. In addition, identify the vertical grooves of the medial aspect of the forearm that separate (in a craniocaudal order) the extensor carpi radialis M. from the radius, the radius from the flexor carpi radialis M., and the flexor carpi radialis M. from the flexor carpi ulnaris M. The grooves of the metacarpal area are similar to those on the lateral side.

In a manner similar to that for the horse, skin the limb down to the metacarposesamophalangeal joint and remove the hoof either with a saw and pliers or by boiling (see this procedure described in the section on the horse). Skin the digital area.

Turn the limb so that the medial side is up, and identify the pectoral muscles, which were previously dissected and transected.

Notice that the subclavius M. in the large ruminants is very small in comparison with the horse. It is inserted on the medial aspect of the brachiocephalicus M., just cranial to the shoulder joint.

Pull up the brachial plexus, and attempt to identify the nerves in conjunction with the muscles. The same nerves, muscles, and vessels encountered in the horse are present in the cow. Identify the axillary ln. of the first rib, if it did not remain attached to the lateral wall of the thorax (first intercostal space). In relation to the thoracic limb, this ln. is found on the medial aspect of the supraspinatus M., close to the shoulder joint. The axillary lnn. are scattered along the thoracodorsal A.V.

Notice that no cubital ln. is present in the large ruminants (Fig. 6-21).

Separate and remove the pectoral Mm. to expose the medial aspect of the arm. Identify and dissect the following muscles, vessels, and nerves of the shoulder and arm (in a craniocaudal order): the supraspinatus, the subscapularis, the teres major and latissimus dorsi Mm., which are visible in the scapular area. In addition, locate the attachment of the serratus ventralis Mm. on the facies serrata of the scapula. The suprascapular A.V.N. penetrate between the supraspinatus and subscapularis Mm. The subscapular N. supplies the muscle with the same name. The axillary N. and subscapular A.V. penetrate between the subscapularis and teres major Mm.

Caution! The axillary N. sends a branch to supply the teres major M., whereas the subscapular A.V. emits the thoracodorsal A.V. These vessels are accompanied by the thoracodorsal N., with which they run between the teres major and latissimus dorsi Mm.

The musculocutaneus N. crosses the lateral side of the axillary A.V., whereas the median, ulnar, and radial Nn. cross the medial side of the same vessels.

In the brachial area, identify and dissect the structures similar to those described in horse (Fig. 6-22).

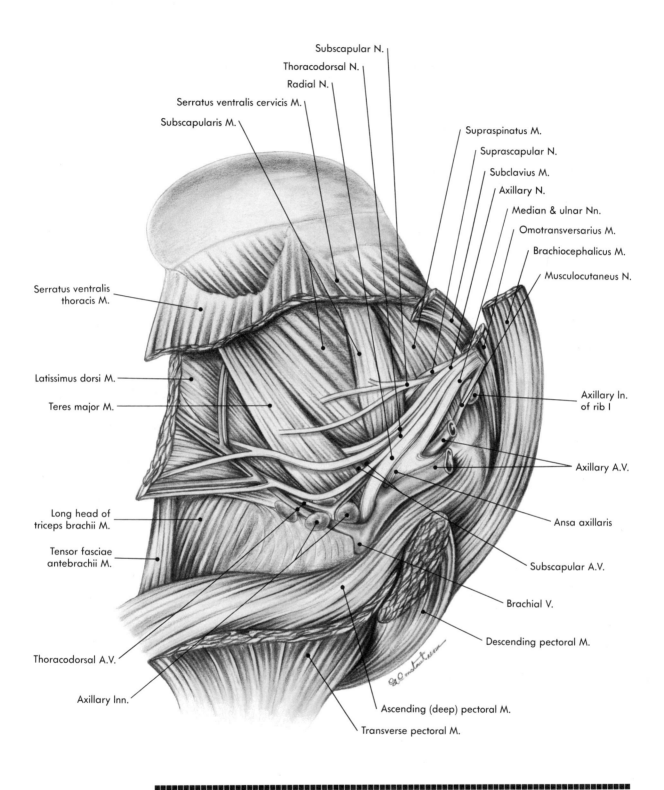

Brachial plexus and adjacent structures of left thoracic limb (medial aspect).

Fig. 6-21

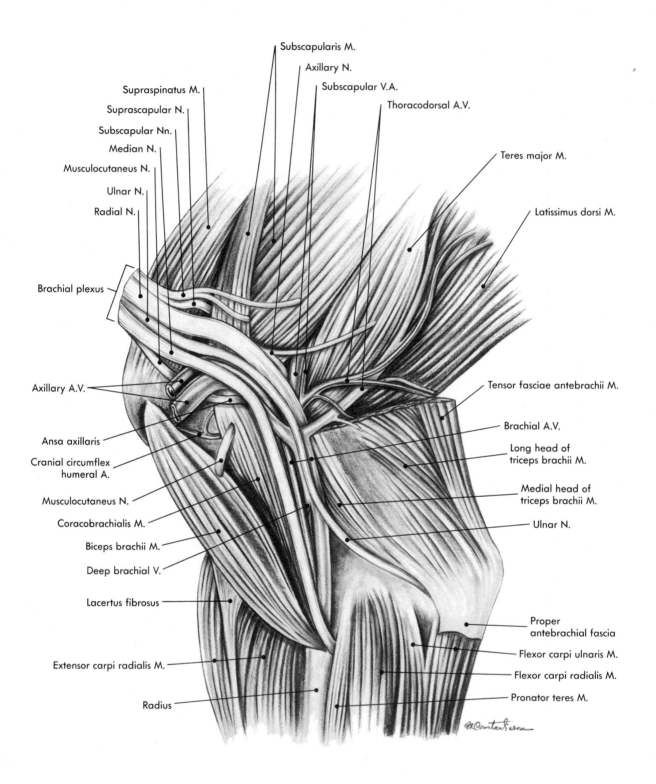

Subscapularis M.

Axillary N.

Subscapular V.A.

Thoracodorsal A.V.

Supraspinatus M.

Suprascapular N.

Subscapular Nn.

Median N.

Musculocutaneus N.

Ulnar N.

Radial N.

Brachial plexus

Axillary A.V.

Ansa axillaris

Cranial circumflex humeral A.

Musculocutaneus N.

Coracobrachialis M.

Biceps brachii M.

Deep brachial V.

Lacertus fibrosus

Extensor carpi radialis M.

Radius

Teres major M.

Latissimus dorsi M.

Tensor fasciae antebrachii M.

Brachial A.V.

Long head of triceps brachii M.

Medial head of triceps brachii M.

Ulnar N.

Proper antebrachial fascia

Flexor carpi ulnaris M.

Flexor carpi radialis M.

Pronator teres M.

Fig. 6-22   Vessels and nerves of medial shoulder and arm and adjacent structures of right thoracic limb.

Notice that the bicipital and transverse cubital Aa. are branches of the brachial A., whereas the bicipital V. is a branch of the transverse cubital V. in all ruminants.

Turn the limb so that the lateral side is up and dissect all the structures shown in Fig. 6-23.

Caution! Do not separate the aponeurosis of the scapular part of the deltoideus M. from the lateral axillary fascia, which fuses with the infraspinatus M.

Caution! Dissect carefully in the area of the lateral head of the triceps brachii and brachialis Mm. to expose the caudal circumflex humeral A.V. and the axillary N.

Caudal to the origin of the cranial cutaneous antebrachial n., identify and dissect the lateral cutaneous antebrachial N. (Fig 6-23).

Remember! The cranial cutaneous antebrachial N. is a branch of the axillary N., whereas the lateral cutaneous antebrachial N. is a branch of the radial N.

Caution! Do not remove the lateral axillary fascia from the supraspinatus and the infraspinatus Mm.

Transect the deltoideus M. from the deltoid tuberosity and from the acromion. Transect the lateral head of the triceps brachii M. in the middle, and partially transect its humeral attachment. Reflect the muscle dorsocaudally, and expose the radial N. and the collateral A.V.

Transect the tendon of the infraspinatus M., leaving a short portion attached to the facies infraspinata and reflect the rest of the muscle. Identify the muscular bursa between the tendon and the caudal part of the greater tubercle of the humerus. Carefully dissect the deep aspect of the reflected tendon and muscle to expose the deep attachment of the infraspinatus and the teres minor Mm. Transect them carefully to expose the lateral aspect of the shoulder joint capsule. Reflect the deltoid M. caudally. The main trunk of the axillary N. and the caudal circumflex humeral A.V. are exposed.

Turn the limb again so the medial side is up, and continue the dissection of the structures from the arm toward the elbow and forearm as shown in Fig. 6-24.

Remember! The medial cutaneous antebrachial N. is a branch of the musculocutaneus N.

Notice that the median cubital V. joins the cephalic venous system to the deep (muscular) venous system (the brachial Vv.).

Continue the dissection of the medial aspect of the forearm. Those structures are illustrated in Figs. 6-25 and 6-26.

Turn the limb so that the lateral side is up again and dissect the superficial and deep structures as shown in Figs. 6-27 and 6-28.

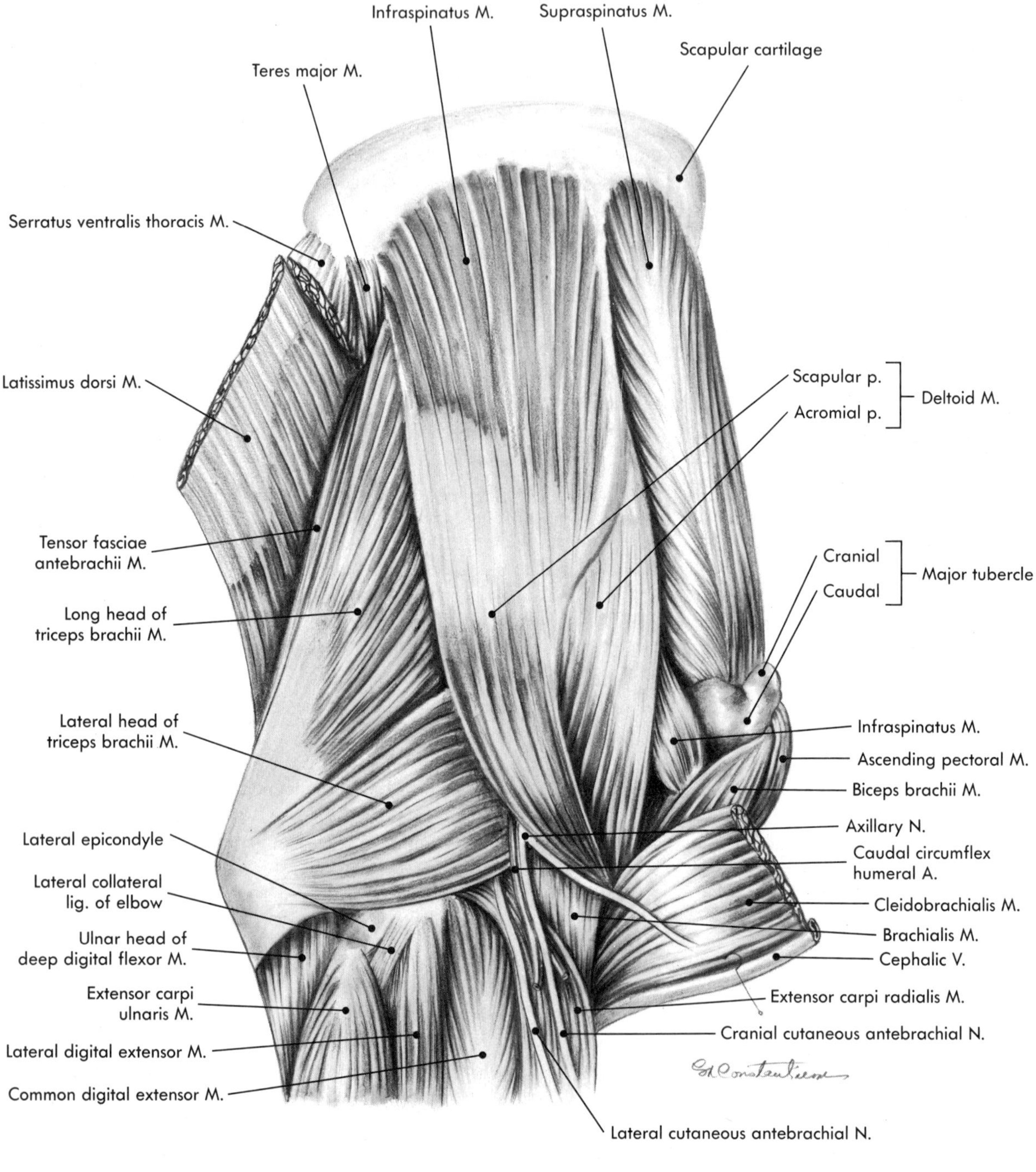

Fig. 6-23

Lateral shoulder of right thoracic limb.

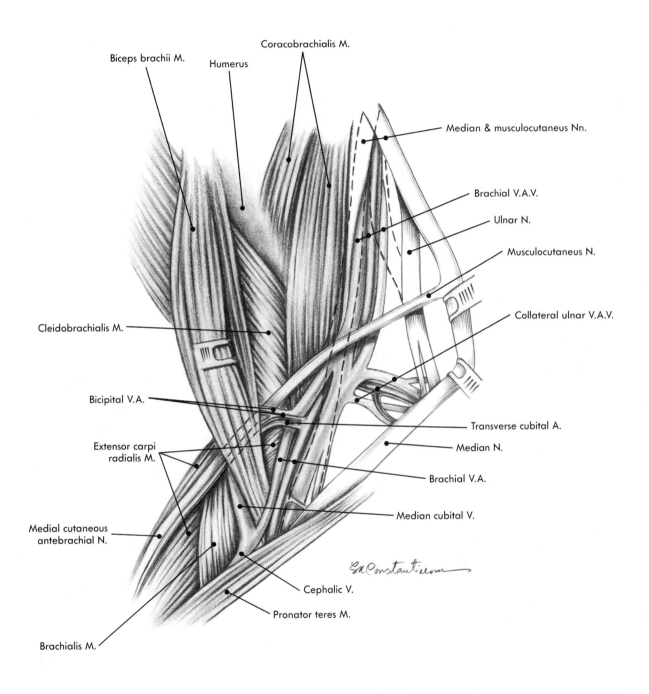

Vessels, nerves, and muscles of elbow of right thoracic limb (medial aspect).

Fig. 6-24

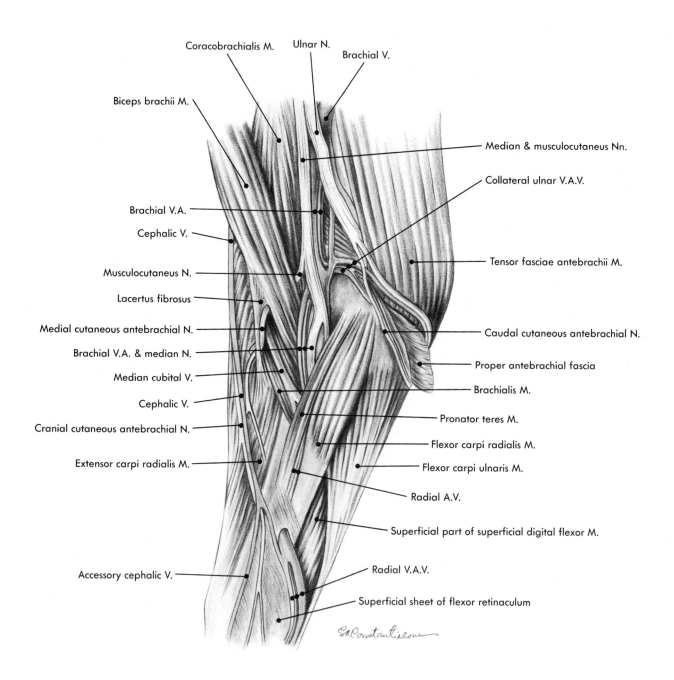

Coracobrachialis M.

Ulnar N.

Brachial V.

Biceps brachii M.

Median & musculocutaneus Nn.

Collateral ulnar V.A.V.

Brachial V.A.

Cephalic V.

Musculocutaneus N.

Tensor fasciae antebrachii M.

Lacertus fibrosus

Medial cutaneous antebrachial N.

Caudal cutaneous antebrachial N.

Brachial V.A. & median N.

Proper antebrachial fascia

Median cubital V.

Brachialis M.

Cephalic V.

Pronator teres M.

Cranial cutaneous antebrachial N.

Flexor carpi radialis M.

Extensor carpi radialis M.

Flexor carpi ulnaris M.

Radial A.V.

Superficial part of superficial digital flexor M.

Accessory cephalic V.

Radial V.A.V.

Superficial sheet of flexor retinaculum

Fig. 6-25   Superficial structures of arm and forearm of right thoracic limb (medial aspect).

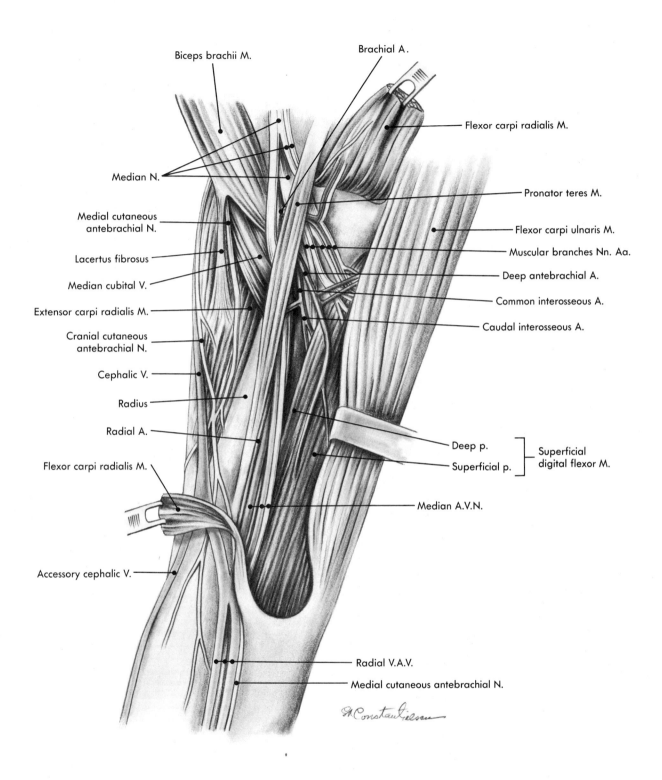

Biceps brachii M.

Brachial A.

Flexor carpi radialis M.

Median N.

Pronator teres M.

Medial cutaneous
antebrachial N.

Flexor carpi ulnaris M.

Muscular branches Nn. Aa.

Lacertus fibrosus

Deep antebrachial A.

Median cubital V.

Common interosseous A.

Extensor carpi radialis M.

Caudal interosseous A.

Cranial cutaneous
antebrachial N.

Cephalic V.

Radius

Deep p.

Radial A.

Superficial p.

Superficial
digital flexor M.

Flexor carpi radialis M.

Median A.V.N.

Accessory cephalic V.

Radial V.A.V.

Medial cutaneous antebrachial N.

Deep structures of elbow and forearm of right thoracic limb (medial aspect).

Fig. 6-26

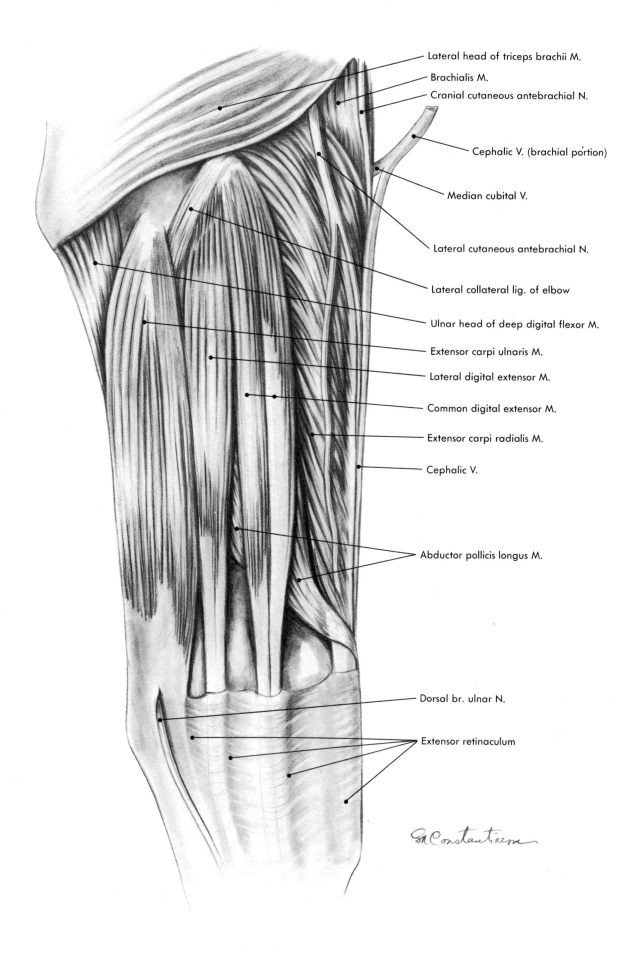

Lateral head of triceps brachii M.

Brachialis M.

Cranial cutaneous antebrachial N.

Cephalic V. (brachial portion)

Median cubital V.

Lateral cutaneous antebrachial N.

Lateral collateral lig. of elbow

Ulnar head of deep digital flexor M.

Extensor carpi ulnaris M.

Lateral digital extensor M.

Common digital extensor M.

Extensor carpi radialis M.

Cephalic V.

Abductor pollicis longus M.

Dorsal br. ulnar N.

Extensor retinaculum

Fig. 6-27   Superficial structures of forearm of right thoracic limb (lateral aspect).

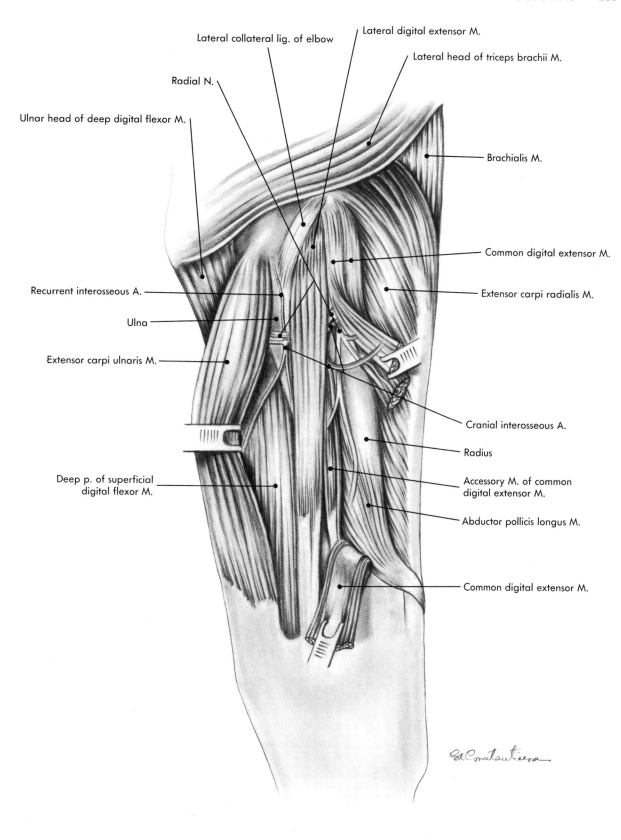

Lateral collateral lig. of elbow

Lateral digital extensor M.

Lateral head of triceps brachii M.

Radial N.

Ulnar head of deep digital flexor M.

Brachialis M.

Common digital extensor M.

Extensor carpi radialis M.

Recurrent interosseous A.

Ulna

Extensor carpi ulnaris M.

Cranial interosseous A.

Radius

Deep p. of superficial digital flexor M.

Accessory M. of common digital extensor M.

Abductor pollicis longus M.

Common digital extensor M.

Deep structures of forearm of right thoracic limb (lateral aspect).

Fig. 6-28

Turn the limb so that the dorsal aspect of the autopodium is up, and identify the tendons bound by the extensor retinaculum (in the carpal area).

Remember! *Dorsal* for the segments of the autopodium means *cranial* for the rest of the body, except the head.

Because the extensor retinaculum continues in the metacarpophalangeal area as loose connective tissue, the dissection of the structures of the dorsal aspect of this area is easily performed. Only the tendons of the digital extensor Mm. continue in the metacarpal and phalangeal areas. They are the tendons of the lateral digital extensor M. and the two tendons of the common digital extensor M.: the lateral one was formerly called the tendon of either the common digital extensor M. or the digital extensor M. III and IV, whereas the medial one was called the tendon of either the medial digital extensor M. or the proper digital extensor M. III. Dissect them along with the vessels and nerves.

According to the *Nomina Anatomica Veterinaria,* the terms for the arteries, veins, and nerves of the autopodium can be explained in the following: The superficial Aa.Vv. of the metapodium are designated *common digital* Aa.Vv.; the deep vessels are termed *metacarpal (metatarsal)* Aa.Vv. Digital vessels that originate from the bifurcation of common digital Aa.Vv. are called *proper digital* Aa.Vv. When abaxial digital vessels are present on the most medial or lateral digits, they come from some other source and are called *abaxial digital* Aa.Vv. The superficial Nn. of the metapodium are designated common digital Nn. and the deep Nn. are designated metacarpal (metatarsal) Nn. Digital Nn. that originate from the bifurcation of the common digital Nn. are called proper digital Nn. Those that originate from some other source are simply digital Nn. Dorsal common digital Nn. that occur in ruminants are II and III. In ruminants, the dorsal branch of the ulnar N. forms the dorsal common digital N. IV. Digital nerves are termed *proper* if they originate by bifurcation of a common digital nerve. If they arise independently the term *proper* is omitted. The dorsal branch of the ulnar N. continues in the metacarpal area as the dorsal common digital N. IV, which in the digital area is called the dorsal proper abaxial digital N. IV. In the interdigital space, the dorsal common digital V. III branches into the dorsal proper digital Vv. (III and IV in an axial position). The deepest structures, the dorsal metacarpal A.V. III, are located in the metacarpal dorsal longitudinal sulcus.

Saving the vessels and nerves of the dorsal aspect in the digital area, dissect the tendons and the interdigital ligaments. The lateral tendon of the common digital extensor M. branches to each digit. They course along the axial border of the third and fourth digits, are surrounded by tendinous sheaths, and insert on P. III. The medial tendon of the common digital extensor M. and the tendon of the lateral digital extensor M. are similar. They course on the dorsal aspect of the fetlock joint with tendinous bursae and fuse with the symmetrical abaxial extensions of the interosseus medius tendon. At the level of P. I, they fuse with the symmetrical axial extensions of the same interosseus medius tendon. Finally, they insert on both P. II and P. III.

The proximal and distal interdigital ligg., which connect the symmetrical P. Is and the symmetrical P. Is and P. IIIs, respectively, are better exposed on the palmar aspect.

Turn the limb so that the palmar aspect of the autopodium is up. Make an incision in the two sheets of flexor retinaculum, and expose, identify, and dissect the tendons (surrounded and protected by their tendinous sheaths) and the vessels and nerves of the palmar carpal area. Proceed in a similar manner in the metacarpal area after making an incision in the palmar fascia in the long axis of the limb (Fig 6-29).

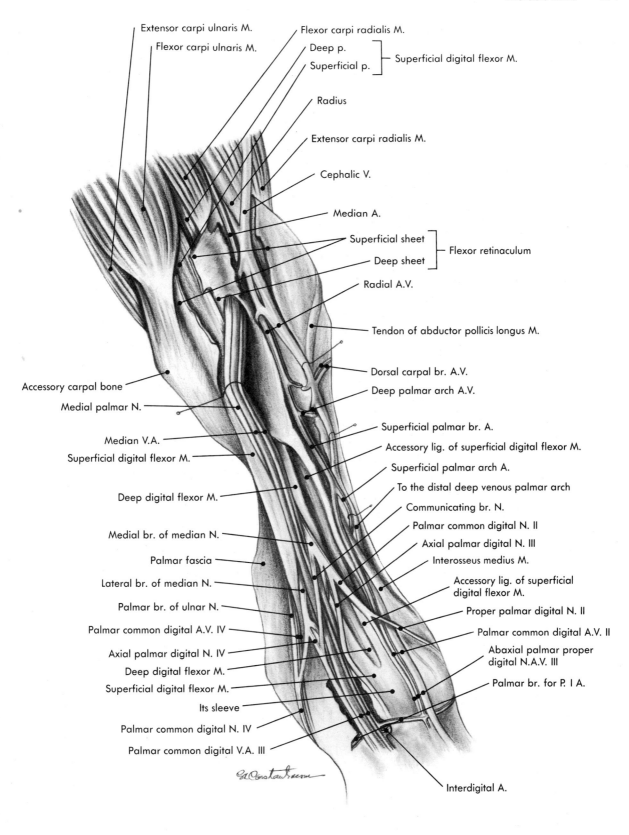

Extensor carpi ulnaris M.

Flexor carpi ulnaris M.

Flexor carpi radialis M.

Deep p.

Superficial p. ⎫ Superficial digital flexor M.

Radius

Extensor carpi radialis M.

Cephalic V.

Median A.

Superficial sheet

Deep sheet ⎫ Flexor retinaculum

Radial A.V.

Tendon of abductor pollicis longus M.

Dorsal carpal br. A.V.

Deep palmar arch A.V.

Accessory carpal bone

Medial palmar N.

Superficial palmar br. A.

Accessory lig. of superficial digital flexor M.

Median V.A.

Superficial digital flexor M.

Superficial palmar arch A.

To the distal deep venous palmar arch

Deep digital flexor M.

Communicating br. N.

Palmar common digital N. II

Medial br. of median N.

Axial palmar digital N. III

Palmar fascia

Interosseus medius M.

Lateral br. of median N.

Accessory lig. of superficial digital flexor M.

Palmar br. of ulnar N.

Proper palmar digital N. II

Palmar common digital A.V. IV

Palmar common digital A.V. II

Axial palmar digital N. IV

Abaxial palmar proper digital N.A.V. III

Deep digital flexor M.

Superficial digital flexor M.

Palmar br. for P. I A.

Its sleeve

Palmar common digital N. IV

Palmar common digital V.A. III

Interdigital A.

Palmar structures of carpus and metacarpus of left thoracic limb (medial aspect).

Fig. 6-29

Disarticulate one fetlock joint after transecting the two interdigital ligaments, and dissect the fascia of the digital cushion between the digital cushion and the corresponding ergot; transect it in the middle and reflect the stumps. Dissect and transect the proximal digital annular lig.

Dissect the distal interdigital lig. with its four attachments (on P. I and P. III), which partially protects the tendons of the superficial and deep digital flexors. Transect the abaxial attachment of the ligament, and reflect it. Dissect and transect the distal digital annular lig. Transect both the superficial and deep digital flexor tendons of one digit proximal to the fetlock. Make a vertical incision through the digital fascia close to its attachment on the abaxial sesamoid bone, and expose the distal stumps of the previously transected digital flexor tendons.

Pull out the tendon of the deep digital flexor M. from between the two insertions of the superficial digital flexor tendon, and reflect it to expose the transverse lamina (similar to that of the horse). Transect the transverse lamina and expose the navicular bone, and the insertion of the deep digital flexor tendon. Transect the proximal interdigital lig. and disarticulate the fetlock joint.

The corresponding tendons, vessels, and nerves of the disarticulated digit are similar to those of the remaining digit. Dissect these structures on the axial aspect (first the vessels and nerves and then the ligaments).

Examine and identify the structures of the digital organ. In this section only the differences between the large ruminants and the horse are discussed.

The two symmetrical hooves (or claws) correspond to the digits and, when in close apposition, look similar to a horse hoof. Each hoof has an axial and an abaxial surface (including a dorsal part, the heel, and the bulb). The hooves bear no bars or frogs. Only the wall and sole are present. The sole is reduced to the area within the angle of inflection of the wall and continues without a distinct limit with the bulb. A very reduced white zone is observed. Inside the hooves, the perioplic and coronary grooves are wider but not as deep as in horse. The horny lamellae are as numerous as the lamellae of the laminar corium (Fig. 6-30).

In the live tissue (the corium), the perioplic and the coronary corium are thicker in the ox. The lamellae of the laminar corium are much shorter but much more numerous (1000 to 1300). No secondary lamellae are observed.

Only the digital cushion is covered by the corium. It is much reduced compared to the digital cushion in the horse and is not protected by a frog (see Fig. 6-30).

Dewclaws are present in the large ruminants. They resemble the main claws and cover one to two vestigial phalanges.

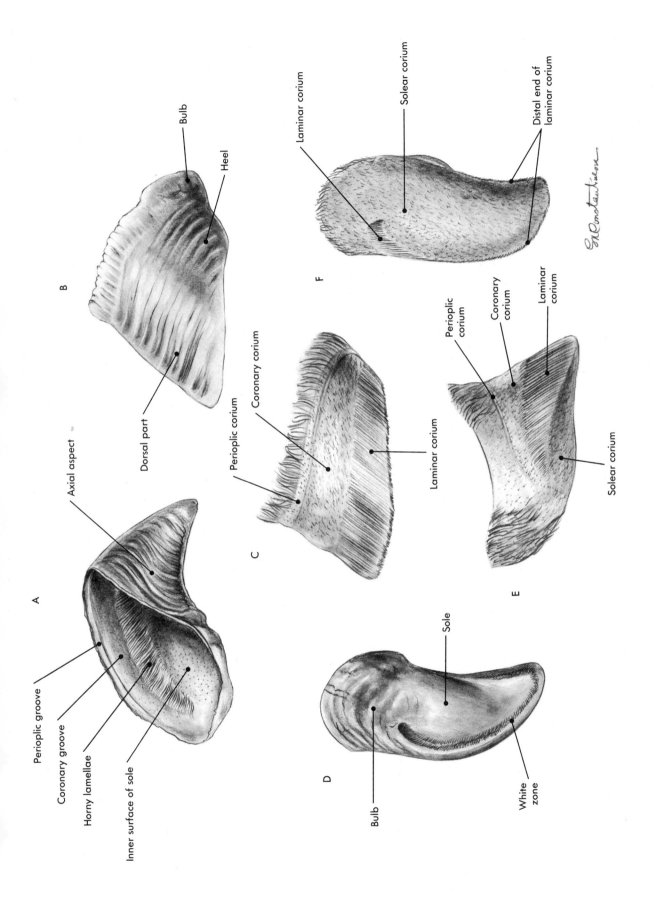

A, Internal aspect of hoof. B, Abaxial aspect of wall of hoof. C, Corium of digit (abaxial aspect). D, Solear aspect of hoof. E, Corium of digit (axial and ventral aspects). F, Solear corium of digit.

Fig. 6-30

Sheep

Step by step, follow the directions suggested for the large ruminants.

(Figs. 6-31 and 6-32)

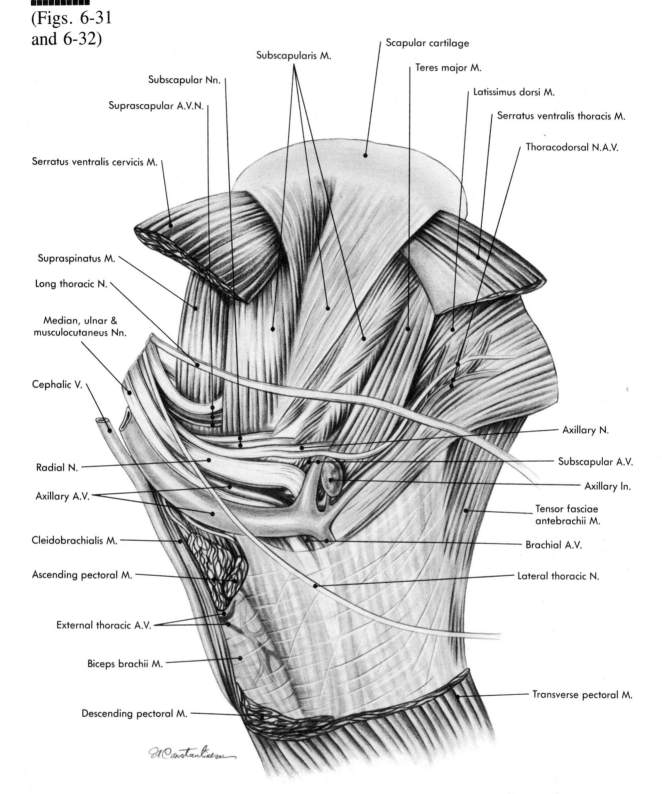

Fig. 6-31 Brachial plexus and adjacent structures of right thoracic limb (medial aspect).

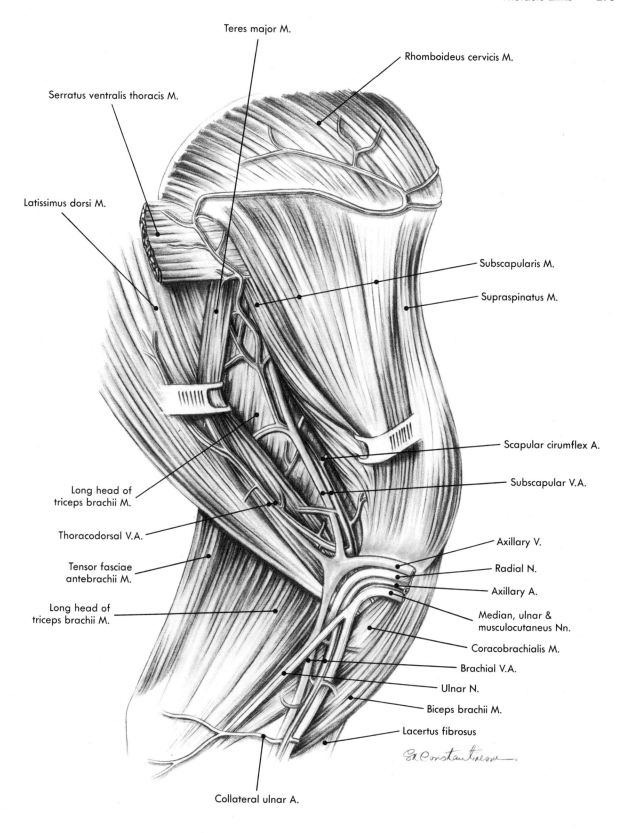

Teres major M.

Rhomboideus cervicis M.

Serratus ventralis thoracis M.

Latissimus dorsi M.

Subscapularis M.

Supraspinatus M.

Scapular cirumflex A.

Subscapular V.A.

Long head of
triceps brachii M.

Thoracodorsal V.A.

Tensor fasciae
antebrachii M.

Long head of
triceps brachii M.

Axillary V.

Radial N.

Axillary A.

Median, ulnar &
musculocutaneus Nn.

Coracobrachialis M.

Brachial V.A.

Ulnar N.

Biceps brachii M.

Lacertus fibrosus

Collateral ulnar A.

Deep structures of shoulder and arm of left thoracic limb (medial aspect).

Fig. 6-32

Goat
■■■■■■■
(Figs. 6-33 and 6-34)

Some of the few differences observed between goat and sheep are shown in Figs. 6-33 and 6-34.

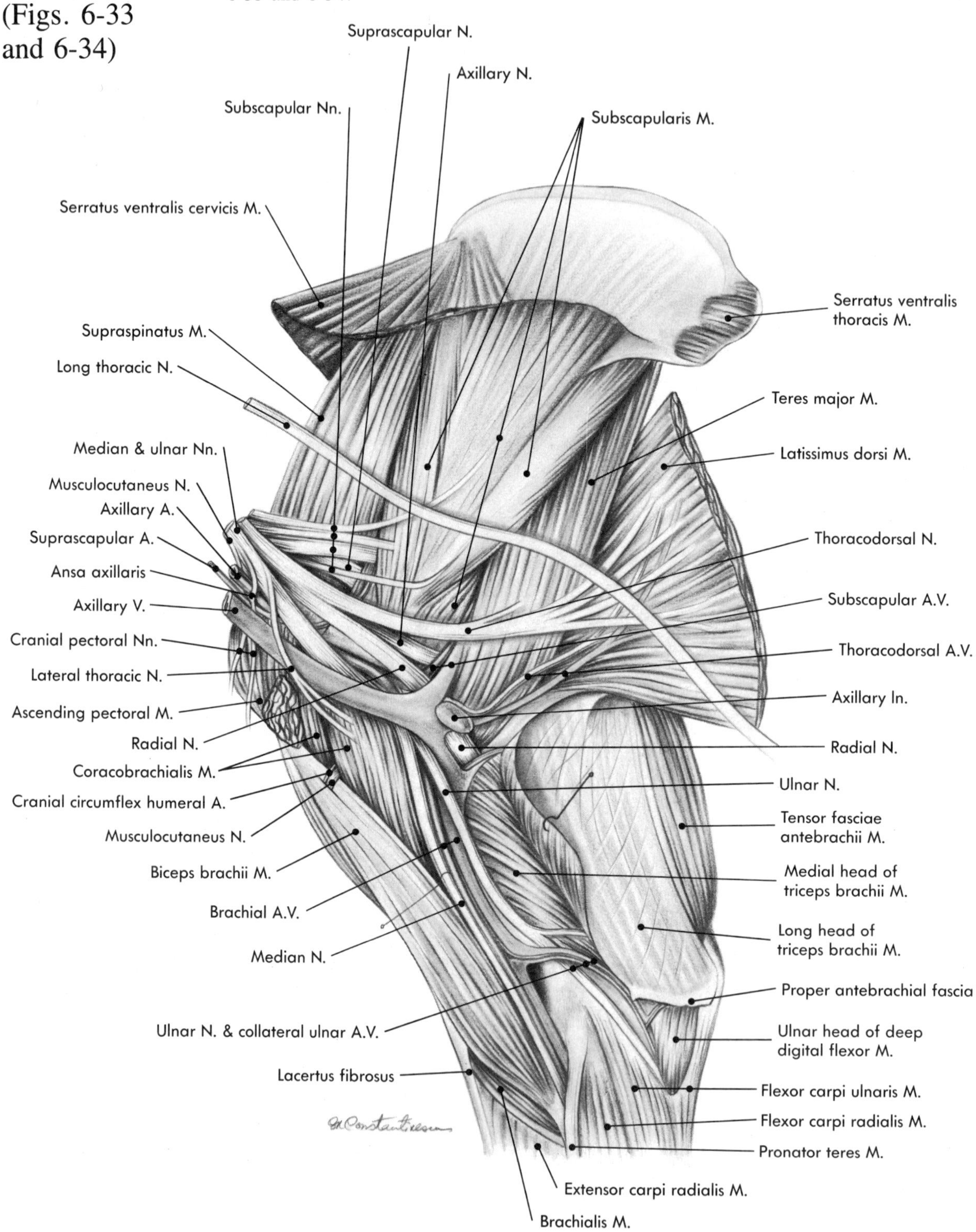

Suprascapular N.

Axillary N.

Subscapular Nn.

Subscapularis M.

Serratus ventralis cervicis M.

Serratus ventralis thoracis M.

Supraspinatus M.

Teres major M.

Long thoracic N.

Latissimus dorsi M.

Median & ulnar Nn.

Musculocutaneus N.

Axillary A.

Suprascapular A.

Thoracodorsal N.

Ansa axillaris

Subscapular A.V.

Axillary V.

Thoracodorsal A.V.

Cranial pectoral Nn.

Axillary ln.

Lateral thoracic N.

Ascending pectoral M.

Radial N.

Radial N.

Coracobrachialis M.

Ulnar N.

Cranial circumflex humeral A.

Tensor fasciae antebrachii M.

Musculocutaneus N.

Medial head of triceps brachii M.

Biceps brachii M.

Brachial A.V.

Long head of triceps brachii M.

Median N.

Proper antebrachial fascia

Ulnar N. & collateral ulnar A.V.

Ulnar head of deep digital flexor M.

Lacertus fibrosus

Flexor carpi ulnaris M.

Flexor carpi radialis M.

Pronator teres M.

Extensor carpi radialis M.

Brachialis M.

Fig. 6-33   Brachial plexus and adjacent structures of shoulder and arm of right thoracic limb (medial aspect).

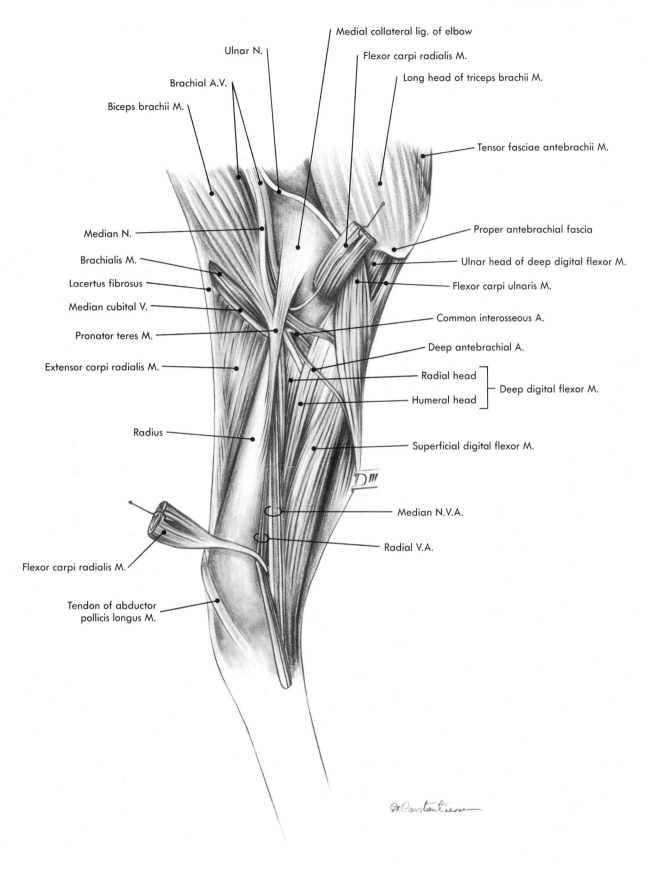

Biceps brachii M.

Brachial A.V.

Ulnar N.

Medial collateral lig. of elbow

Flexor carpi radialis M.

Long head of triceps brachii M.

Tensor fasciae antebrachii M.

Median N.

Brachialis M.

Lacertus fibrosus

Median cubital V.

Pronator teres M.

Extensor carpi radialis M.

Radius

Flexor carpi radialis M.

Tendon of abductor
pollicis longus M.

Proper antebrachial fascia

Ulnar head of deep digital flexor M.

Flexor carpi ulnaris M.

Common interosseous A.

Deep antebrachial A.

Radial head

Humeral head

Deep digital flexor M.

Superficial digital flexor M.

Median N.V.A.

Radial V.A.

Deep structures of forearm of right thoracic limb (medial aspect).

Fig. 6-34

Pig
■■■■■
(Figs. 6-35
to 6-40)

Due to the excessive amount of fat, only the landmarks provided by the bones and joints can be seen and palpated—and this can sometimes only be approximated. Checking the landmarks on a skeleton is highly recommended.

In the scapular region, the most visible structure is the tuber of the scapular spine.

Notice that the tuber of the scapular spine is encountered only in the horse and the pig.

Notice that, in the pig, the second digit is the most medial, whereas the fifth digit is the most lateral. The main digits are the third and fourth and are the only digits to come in contact with the ground surface, that is, the only weight-bearing digits.

Figs. 6-35 to 6-40 show the dissected structures of the thoracic limb in the pig.

Notice that, in the pig (as in the sheep), the tendon of the biceps brachii M. penetrates the shoulder joint and perforates the joint capsule.

Reflect the coracobrachialis M. caudally to expose the articularis M.

Remember! The articularis M. is also present in the horse.

Caution! The nomenclature of the digital extensor muscles is often confused in the literature. The only footnote in the *Nomina Anatomica Veterinaria* regarding the common digital extensor is: "This also applies to the parts of the muscle in pig." No statement is made regarding the lateral digital extensor M. In this guide, the names of the muscles are taken from the *Nomina Anatomica Veterinaria*. However, the heads and tendons of these muscles are named according to the digits they act upon. Three heads of the common digital extensor M. act distinctly upon digits II and III, III and IV, and IV and V. Two heads of the lateral digital extensor M. act distinctly upon digit IV and digit V. A digital extensor M. also acts upon the second digit.

Caution! There are numerous variations in the branching of the tendons of the digital extensor muscles in pig.

Caution! Do not dissect these tendons separately. Identify them and focus your attention on the vessels and nerves. Dissect the vessels and nerves along with the tendons.

Caution! For an easier dissection, keep the secondary digits in abduction at all times.

Caution! Only the deep digital flexor M. sends tendons to the secondary digits. The superficial digital flexor M. gives off only two tendons, one for each of the main digits.

The digital adductor Mm. II and V are overlapped by the digital flexor tendons.

Explore each joint of the thoracic limb. The approach for the shoulder and the elbow joints is from the medial aspect. In the shoulder area, examine the relationship between the joint capsule and the tendon of the biceps brachii M. At the elbow, examine the relationship between the pronator teres M. and the medial collateral lig. The antebrachial interosseous membrane firmly connects the two bones of the forearm with one another. Examine the carpus with its two rows of four bones each and the numerous ligaments connecting them. Dorsal and palmar carpometacarpal ligg. are present, as are proximal and distal interdigital ligg. The deep transverse metacarpal lig. consists of branches with complex attachments.

Remove a main hoof; examine it and the living tissue underneath. The main components of the ruminants' hoof are present. The only significant difference is the prominent bulbar segment of the hoof, which protects the corresponding prominent bulbar cushion.

The secondary, or accessory, digits are equipped with similar structures.

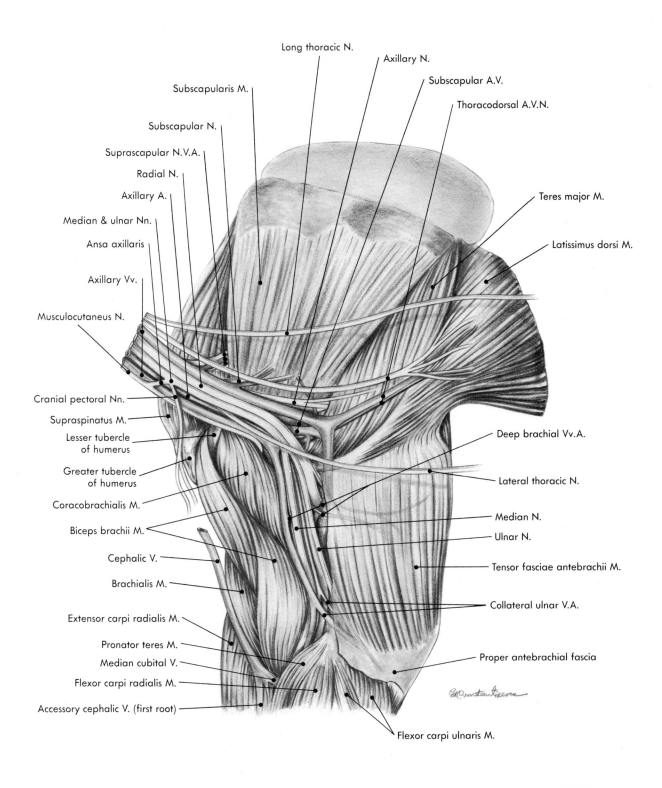

Long thoracic N.

Axillary N.

Subscapular A.V.

Thoracodorsal A.V.N.

Subscapularis M.

Subscapular N.

Suprascapular N.V.A.

Radial N.

Axillary A.

Median & ulnar Nn.

Ansa axillaris

Axillary Vv.

Musculocutaneus N.

Cranial pectoral Nn.

Supraspinatus M.

Lesser tubercle
of humerus

Greater tubercle
of humerus

Coracobrachialis M.

Biceps brachii M.

Cephalic V.

Brachialis M.

Extensor carpi radialis M.

Pronator teres M.

Median cubital V.

Flexor carpi radialis M.

Accessory cephalic V. (first root)

Teres major M.

Latissimus dorsi M.

Deep brachial Vv.A.

Lateral thoracic N.

Median N.

Ulnar N.

Tensor fasciae antebrachii M.

Collateral ulnar V.A.

Proper antebrachial fascia

Flexor carpi ulnaris M.

Brachial plexus and adjacent structures of shoulder, arm, and elbow of right thoracic limb (medial aspect).

Fig. 6-35

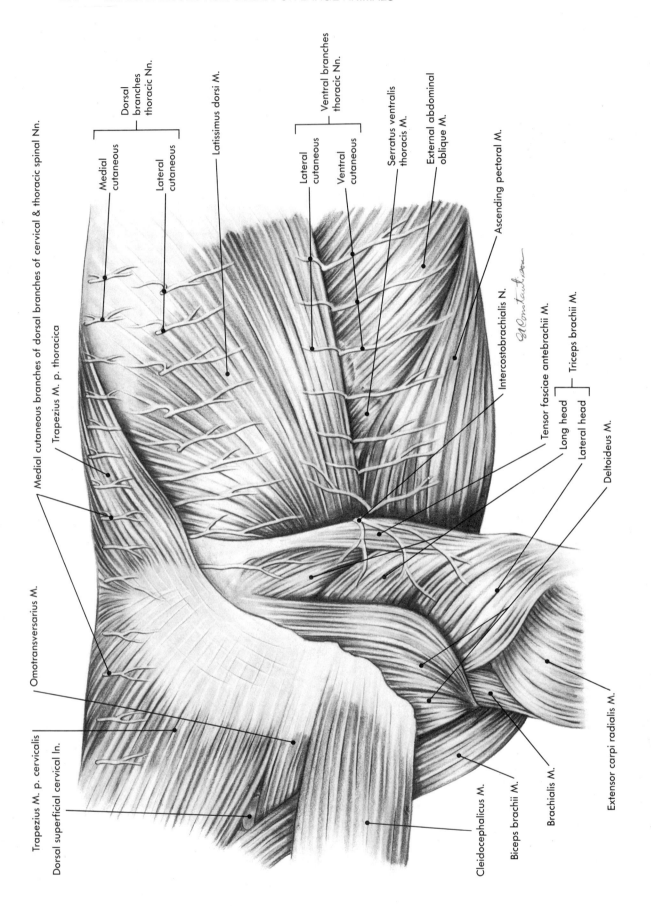

Fig. 6-36 Superficial structures of shoulder, arm, and thorax of left side (lateral aspect).

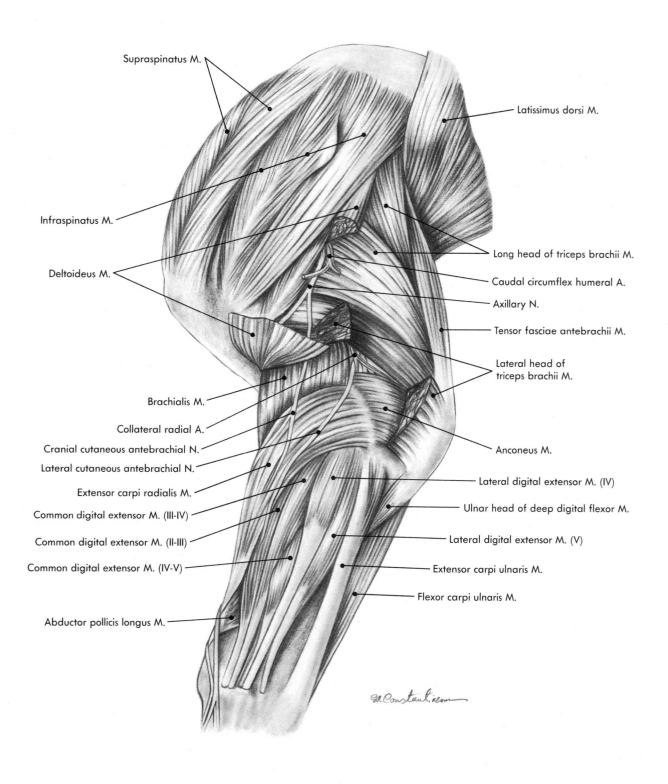

Supraspinatus M.

Latissimus dorsi M.

Infraspinatus M.

Long head of triceps brachii M.

Caudal circumflex humeral A.

Deltoideus M.

Axillary N.

Tensor fasciae antebrachii M.

Lateral head of
triceps brachii M.

Brachialis M.

Collateral radial A.

Cranial cutaneous antebrachial N.

Anconeus M.

Lateral cutaneous antebrachial N.

Extensor carpi radialis M.

Lateral digital extensor M. (IV)

Common digital extensor M. (III-IV)

Ulnar head of deep digital flexor M.

Common digital extensor M. (II-III)

Lateral digital extensor M. (V)

Common digital extensor M. (IV-V)

Extensor carpi ulnaris M.

Flexor carpi ulnaris M.

Abductor pollicis longus M.

Shoulder, arm, elbow, and forearm of left thoracic limb (lateral aspect).

Fig. 6-37

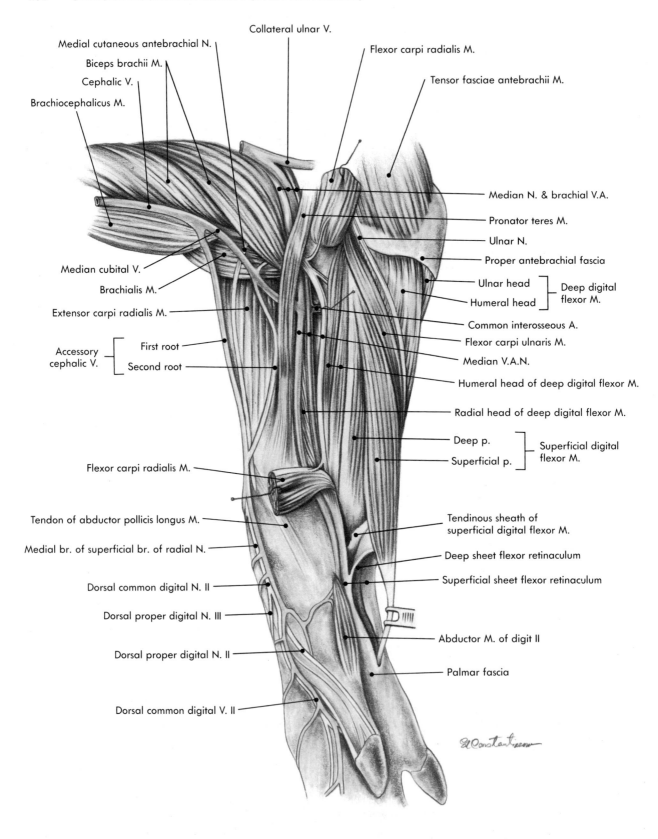

Collateral ulnar V.

Medial cutaneous antebrachial N.

Biceps brachii M.

Cephalic V.

Brachiocephalicus M.

Flexor carpi radialis M.

Tensor fasciae antebrachii M.

Median N. & brachial V.A.

Pronator teres M.

Ulnar N.

Proper antebrachial fascia

Median cubital V.

Brachialis M.

Extensor carpi radialis M.

Ulnar head ⎤
Humeral head ⎦ Deep digital flexor M.

Common interosseous A.

Accessory cephalic V. ⎡ First root
⎣ Second root

Flexor carpi ulnaris M.

Median V.A.N.

Humeral head of deep digital flexor M.

Radial head of deep digital flexor M.

Deep p. ⎤
Superficial p. ⎦ Superficial digital flexor M.

Flexor carpi radialis M.

Tendon of abductor pollicis longus M.

Tendinous sheath of superficial digital flexor M.

Medial br. of superficial br. of radial N.

Deep sheet flexor retinaculum

Superficial sheet flexor retinaculum

Dorsal common digital N. II

Dorsal proper digital N. III

Abductor M. of digit II

Dorsal proper digital N. II

Palmar fascia

Dorsal common digital V. II

Fig. 6-38   Forearm, carpus, and metacarpus of right thoracic limb (mediopalmar aspect).

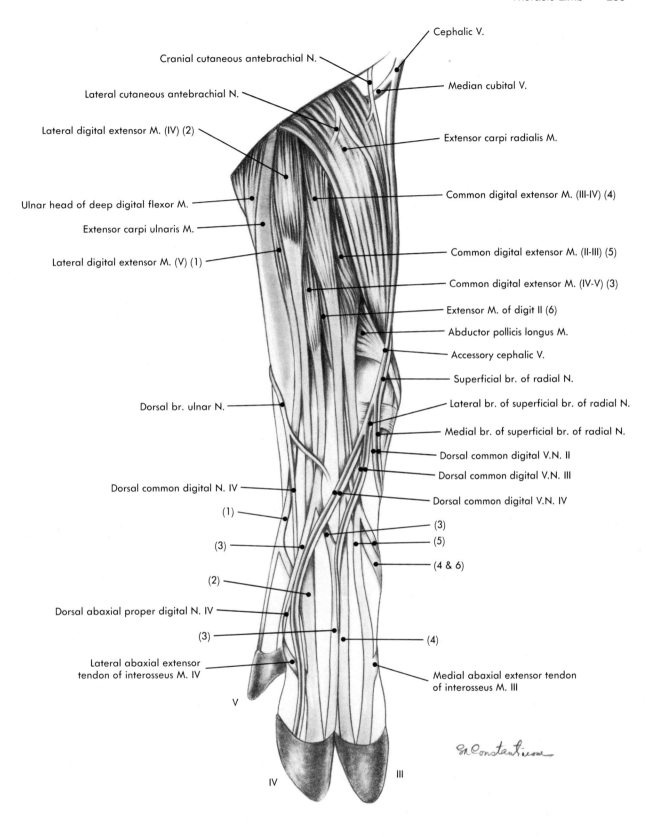

Cephalic V.

Cranial cutaneous antebrachial N.

Lateral cutaneous antebrachial N.

Median cubital V.

Lateral digital extensor M. (IV) (2)

Extensor carpi radialis M.

Ulnar head of deep digital flexor M.

Common digital extensor M. (III-IV) (4)

Extensor carpi ulnaris M.

Lateral digital extensor M. (V) (1)

Common digital extensor M. (II-III) (5)

Common digital extensor M. (IV-V) (3)

Extensor M. of digit II (6)

Abductor pollicis longus M.

Accessory cephalic V.

Superficial br. of radial N.

Dorsal br. ulnar N.

Lateral br. of superficial br. of radial N.

Medial br. of superficial br. of radial N.

Dorsal common digital V.N. II

Dorsal common digital V.N. III

Dorsal common digital N. IV

Dorsal common digital V.N. IV

(1)

(3)

(3)

(5)

(4 & 6)

(2)

Dorsal abaxial proper digital N. IV

(3)

(4)

Lateral abaxial extensor
tendon of interosseus M. IV

Medial abaxial extensor tendon
of interosseus M. III

V

IV

III

Forearm and autopodium of right thoracic limb (dorsolateral aspect).

Fig. 6-39

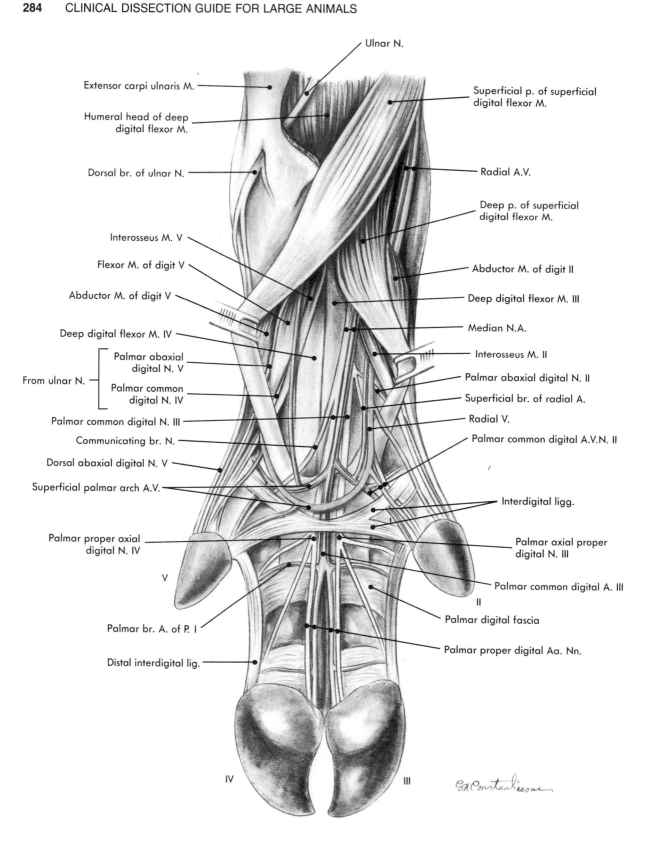

Ulnar N.

Extensor carpi ulnaris M.

Humeral head of deep
digital flexor M.

Superficial p. of superficial
digital flexor M.

Dorsal br. of ulnar N.

Radial A.V.

Deep p. of superficial
digital flexor M.

Interosseus M. V

Flexor M. of digit V

Abductor M. of digit II

Abductor M. of digit V

Deep digital flexor M. III

Deep digital flexor M. IV

Median N.A.

From ulnar N. { Palmar abaxial
digital N. V

Interosseus M. II

Palmar abaxial digital N. II

Palmar common
digital N. IV

Superficial br. of radial A.

Palmar common digital N. III

Radial V.

Communicating br. N.

Palmar common digital A.V.N. II

Dorsal abaxial digital N. V

Superficial palmar arch A.V.

Interdigital ligg.

Palmar proper axial
digital N. IV

Palmar axial proper
digital N. III

V

Palmar common digital A. III

II

Palmar digital fascia

Palmar br. A. of P. I

Palmar proper digital Aa. Nn.

Distal interdigital lig.

IV

III

Fig. 6-40   Autopodium of left thoracic limb (deep palmar aspect).

Before starting the dissection, identify the following landmarks and structures.

On the lateral side of *the head:* the upper lip, the nostril, the alar cartilage (at the medial border of the nostril), the nasal process, the nasoincisive incisure, the osseous nasal aperture, the superior incisors, the diastema, the superior premolars and molars, the facial tubercle and crest, the infraorbital foramen, the orbita, the supraorbital foramen, the zygomatic arch, the temporal fossa, the ear, the temporomandibular joint, the jugular (paracondylar) process of the occipital bone, the caudal border of the mandible with the angle of the mandible, the stylohyoid angle of the stylohyoid bone, the vascular incisure of the mandible, the rostral border of the masseter M., the linguofacial V., the tendon of the sternomandibularis M. and the occipitomandibular part of the digastric M., the lower lip and the chin, the inferior molars, premolars, and incisors, the mental foramen, the oral cleft, and the commissure of the lips.

On the frontal aspect, locate the following: the philtrum of the upper lip, the incisive bones, the nasal process, the osseous nasal aperture, the alar cartilages (of the nostrils), the orbitae, the supraorbital foramina, the zygomatic arches, the temporal lines, the temporal fossae, and the external occipital protuberance.

On the ventral aspect, identify the following: the body of the mandible, the two rami of the mandible, the mylohyoideus Mm., the basihyoid bone, and the larynx.

Identify the following structures of the ventral aspect of *a skull* (in rostrocaudal order) (Fig. 7-1, *A*).

At the incisive region, examine the corresponding alveoli and teeth, the interincisive canal, the palatine processes of the incisive and maxillary bones, the horizontal laminae of the palatine bones, and the major palatine foramina. Also identify the dental alveoli of the superior premolars and molars. Compare the superior and the inferior alveoli and teeth. Examine the corpus mandibulae up to the mandibular angles.

In the guttural region, locate the choanae, the vomer, the pterygoid processes, and the pterygoid hamulus (paired structure).

Identify the articular tubercles, the mandibular fossae, and the retroarticular processes of the squamous part of the temporal bones; the muscular processes, the tympanic bullae, and the external acoustic meatus (from the tympanic part); and the styloid processes, the mastoid processes, and the stylomastoid foramina (from the petrosal part of the temporal bones).

On the base of the skull, the basisphenoid bone and the basal part of the occipital bone (at the midline) and, on both sides of these, the petrosal parts of the symmetrical temporal bones surround the two symmetrical spaces named the foramen lacerum. These foramina communicate caudally with the jugular foramina through the petrooccipital fissure. At the rostral border of the foramen lacerum, the carotid, oval, and spinous incisures are visible (in a mediolateral direction) (see Fig. 7-5). Identify the jugular processes with the corresponding hypoglossal foramina, then the foramen magnum and the two occipital condyles.

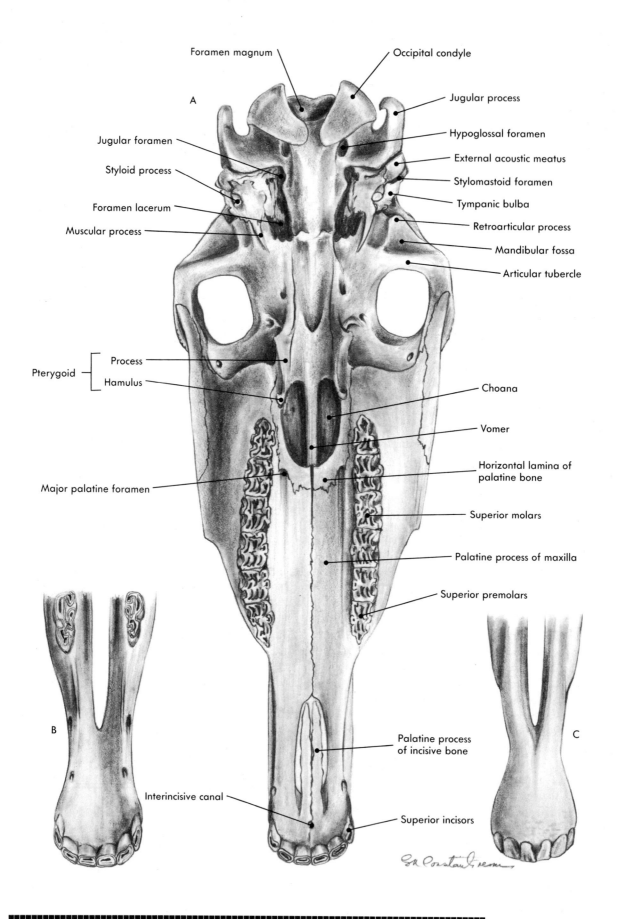

Foramen magnum

Occipital condyle

Jugular process

Hypoglossal foramen

External acoustic meatus

Stylomastoid foramen

Tympanic bulba

Retroarticular process

Mandibular fossa

Articular tubercle

Jugular foramen

Styloid process

Foramen lacerum

Muscular process

A

Pterygoid { Process / Hamulus }

Choana

Vomer

Horizontal lamina of palatine bone

Major palatine foramen

Superior molars

Palatine process of maxilla

Superior premolars

B

C

Palatine process of incisive bone

Interincisive canal

Superior incisors

Fig. 7-1   **A,** Skull (ventral aspect). **B,** Rostral end of mandible (dorsal aspect). **C,** Rostral end of mandible (ventral aspect).

On the lateral aspect of the skull (Fig. 7-2, *A*), identify the following foramina within the bony orbita (Fig. 7-3, *A*): the maxillary, the sphenopalatine and caudal palatine foramina (located dorsoventrally in the pterygopalatine fossa); the ethmoidal foramen; the optic foramen; the trochlear foramen (not mentioned in the *Nomina Anatomica Veterinaria*); the orbital fissure; the rostral alar foramen and foramen rotundum (located rostral to the pterygoid crest and within the pterygoid fossa but not the ethmoidal foramen); the caudal alar foramen; and the foramen alare parvum (caudal to the pterygoid crest). In addition, note the lacrimal process, the fossa of the lacrimal sac, and the fossa of the ventral oblique M.

At the articular area of the mandible, identify the coronoid process, the mandibular incisure, and the condylar process. On the medial aspect of the ramus mandibulae, identify the pterygoid fossa* and the mandibular foramen (Fig. 7-3, *B*).

---

*There is both a pterygoid fossa of the pterygoid bone and a pterygoid fossa of the mandible.

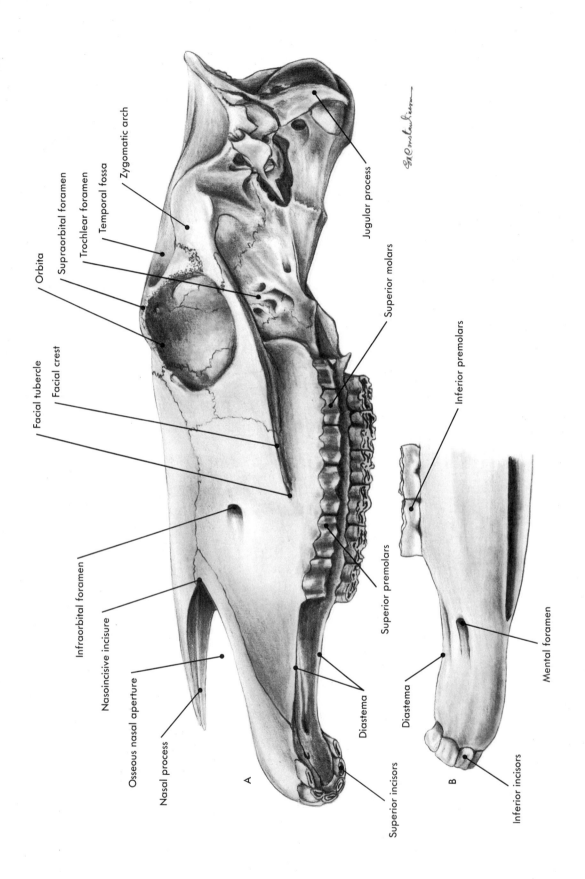

Zygomatic arch

Temporal fossa

Trochlear foramen

Supraorbital foramen

Orbita

Facial tubercle

Facial crest

Infraorbital foramen

Nasoincisive incisure

Osseous nasal aperture

Nasal process

Jugular process

Superior molars

Superior premolars

Superior incisors

Diastema

Inferior premolars

Diastema

Mental foramen

Inferior incisors

A

B

Fig. 7-2   **A,** Skull (lateral aspect). **B,** Rostral end of mandible (lateral aspect).

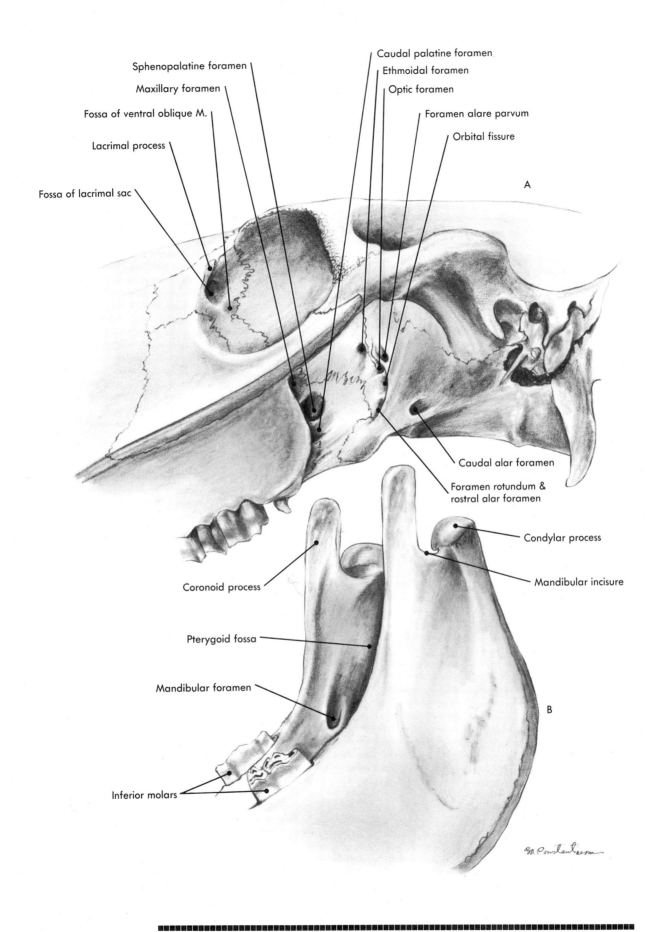

Sphenopalatine foramen

Maxillary foramen

Fossa of ventral oblique M.

Lacrimal process

Fossa of lacrimal sac

Caudal palatine foramen

Ethmoidal foramen

Optic foramen

Foramen alare parvum

Orbital fissure

A

Caudal alar foramen

Foramen rotundum &
rostral alar foramen

Condylar process

Coronoid process

Mandibular incisure

Pterygoid fossa

Mandibular foramen

B

Inferior molars

A, Orbital area of skull (lateroventral aspect). **B,** Caudal half of mandible (rostrolateral aspect).

Fig. 7-3

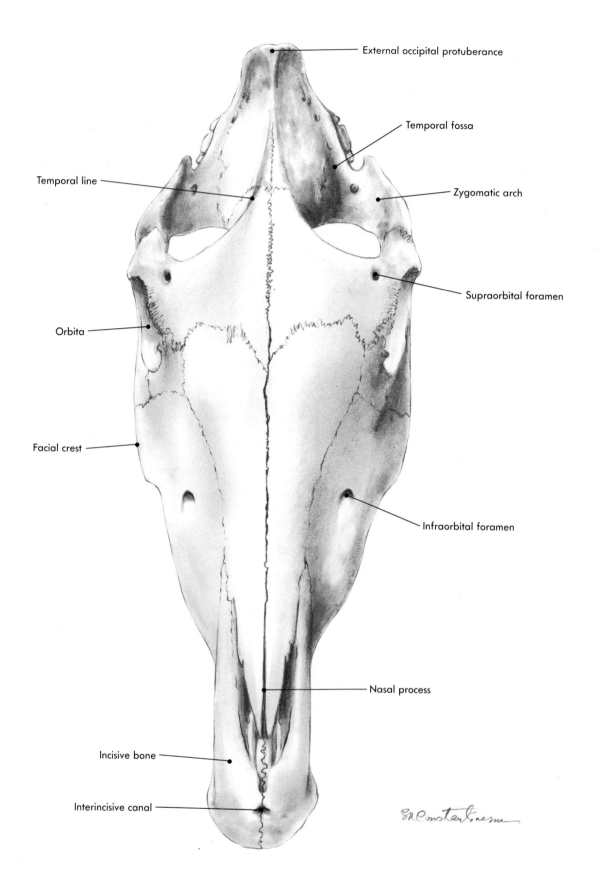

External occipital protuberance

Temporal fossa

Temporal line

Zygomatic arch

Supraorbital foramen

Orbita

Facial crest

Infraorbital foramen

Nasal process

Incisive bone

Interincisive canal

Fig. 7-4

Skull (frontal aspect)

Identify the structures of the skull from the frontal perspective (Fig. 7-4).

Explore the cranial cavity and identify (in rostrocaudal order) the cribriform lamina (plate) and crista galli (of the ethmoid bone); the pre- and basisphenoid, the sulcus chiasmatis, the sella turcica with the hypophyseal fossa, the carotid, ophthalmic, maxillary, and trochlear grooves (of the sphenoid bone); and the petrosal crest and internal acoustic meatus (of the temporal bone).

Notice that the trochlear groove is not mentioned in the *Nomina Anatomica Veterinaria,* although it is an obvious structure (at least in the horse) (Fig. 7-5).

Examine the mandible in Figs. 7-1, *B* and *C*; 7-2, *B*, and 7-3, *B*).

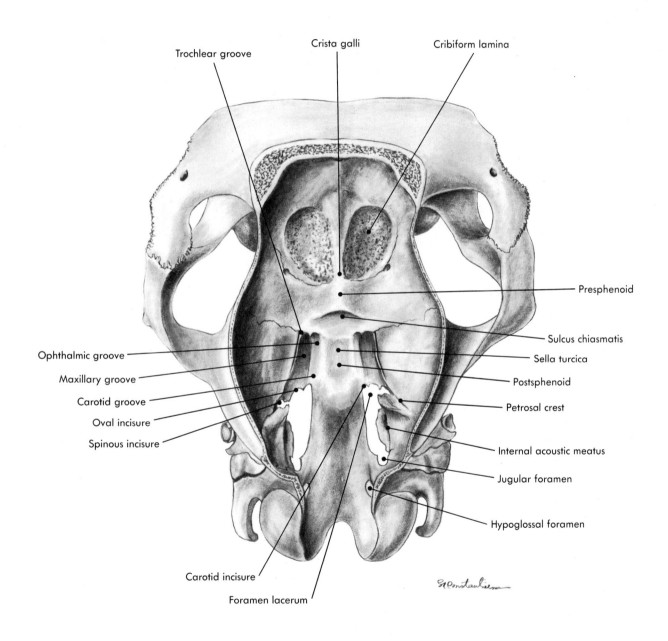

Cranial cavity (rostral wall and floor).   **Fig. 7-5**

Return to your specimen and prepare it for splitting into two halves. With a knife, make a precise midline incision in the lips, nasal area, and ventral aspect of the head. Deepen the incision on the ventral aspect through the tongue, larynx, pharynx, and esophagus; do the same on the dorsal midline of the neck. Saw the head carefully to obtain two symmetrical halves.

Make an incision in the skin from the chin around the commissure of the lips, caudal to the nostrils, and up to the dorsal midline. Make a 1 cm skin incision around the eyelids and another around the base of the ear. Begin skinning in a ventrodorsal direction.

Caution! The more carefully you skin, the better you will expose the underlying structures.

The cutaneus faciei M., in continuation with the cutaneus colli M., is the first muscle exposed within the superficial fascia of the head. The cutaneus faciei M. is inserted on the orbicularis oris M. at the level of the commissure of the lips. Dissect and reflect the platysma rostrally up to the commissure of the lips, saving all the underlying structures. Keep it only a few cm long; transect and remove the rest. In a caudorostral direction, the sternomandibularis and omohyoideus Mm., the external jugular and linguofacial Vv., the parotid salivary gland partially overlapped by the parotidoauricular M., the masseteric br. (A.V.), the masseter, and the buccinator and depressor labii inferioris Mm. will be exposed, as will the ventral buccal br. of the facial N., the facial A.V., the parotid duct, and the inferior labial A.V. Dissect the orbicularis oris M. around the commissure of the lips.

Remember! The platysma is represented by the cutaneus faciei and the cutaneus colli Mm.

With firm but gentle incisions, outline the zygomatic M. as a 1.5 to 2 cm strip running from the facial crest to the orbicularis oris M. Then outline the levator nasolabialis M., starting with its ventral border, from the medial canthus (of the eyelids) to the orbicularis oris M. and dorsal to the commissure of the lips; the dorsal border of the muscle is parallel with and 5 cm dorsal to the previous border, close to the dorsal commissure of the nostril.

Caution! Save the dorsal nasal V. which crosses the levator nasolabialis M.

Remember! The levator nasolabialis M. has two parts (nasal and labial—the name of the muscle is suggestive). Between the two parts, another muscle passes: the caninus M.

To differentiate the two muscles, it is necessary to tense the lateral wall of the nostril, pulling it rostrally. A triangular-shaped muscle (the caninus M.) becomes apparent, with a delicate tendon pointed toward the facial tubercle. The fan-shaped muscular portion represents the superficial layer of the lateral wall of the nostril. The muscle passes under the labial part of the levator nasolabialis M. and overlaps the nasal part of the same muscle.

Notice that the nasal part of the levator nasolabialis M. represents the deep layer of the lateral wall of the nostril.

Outline the caninus M., saving the vessels crossing its tendon.

From the medial canthus, trace the rostral border of the malaris M. toward the facial tubercle (1 to 2 cm caudal to it), taking care with the vessels of the area. Between the levator nasolabialis and the malaris Mm., the caudal extent of the levator labii superioris is visible. It continues rostrally and deep to the levator nasolabialis toward the nasal process with a thin tendon. The tendons of the two symmetrical muscles fuse in an aponeurosis, which overlaps the dilatator naris apicalis M. and blends its fibers with the orbicularis oris M.

Move to the area of the temporomandibular joint. From under the rostral border of the parotid (salivary) gland, the following vessels and nerves become superficial on the lateral aspect of the masseter M. and ventral to the joint: the transverse facial A.V. and the auriculotemporal and facial Nn.

Remember! The transverse facial vessels are branches of the superficial temporal vessels. The auriculotemporal N. is a branch of the mandibular N. (V-trigeminal), whereas the facial N. is the cranial N. VII. These two nerves communicate with each other; the dorsal and ventral buccal branches of the facial N. continue rostrally, making multiple communications in a sort of plexus.

The nervous branches that supply the facial muscles cross the structures that run parallel to the rostral border of the masseter M.—that is, the facial A.V. and the parotid duct (in rostrocaudal order). Dissect them and the branches of the vessels (inferior labial and superior labial Aa.Vv.).

Caution! Before giving off the superior labial A.V., the facial vessels travel on the deep surface of the zygomaticus M. (Fig. 7-6).

Examine the parotid duct, which changes direction by obliquely crossing the deep aspect of the vessels between them and the buccinator M. The parotid duct finally pierces the buccinator M. at the level of the third superior premolar.

Continue to dissect the facial vessels and expose the lateral nasal, dorsal nasal, and angularis oculi Aa.Vv.

Introduce your finger into the nasal cavity through the dorsal commissure of the nostril, which leads into the nasal diverticulum. This structure, which is the size and shape of a glove finger, lies against the membrane that fills the osseous nasal aperture. Palpate the alar cartilage of the nostril, its lamina (dorsally), and cornu (ventrolaterally). Make an incision in the orbicularis oris M. at the level of the end of the cornu, and expose the rostral part of the lateral nasal M.

The incisive Mm. (superior and inferior) are inserted on the incisive bones and on the incisive part of the body of the mandible, respectively, and on the two corresponding lips under their mucous membrane (mucosa). Dissect the mucosa and expose the muscles. Within the chin, another muscle is encountered: the mentalis M. Relate the mentalis M. and the chin to the insertion of the tendon of the depressor labii inferioris M. In its route, this tendon overlaps the mental foramen, nerve, and vessels.

To fully expose the deep vessels and nerves of the facial area, transect the levator nasolabialis M. parallel to the long axis of the levator labii superioris M., and carefully dissect and reflect the stumps. Free the caninus M. from its tendinous attachment, and reflect it rostrally with the rostral stump of the levator labii superioris M. Do not transect the levator labii superioris M., but reflect it dorsally to expose the infraorbital foramen, nerve, and vessels. Transect the zygomatic M. and reflect the stumps. Dissect all the vessels and nerves and their branches within the area.

If your instructor decides that you should dissect the parotid region as a separate area, refer to the chapter on the parotid region in this guide.

Focus your attention on the parotid region. Follow the course of the auriculotemporal and facial Nn. and of the transverse facial A.V., which are under the parotid gland. Reflect the rostral border of the parotid gland caudally and look for the parotid ln., a small structure located on the subparotid route of the transverse facial A. Extend the reflection of the gland to the entire rostral border, and identify the origin of the parotid duct. The most superficial structure within the gland is the maxillary V. Dissect it in a dorsal direction and continue to dissect the superficial temporal V. parallel to the caudal border of the mandible and the temporomandibular joint. Isolate, double ligate, and transect the superficial temporal V. between the two ligatures. Continue to dissect the subparotid route of the facial N. to expose and dissect its branches.

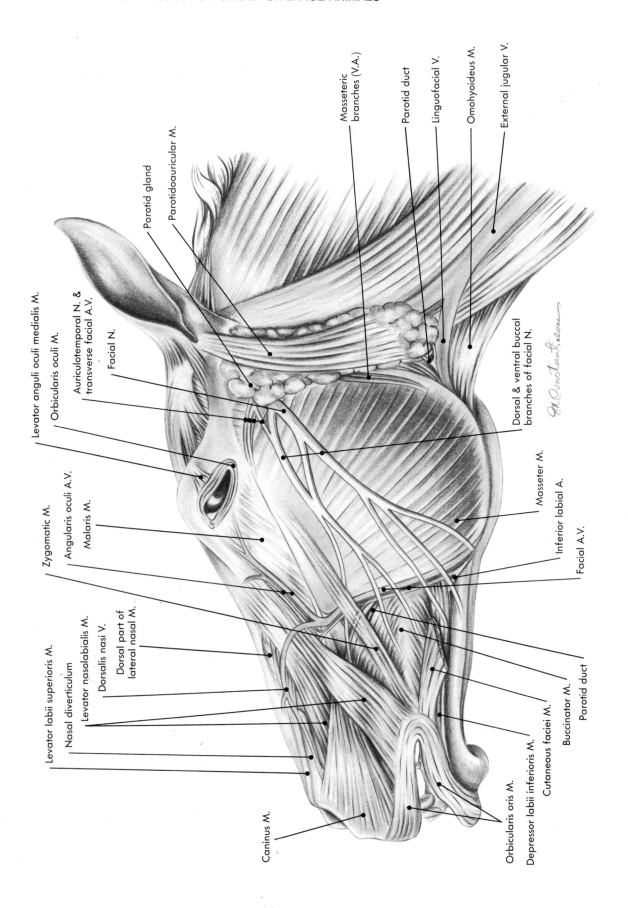

Masseteric branches (V.A.)

Parotid duct

Linguofacial V.

Omohyoideus M.

External jugular V.

Parotid gland

Parotidoauricular M.

Levator anguli oculi medialis M.

Orbicularis oculi M.

Auriculotemporal N. & transverse facial A.V.

Facial N.

Dorsal & ventral buccal branches of facial N.

Masseter M.

Inferior labial A.

Facial A.V.

Zygomatic M.

Angularis oculi A.V.

Malaris M.

Levator labii superioris M.

Nasal diverticulum

Levator nasolabialis M.

Dorsalis nasi V.

Dorsal part of lateral nasal M.

Parotid duct

Buccinator M.

Cutaneous faciei M.

Orbicularis oris M.

Depressor labii inferioris M.

Caninus M.

Fig. 7-6   Superficial structures of head (lateral aspect).

Caution! The facial N. issues three branches in a dorsal direction and three branches and an ansa in a ventral direction; all of them run through the parotid glandular tissue. In rostrocaudal order, the three dorsal branches are the auriculopalpebral N., the internal auricular br., and the caudal auricular N. In the same order, the ventral branches are the cervical br., the ansa around the caudal auricular A., the parotid plexus, and the muscular br., which supplies the caudal belly and occipitomandibular part of the digastric M. (ramus digastricus) and the stylohyoideus M. (ramus stylohyoideus).

Caution! The cervical br. of the facial N. is a delicate nerve, the route of which is not the same in all specimens. At the ventral extent of the parotid gland, the nerve parallels the external jugular V. on its lateral side. Carefully dissect the nerve and the maxillary V., which pass through the glandular tissue.

Double ligate the maxillary V., transect it between the two ligatures, and reflect the stumps to expose the external carotid A. Remove the rest of the parotid gland and expose the subparotid aponeurosis, which separates the parotid from the mandibular gland.

Remember! The subparotid aponeurosis is the 2- to 3-cm-wide connection between the aponeurosis of the cleidomastoideus M. and the tendon of the sternomandibularis M.

Retain the head in maximum extension to expose the following structures (in caudorostral order): the jugular process of the occipital bone, the occipitohyoideus M., the stylohyoid angle of the stylohyoid bone, the occipitomandibular part and caudal belly of the digastric M., and the stylohyoideus M.

To expose, dissect, and identify the deep structures of the head (Fig. 7-7), it is necessary to remove the mandible. First, transect the dorsal and ventral buccal branches of the facial N. Free and reflect them from the masseter M. Transect the masseter M. parallel and 1 cm ventral to the facial crest and reflect it caudally, saving the three veins that join the facial V; in dorsoventral order, they are the transverse facial, deep facial, and buccal Vv.

Notice that, in the horse, these three veins form dilations named sinuses.

Notice that the transverse facial V. connects the facial V. to the superficial temporal V.; the deep facial V. discharges blood from the orbital and pterygopalatine regions; and the buccal V. connects the facial V. to the maxillary V.

The molar (caudal) part of the buccinator M. and the caudal extent of the depressor labii inferioris M. are implicitly exposed, as are the dorsal buccal (salivary) glands.

Transect the facial A.V. and the parotid duct at the level of the mandibular (vascular) incisure, and reflect them dorsally. Make an incision in the mucosa of the oral cavity, following the ventral border of the depressor labii inferioris M. and continuing up to the mental foramen. Transect the occipitomandibular part of the digastric M., the tendon of the sternomandibularis M., and the masseteric and transverse facial vessels close to the caudal border of the mandible.

Using a chisel and a hammer, make four sections in the skull: (1) a transverse section of the zygomatic process of the frontal bone (as close as possible to the base of the process); (2) a transverse section of the temporal process of the zygomatic bone (rostral to the medial canthus); (3) an oblique section of the zygomatic arch (caudal to the temporomandibular joint and in a caudorostral direction); and (4) a transverse section of the mandible (between the mental foramen and the first inferior premolar).

Turn the head so that the ventral intermandibular aspect is up. Dissect the mylohyoideus M. and the rostral belly of the digastric M., the mandibular lnn., the omohyoideus and sternohyoideus Mm., the facial A.V., and the parotid duct. Reflect the vessels and the parotid duct caudally. Free the mylohyoideus M. and the rostral belly of the digastric M. from the medial aspect of the mandible. Gently pull the mandible and free it from the mylohyoideus to detach the muscle from its insertion. Toward the caudal extent of the mylohyoideus M., remove the periosteum from the medial aspect of the mandible to protect the mylohyhoid N.

Turn the head so that the lateral side is up, and remove the zygomatic arch. To accomplish this, it is necessary to free the zygomatic arch from both the temporal and masseter Mm. Leave the articular disc connected to the mandible for the moment, after sectioning the joint capsule around the disc. Pulling the mandible higher and higher, section the insertion of the medial pterygoid M. from the pterygoid fossa (on the medial aspect of the ramus of the mandible), the insertion of the temporal M. from the medial aspect of the coronoid process, and the insertion of the lateral pterygoid M. from the medial aspect of the neck and condyle of the mandible.

Continue to carefully detach the periosteum from the medial aspect of the mandible up to the mandibular foramen.

Caution! The slender mylohyoid N., which supplies the muscle of the same name and the rostral belly of the digastric M., runs between the bone and the periosteum.

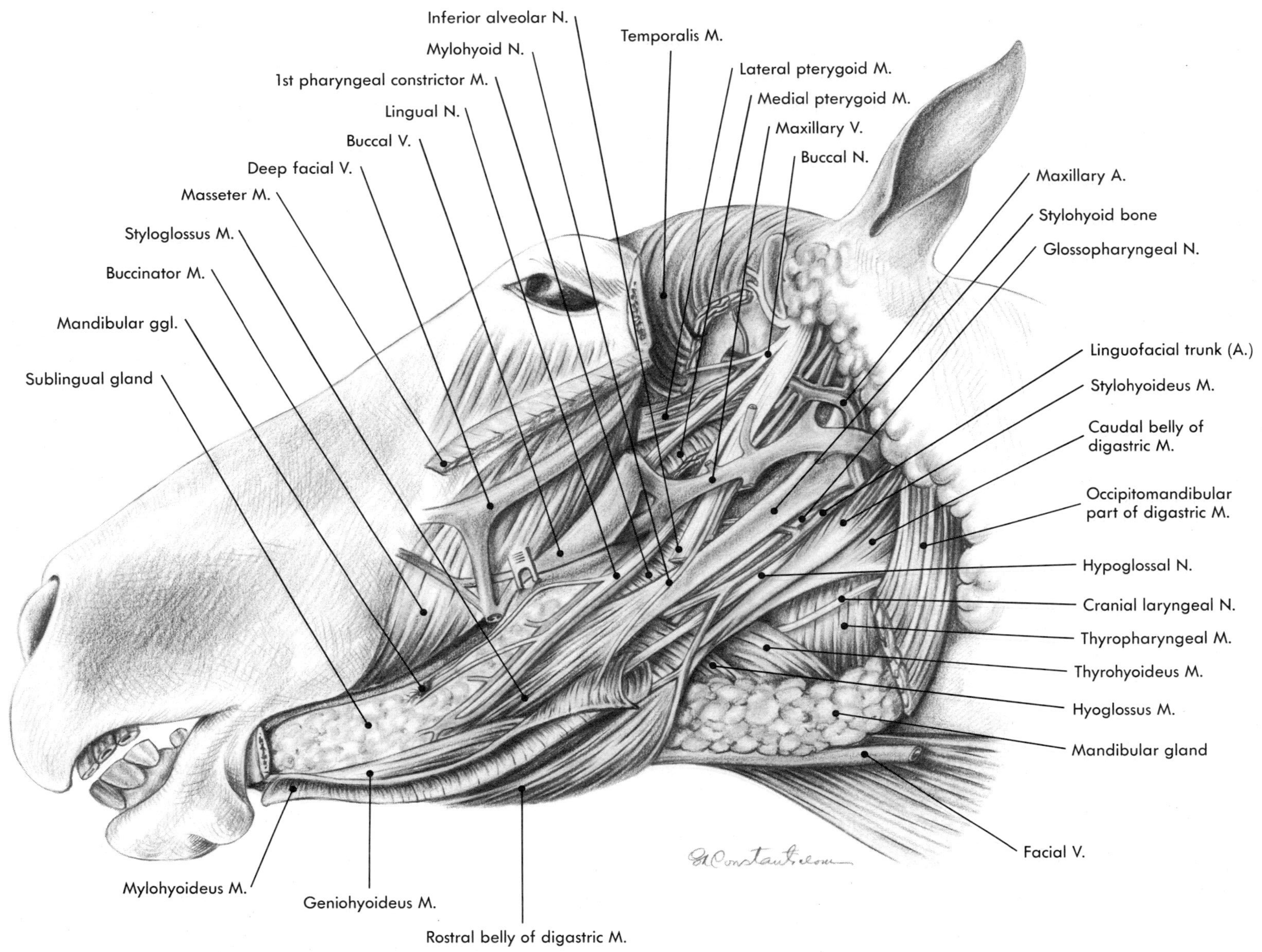

Inferior alveolar N.

Mylohyoid N.

1st pharyngeal constrictor M.

Lingual N.

Buccal V.

Deep facial V.

Masseter M.

Styloglossus M.

Buccinator M.

Mandibular ggl.

Sublingual gland

Temporalis M.

Lateral pterygoid M.

Medial pterygoid M.

Maxillary V.

Buccal N.

Maxillary A.

Stylohyoid bone

Glossopharyngeal N.

Linguofacial trunk (A.)

Stylohyoideus M.

Caudal belly of digastric M.

Occipitomandibular part of digastric M.

Hypoglossal N.

Cranial laryngeal N.

Thyropharyngeal M.

Thyrohyoideus M.

Hyoglossus M.

Mandibular gland

Facial V.

Rostral belly of digastric M.

Geniohyoideus M.

Mylohyoideus M.

Fig. 7-7

Deep structures of head (lateral aspect). (Mandible has been removed.)

Transect the inferior alveolar A.V.N. before entering the mandibular foramen.

Caution! While removing the mandible, detach the articular disc of the temporomandibular joint and leave it connected to the masseteric N., which runs parallel to the rostral border of the disc.

At the conclusion of the procedure, the mandible is removed.

The following muscles, vessels, and nerves are exposed: the temporalis, lateral pterygoid, medial pterygoid, rostral belly of the digastric, and mylohyoideus Mm. Crossing the two pterygoid Mm., the branches of the mandibular N. are visible. In addition and as the most superficial structure, the maxillary V. is also exposed.

Carefully remove the two pterygoid Mm., saving all the nerves and vessels already exposed and the deep structures, including the guttural pouch with its thin and delicate wall.

Carefully dissect the branches of the mandibular N., and identify them one by one (in rostrocaudal order): the masticatory N., giving off the deep (caudal) temporal Nn. and continuing as the masseter N.; the buccal N., giving off the deep (rostral) temporal N. and the lateral pterygoid N.; the medial pterygoid N.; the lingual N.; the inferior alveolar N.; the mylohyoid N.; and the auriculotemporal N.

Notice that, in the *Nomina Anatomica Veterinaria,* third edition, 1983, footnote no. 469 (p. A128) is wrong: "N. buccalis is the sensory nerve to the mucous membrane . . . of the cheek, lips, and nose." In fact, the buccal N. is a mixed nerve, supplying the rostral part of the temporalis M. by the deep rostral temporal N. and the lateral pterygoid M. by the nerve of the same name. In addition, it gives off the Nn. tensoris veli palatini and tensoris tympani.

Remember! The otic ggl. (parasympathetic, carrying fibers from the glossopharyngeal N.) is located at the origin of the buccal N. The chorda tympani N., carrying parasympathetic fibers from the intermediate N. (br. of the intermediofacial N.), joins the lingual N. and ends in the mandibular salivary gland. The auriculotemporal N. surrounds the temporomandibular joint, runs on the lateral aspect of the masseter M., and emits one communicating branch to the dorsal buccal branch of the facial N.

Notice that the chorda tympani N. crosses the deep aspect and the lingual N. crosses the superficial aspect of the maxillary A. before these nerves join one another.

The only hard structure within the area is the hyoid apparatus, the stylohyoid bone of which is visible and surrounded by the guttural pouch. All the deep vessels and nerves run on the surface of the pouch. Identify the stylohyoid M. and its split tendon, which allows for the passage of the tendon of the digastric M. Identify the external carotid A., which comes from under the stylohyoid M., runs over the lateral aspect of the stylohyoid angle (of the stylohyoid bone), and branches into the superficial temporal and maxillary Aa. as terminal branches. Between the stylohyoid bone and the stylohyoid M., the largest branch of the external carotid A. is visible: the linguofacial trunk. This artery is paralleled by the glossopharyngeal N. (rostrally) and the hypoglossal N. (caudally). Identify the ascending palatine, lingual, facial, and sublingual Aa.

Reflect the lingual mucosa and continue the dissection of those structures that continue into the tongue, such as the lingual N.Vv.; the glossopharyngeal and hypoglossal Nn.; the lingual A.; and the styloglossus, hyoglossus, genioglossus, and geniohyoideus Mm.

The deepest structures of the area are the muscles of the soft palate and the muscles of the pharynx. Before examining them, identify the branches of the maxillary A.: the rostral tympanic, middle meningeal, and caudal deep temporal Aa. (in a dorsal direction) and the inferior alveolar A. and pterygoid branches (in a ventral direction).

Deep to the medial pterygoid M. and in an oblique position from the base of the ear to the pterygoid crest, identify the tensor (laterally) and levator (medially) veli palatini

Mm. They are parallel with one another and with the auditory tube, which is located medial to the levator veli palatini M.

Notice that the guttural pouch represents the herniation of the mucosa of the auditory tube through a fissure (split) in the tube and is a specific structure for the horse.

Starting from the pterygoid process and looking in a caudal direction, identify the pterygopharyngeus and palatopharyngeus Mm. (the first constrictor of the pharynx), the hyopharyngeus M. (the second constrictor of the pharynx), and the thyropharyngeus and cricopharyngeus Mm. (the third constrictor of the pharynx). On the medial aspect of the stylohyoid bone, identify the stylopharyngeus M. as the only dilator of the pharynx.

Identify the cranial laryngeal N. (branch of the vagus N.), entering the larynx through the thyroid fissure, turned into a foramen (foramen thyroideum).

Reflect the mandibular gland ventrally. Transect the following muscles transversely: the hyoglossus, hyopharyngeus, thyropharyngeus, and cricopharyngeus Mm. Reflect the stumps to expose the hyoid and laryngeal muscles and joints.

The following structures will be exposed: the lingual A., which runs parallel to the dorsal border of the ceratohyoideus M. and crosses the ceratohyoid bone on its way toward the tongue and is accompanied by the lingual br. of the glossopharyngeal N. and by the hypoglossal N.; the ceratohyoideus M.; the thyrohyoideus M.; the cricothyroideus M.; the sternothyroideus M.; the cricoarytenoideus dorsalis M.; and the arytenoideus transversus M.

The recurrent laryngeal N. was previously exposed during the dissection of the neck. Follow its stump in a cranial direction, and observe that the nerve penetrates the larynx between the cricopharyngeus and cricoarytenoideus dorsalis Mm.

Remember! The recurrent (caudal) laryngeal N. is a mixed nerve, carrying motor fibers for all the laryngeal muscles except the cricothyroideus M. (which is supplied by the cranial laryngeal N.). The two laryngeal nerves form a plexus under the laryngeal mucosa, which they supply.

Dissect the branch of the accessory N. that supplies the sternomandibularis M. This branch usually parallels the occipital V. Dissect the vein, as well. Transect the occipital V., then dissect and reflect the origin of the external jugular V. cranially. Dissect the omohyoideus M. and reflect it rostrally, exposing the origin of the trachea and esophagus and the cranial extent of the common carotid A.

Retain the head in extension and pull the combined common carotid A. and sympathetic trunk backward. The cranial thyroid A. and the inconstant caudal thyroid A., which surround the thyroid gland, are visible.

Remember! The thyroid and parathyroid glands lie on the lateral aspect of the first 3 to 4 tracheal rings. The symmetrical thyroid lobes are connected through an isthmus, which surrounds the trachea ventrally.

Identify the ascending pharyngeal and cranial laryngeal Aa. and dissect them. Identify the trifurcation of the common carotid A.: the occipital A. runs dorsally toward the occipitoatlantal area; the internal carotid A. runs rostrodorsally toward the carotid incisure of the sphenoid bone (within the foramen lacerum); the direct continuation of the common carotid A. is the external carotid A., the location, topography, and branches of which have already been dissected and are known (Fig. 7-8).

Caution! The following structures lie on or come in intimate contact with the lateral wall of the guttural pouch. Dissect them carefully.

Follow the route of the vagosympathetic trunk toward the atlantal fossa, and observe the separation of the cervical sympathetic trunk from the vagus N., which leads to the cranial cervical ganglion (sympathetic). The latter is located along the route of the internal carotid A. within the atlantal fossa and overlapped by the lateral retropharyngeal lnn.

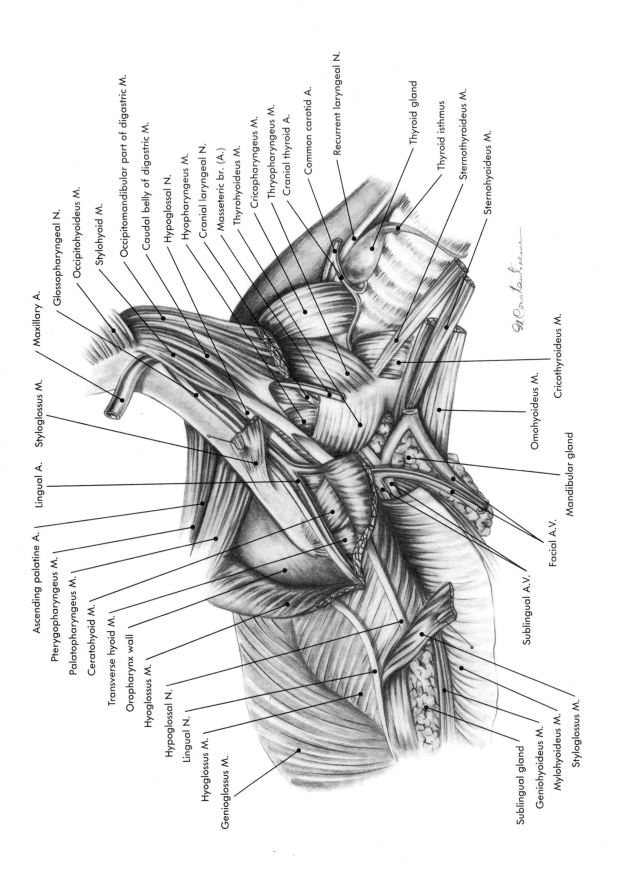

Maxillary A.

Glossopharyngeal N.

Occipitohyoideus M.

Stylohyoid M.

Occipitomandibular part of digastric M.

Caudal belly of digastric M.

Hypoglossal N.

Hyopharyngeus M.

Cranial laryngeal N.

Masseteric br. (A.)

Thyrohyoideus M.

Cricopharyngeus M.

Thryopharyngeus M.

Cranial thyroid A.

Common carotid A.

Recurrent laryngeal N.

Thyroid gland

Thyroid isthmus

Sternothyroideus M.

Sternohyoideus M.

Styloglossus M.

Lingual A.

Ascending palatine A.

Pterygopharyngeus M.

Palatopharyngeus M.

Ceratohyoid M.

Transverse hyoid M.

Oropharynx wall

Hyoglossus M.

Hypoglossal N.

Lingual N.

Hyoglossus M.

Genioglossus M.

Omohyoideus M.

Cricothyroideus M.

Mandibular gland

Facial A.V.

Sublingual A.V.

Sublingual gland

Geniohyoideus M.

Mylohyoideus M.

Styloglossus M.

Structures of tongue, hyoid apparatus, pharynx, and larynx (lateral aspect).

Fig. 7-8

Transect the caudal belly of the digastric M. Reflect the caudal stump, widely exposing the occipitohyoideus M., as well as the glossopharyngeal and hypoglossal Nn. Identify the following structures: the carotid sinus branch of the glossopharyngeal N.; the ansa cervicalis between the hypoglossal N. and ventral branch of C_1; the pharyngeal branches of the vagus N. and the cranial cervical ggl.; the origin of the cranial laryngeal N.; the external carotid N. (from the cranial cervical ggl. to the carotid sinus and glomus); and the external br. of the accessory N., which splits into ventral and dorsal branches.

Remember! The carotid sinus is a baroreceptor zone, a dilatation of the origin of the internal carotid A., whereas the carotid glomus is a chemoreceptor and is located close to the carotid sinus.

Turn the head so that the median (split) side is up. It is recommended that only the dissection of the vessels and nerves described in the previous section should be performed. The dissection of the ear and eye is the next step; this is followed by the dissection of the nasal and oral cavities, the soft palate, the pharynx, the larynx, and the brain (from their median [medial] perspective); and finally, the dissection of the paranasal sinuses is performed. The brain will be described in its entirety. Following these directions, each compartment of the head is dissected and described in this section.

Use the half of the head with the attached mandible. Dissect and reflect the guttural pouch rostrally to free it from the adjacent deep (lateral) structures. Remove the mucosa of nasopharynx to expose the tensor and levator veli palatini Mm. and the first constrictor M. of the pharynx. Reflect the tendon of the genioglossus M. dorsally to expose the geniohyoideus, mylohyoideus, hyoglossus Mm., as well as the other structures shown in Fig. 7-9. Take this opportunity to identify the structures in Figs. 7-9 and 7-10 that are detailed on pp. 311-314 and 317 and that belong to the nasal and oral cavities, pharynx, and larynx.

Pull the common carotid A. and the vagosympathetic trunk backward, and complete the dissection of the occipital, internal carotid, and external carotid Aa. and the linguofacial trunk. Complete the dissection of the nerves of the area and their branches (see Fig. 7-10). Identify the stylopharyngeus M. on the medial aspect of the stylohyoid bone and the occipitohyoideus, caudal belly of the digastric, and stylohyoideus Mm. from their medial perspective.

Remember! The stylopharyngeus M. is the only dilator of the pharynx.

Crossing the lateral aspect of the stylohyoid bone, identify the medial and lateral pterygoid Mm. Parallel to the rostral border of the stylohyoid bone and running between the two pterygoid muscles, identify the mandibular N. of the trigeminal.

To widely expose the last four cranial nerves, transect and reflect the longus capitis and rectus capitis ventralis Mm.

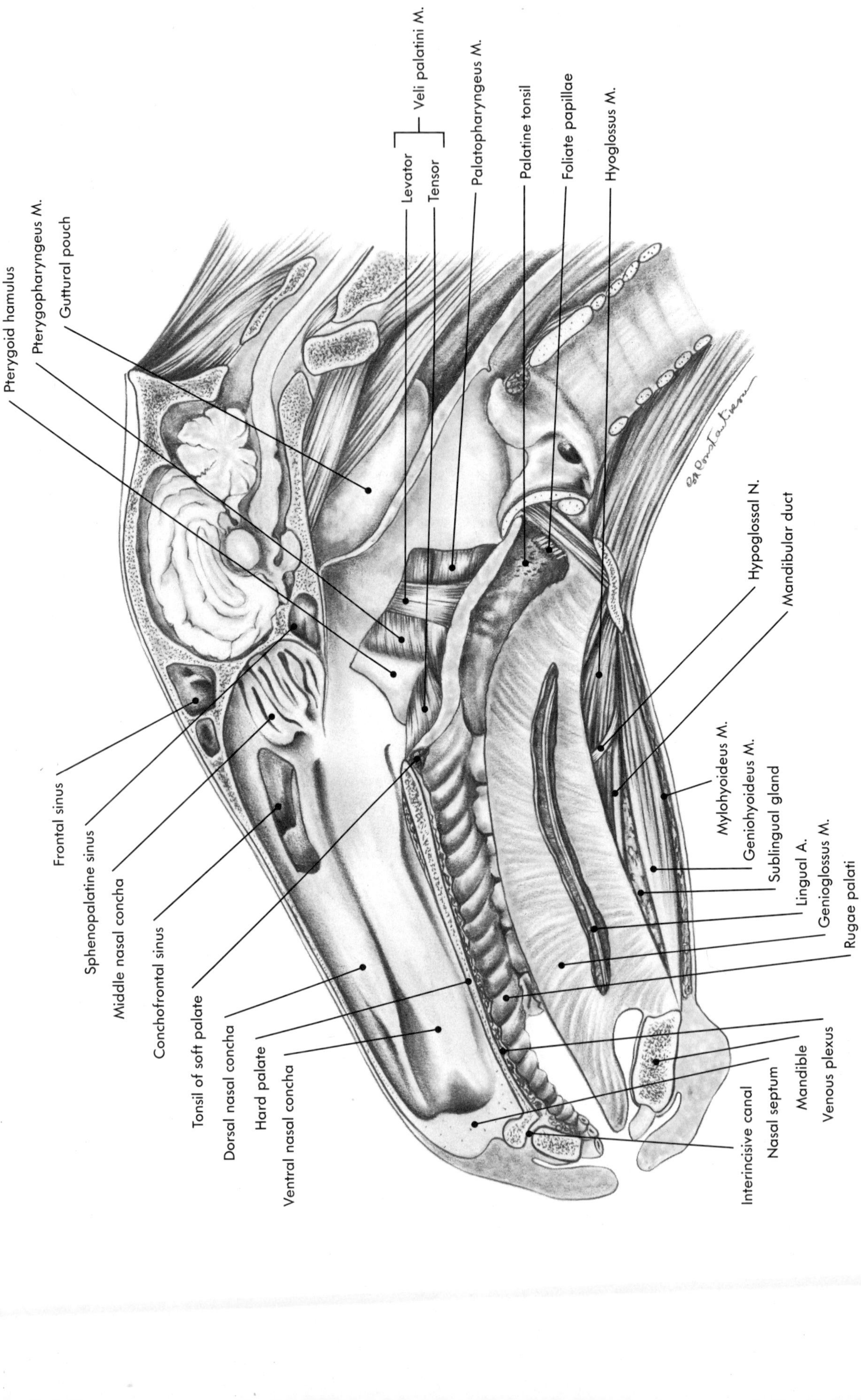

Structures of median aspect.

Fig. 7-9

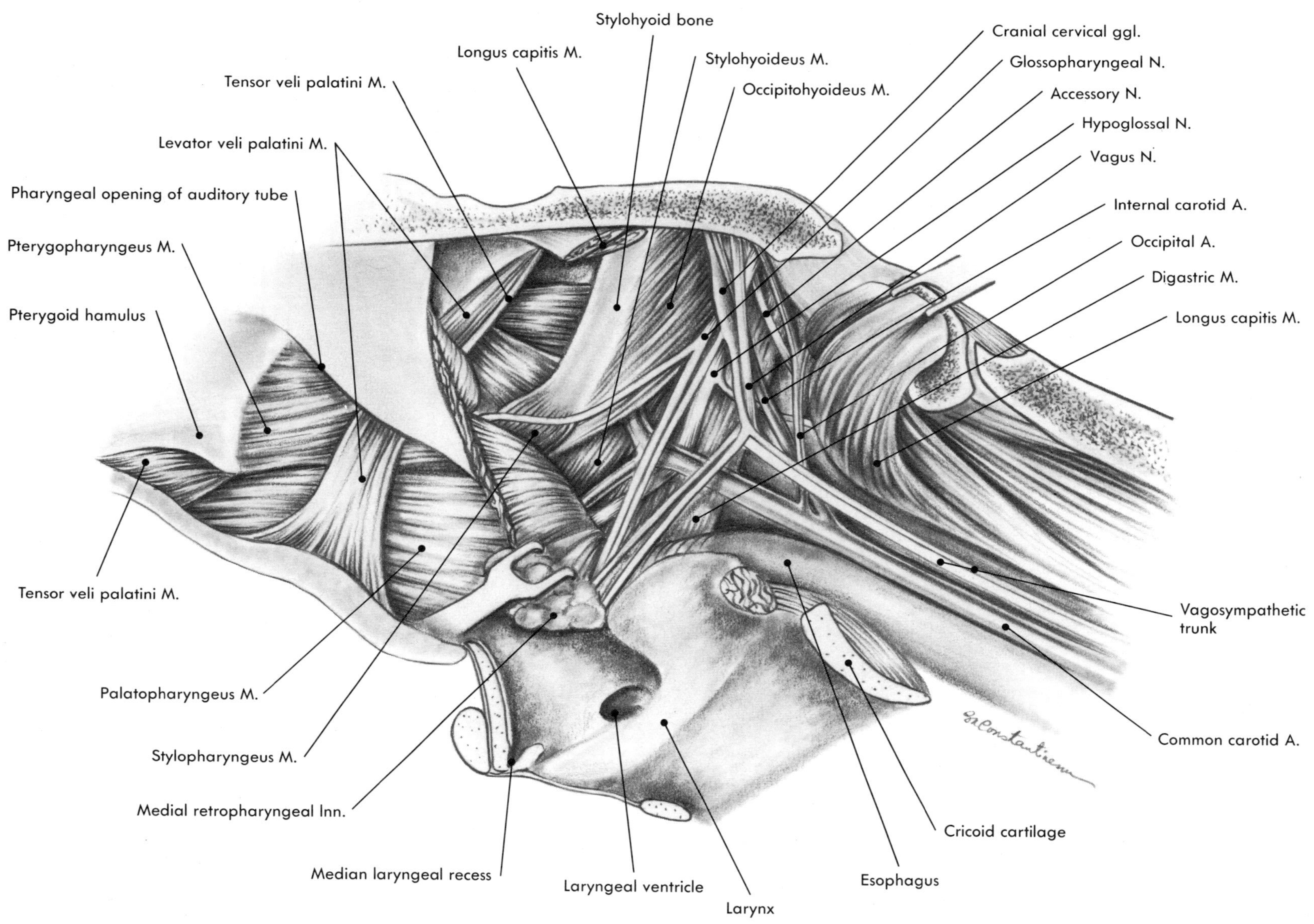

Cranial cervical ggl.

Glossopharyngeal N.

Accessory N.

Hypoglossal N.

Vagus N.

Internal carotid A.

Occipital A.

Digastric M.

Longus capitis M.

Vagosympathetic trunk

Common carotid A.

Cricoid cartilage

Esophagus

Larynx

Laryngeal ventricle

Median laryngeal recess

Stylohyoid bone

Stylohyoideus M.

Occipitohyoideus M.

Longus capitis M.

Tensor veli palatini M.

Levator veli palatini M.

Pharyngeal opening of auditory tube

Pterygopharyngeus M.

Pterygoid hamulus

Tensor veli palatini M.

Palatopharyngeus M.

Stylopharyngeus M.

Medial retropharyngeal lnn.

Fig. 7-10

Structures of pharynx, larynx, and hyoid apparatus (medial aspect).

Turn the head so that the lateral side is up. Focus on the *ear,* and skin it carefully. Nerve fibers from the auriculopalpebral, caudal auricular, occipitalis major, and auricularis magnus Nn. are visible, as are the terminal branches from the rostral auricular and caudal auricular Aa.Vv.

Remember! The auriculopalpebral and caudal auricular Nn. are branches of the facial N. The occipitalis major N. is given off by the dorsal branch of C_1, whereas the auricularis magnus N. is given off by the ventral branch of C_2. The rostral auricular vessels supply the rostral aspect of the ear, whereas the caudal auricular vessels supply the other three aspects of the ear with the medial, intermediate, and lateral branches.

Remember! The muscles of the ear are systematized into rostral, dorsal, caudal, and ventral groups. They originate from the surrounding bones, the ligamentum nuchae, or the scutiform cartilage, and they insert on the scapha (the auricular cartilage) or on the scutiform cartilage (this structure transmits the action of some muscles to the scapha).

Remember! The additional annular cartilage of the ear is interposed between the scapha and the external acoustic meatus.

First, dissect the superficial scutuloauriculares Mm. (which originate from the scutiform cartilage and insert on the scapha) in an order corresponding to their names: dorsal, middle, ventral, and accessory. The dorsal muscle partially overlaps the accessory (oriented medially) and the middle (oriented laterally) scutuloauriculares Mm. Identify the superficial temporal A. between the zygomatoauricularis, frontoscutularis, and ventral part of the scutuloauricularis superficialis Mm. Identify the rostral auricular A. between the middle and ventral parts of the scutuloauricularis superficialis M. Identify the medial br. of the caudal auricular A., which becomes a superficial structure between the accessory part of the scutuloauricularis superficialis M. and the cervicoauricularis superficialis M. Identify the occipitalis major N., which parallels the medial br. of the caudal auricular A. Identify and dissect the zygomatoauricularis, frontoscutularis, interscutularis, cervicoscutularis, and cervicoauricularis superficialis Mm.

To dissect the deep muscles, transect the interscutularis and cervicoscutularis Mm. transversely, and reflect the stumps. Reflect the scutiform cartilage rostrally, and identify the following structures: the parietoauricularis, scutuloauricularis profundus major, and scutuloauricularis profundus minor Mm., the branch of the superficial temporal A., and the medial br. of the caudal auricular A. The superficial temporal A. runs over the temporal M. between the parietoauricularis and scutuloauricularis profundus major Mm., whereas the medial br. of the caudal auricular A. runs parallel to the occipitalis major N. at the rostral border of the cervicoauricularis superficialis M. Dissect the cervicoauricularis medius and profundus Mm. Identify the intermediate br. of the caudal auricular A. It becomes superficial at the caudal border of the insertion of the cervicoauricularis superficialis M. *Note* the auricularis magnus N., which parallels the ventral border of the cervicoauricularis profundus M. (Fig. 7-11).

The parotidoauricular M. connects the base of the auricular cartilage to the parotid gland; it was previously identified when the gland was reflected and removed. To expose the styloauricularis M., rotate the ear laterocaudally and palpate the annular cartilage of the ear and the external acoustic meatus. Dissect this spindle-shaped muscle, which is as thin as a wooden match, 2 to 3 cm long, and located on the rostromedial aspect of the annular cartilage.

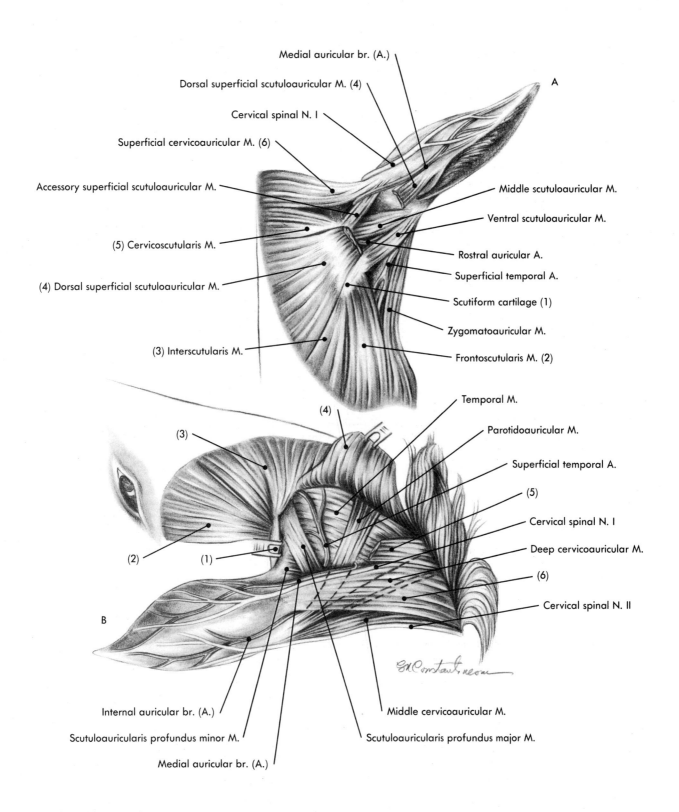

Medial auricular br. (A.)

Dorsal superficial scutuloauricular M. (4)

Cervical spinal N. I

Superficial cervicoauricular M. (6)

Accessory superficial scutuloauricular M.

(5) Cervicoscutularis M.

(4) Dorsal superficial scutuloauricular M.

(3) Interscutularis M.

Middle scutuloauricular M.

Ventral scutuloauricular M.

Rostral auricular A.

Superficial temporal A.

Scutiform cartilage (1)

Zygomatoauricular M.

Frontoscutularis M. (2)

A

(4)

(3)

(2)   (1)

B

Temporal M.

Parotidoauricular M.

Superficial temporal A.

(5)

Cervical spinal N. I

Deep cervicoauricular M.

(6)

Cervical spinal N. II

Internal auricular br. (A.)

Scutuloauricularis profundus minor M.

Medial auricular br. (A.)

Middle cervicoauricular M.

Scutuloauricularis profundus major M.

Fig. 7-11   **A,** Auricular structures (dorsal aspect). **B,** Deep auricular structures (dorsal aspect).

Following the dissection of the ear, remove the three cartilages. Remove all the muscles, vessels, nerves, and remnant skin from the scapha. Identify the prominences, depressions, incisures, and other details of the scapha.

Concentrating on the *eye,* examine the two eyelids and their two commissures, or canthi (medial or nasal and lateral, or temporal). Inside of the medial canthus, identify the lacrimal caruncle and the two puncta lacrimalia, which are located dorsal and ventral to the caruncle on the free borders of the eyelids (Fig. 7-12, *A*).

Remember! The puncta lacrimalia lead into the lacrimal canaliculi, which open into the lacrimal sac, the origin of the nasolacrimal duct.

Evert each eyelid, and examine the palpebral conjunctiva, bulbar conjunctiva, conjunctival fornix, and conjunctival sac. Make an incision through the eyelids from the free border toward the conjunctival fornix; on the cut surface, examine the tarsus, the tarsal (Meibomian) glands, and the cross-section of the orbicularis oculi M. In the superior eyelid, also inspect the fibers from the levator palpebrae superioris M.

The eyelids are "inserted" on the border of the orbital cavity by means of the orbital septum. Examine the medial and lateral palpebral ligg., which anchor the angles of the eyelids on the border of the orbital cavity.

Examine the third eyelid at the medial angle of the bulbus oculi.

The zygomatic process of the frontal bone and the temporal process of the zygomatic bone have already been sectioned and the zygomatic arch removed. Manually remove the large amount of fat located between the orbital area and the temporal M. This is part of the extraperiorbital fat. After removal of the fat, the eyeball, muscles, vessels, and nerves, which are surrounded and protected by the periorbita, are ready for dissection.

Before dissecting the structures within the periorbita identify on your specimen some of the structures already identified on the skull, such as the following: the rostral alar foramen (the maxillary A. comes from it); the foramen rotundum (the maxillary and zygomatic Nn. come from it); the pterygoid crest; the maxillary foramen (the infraorbital A.N. enter it); the sphenopalatine foramen (for the sphenopalatine A. and the caudal nasal N.); and the caudal palatine foramen (for the major palatine A.N.).

Identify the malar A., which runs within the bony orbita outside the periorbita toward the medial angle of the eyeball to supply the eyelids. Identify the buccal A. given off by the maxillary A. after exiting the rostral alar foramen and the minor palatine A., which is a branch of the descending palatine A.

Remember! The descending palatine A. branches into the minor palatine, major palatine, and sphenopalatine Aa.

Also identify the minor palatine and accessory palatine Nn.

Remember! The pterygopalatine N. (from the maxillary) branches into the minor palatine, major palatine, and caudal nasal Nn. The accessory palatine N. is a branch of the major palatine N.

Carefully examine the periorbita. Between the dorsal and lateral rectus Mm., a yellow strip is visible through the transparency. Here, make an incision in the long axis of the periorbita, separate the periorbita from the orbital septum, and carefully transect the periorbita from the attachment of the pterygoid crest. The yellow strip is part of the intraperiorbital fat. Reflect the periorbita dorsally and ventrally as much as possible, saving the lacrimal gland that lies against the lacrimal fossa of the frontal bone. In dorsoventral order, identify the frontal, lacrimal, and zygomatic Nn. between the periorbita and the dorsal and lateral rectus Mm.

Remember! The frontal N. runs toward the supraorbital foramen, passes through it, and continues on as the supraorbital N. The lacrimal N. has two or more branches. Both frontal and lacrimal nerves belong to the ophthalmic N. (trigeminal) and come from the skull through the orbital fissure. The zygomatic N. (branch of the maxillary N. [trigeminal]) emits the zygomaticofacial and zygomaticotemporal Nn. and sends the communicating br. to the lacrimal N. (There are various possible branchings of the zygomatic N.) (Fig. 7-12, *B*).

Reflect the muscles of the eyeball as far laterally as possible to identify the dorsal oblique M. and the trochlear N. above it.

Evert the lateral rectus M., and identify the branch of the abducent N. supplying it. Enlarge the space between the lateral and dorsal rectus Mm., and identify the oculomotor N., which divides into the dorsal and ventral branches. The dorsal br. remains deep to supply the dorsal rectus and levator palpebrae superioris Mm., whereas the ventral br. supplies the medial and ventral rectus Mm. It runs on the superficial aspect of the ventral rectus M. toward the ventral oblique M., which it also supplies. At the origin of the ventral br. of the oculomotor N., observe the tiny ciliary ggl. (parasympathetic).

Very deep in the same space between the dorsal and lateral rectus Mm. and exiting from the orbital fissure is the nasociliary N. (the third branch of the ophthalmic N.), which gives off the ethmoidal and infratrochlear Nn. The first penetrates the foramen of the same name along with the external ethmoidal A., whereas the infratrochlear N. travels toward the lacrimal caruncle and sac and the third eyelid.

Identify the external ophthalmic A., exiting the orbital fissure and branching to supply the structures protected within the periorbita. Its branches include the external ethmoidal A.; muscular branches; and the conjunctival, lacrimal and palpebral Aa., including the A. for the third eyelid, the central A. of the retina, and the long and short posterior ciliary Aa.

The muscles of the eyeball are surrounded and protected by muscular fasciae. To expose all the main arteries and veins, it is necessary to transect the lateral and dorsal rectus Mm. and to reflect the stumps. The vorticose Vv., which perforate the sclera close to the equator of the eyeball, are exposed. Next, transect the retractor bulbi M. from its scleral insertion and reflect the muscle caudally. The long ciliary Aa.Nn., paralleling the optic N., are now exposed.

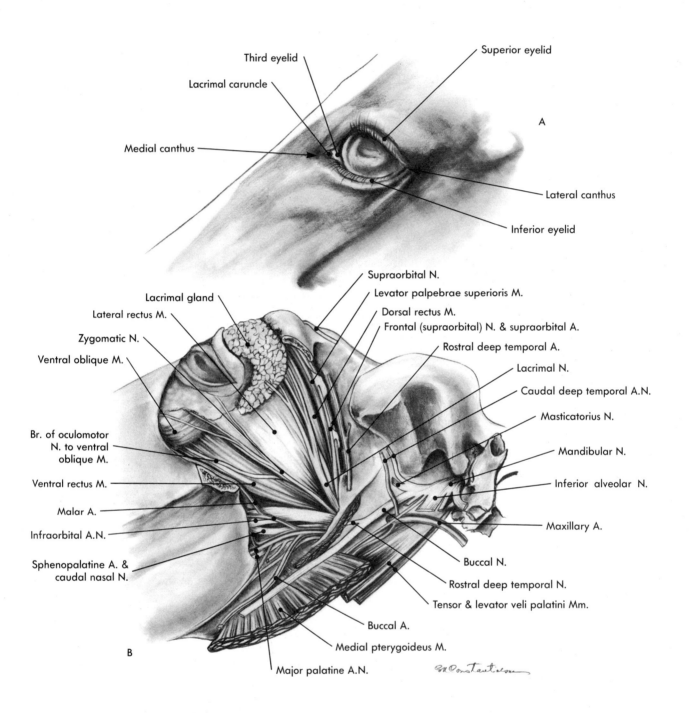

Third eyelid

Lacrimal caruncle

Superior eyelid

A

Medial canthus

Lateral canthus

Inferior eyelid

Supraorbital N.

Levator palpebrae superioris M.

Lacrimal gland

Dorsal rectus M.

Lateral rectus M.

Frontal (supraorbital) N. & supraorbital A.

Zygomatic N.

Rostral deep temporal A.

Ventral oblique M.

Lacrimal N.

Caudal deep temporal A.N.

Masticatorius N.

Br. of oculomotor N. to ventral oblique M.

Mandibular N.

Ventral rectus M.

Inferior alveolar N.

Malar A.

Maxillary A.

Infraorbital A.N.

Buccal N.

Sphenopalatine A. & caudal nasal N.

Rostral deep temporal N.

Tensor & levator veli palatini Mm.

Buccal A.

B

Medial pterygoideus M.

Major palatine A.N.

Fig. 7-12

A, Eye (external aspect). **B,** Structures of eye and orbital area.

Also transect the levator palpebrae superioris M., reflect the stumps, and examine the route of the infratrochlearis N. and the dorsal oblique M. and its trochlea. Reflect the rostral stump of the lateral rectus M., and expose the scleral insertion of the ventral oblique M., which overlaps the scleral insertion of the retractor bulbi M. (Fig. 7-13, *A*).

With a chisel and a hammer, perforate the lateral wall of the alar canal and remove it. Carefully transect the maxillary N. within the alar canal, and reflect the stumps to expose the pterygopalatine ggl. (parasympathetic).

To remove the eye from the bony orbita, transect the muscles, arteries, and nerves from their attachments/origins on the pterygoid crest; free the trochlea of the dorsal oblique M.; free the ventral oblique M. from its fossa of origin; free the eyelids from the orbital septum and the palpebral ligg.; transect the zygomatic N. and the malar A. at their origin; and transect the frontal N. below the supraorbital foramen. Remove the eyeball with its accessory structures and place it on a tray.

Identify the anterior and posterior poles and the equator of the eyeball. An infinite number of parallels and meridians can be traced.

To examine the eyeball tunics and the structures within the eyeball, it is necessary to make a vertical incision through the midline of the eyeball tunics, the lens, the two chambers, and the vitreous cavity and body of the eyeball.

Examine the fibrous tunic represented by the sclera, the cornea, and the sclerocorneal limbus (angle). Identify the area cribrosa for the penetration of the optic N.

Examine the vascular tunic (represented by the choroid, including the tapetum lucidum), the ciliary body (represented by the ciliary processes and muscles), and the iris (perforated by the pupil). Identify the iridocorneal angle with the lig. pectinatum and the iridocorneal spaces.

Examine the lens, which is surrounded and supported by the ciliary zonule, also called the suspensory apparatus of the lens. The zonule originates on the ciliary segment of the retina.

Examine the two chambers of the eyeball: the anterior between the cornea and the iris and the posterior between the iris and the lens. They communicate through the pupil and are filled with aqueous humor.

Examine the inner tunic of the eyeball, which is the retina, a translucent structure easily removed from the vascular tunic (especially from the choroid). Identify the ora serrata, which separates the optic (visual) from the blind (nonvisual) part of the retina. Note the optic disc, which marks the origin of the optic N. The nonvisual part of the retina has two segments that cover the ciliary body and the iris. Examine the arborization of the central A. of the retina, the design of which is specific for each species.

Examine the vitreous body; this is the gel-like structure filling the vitreous cavity (chamber) between the retina and the posterior pole of the lens (Fig. 7-13, *B*).

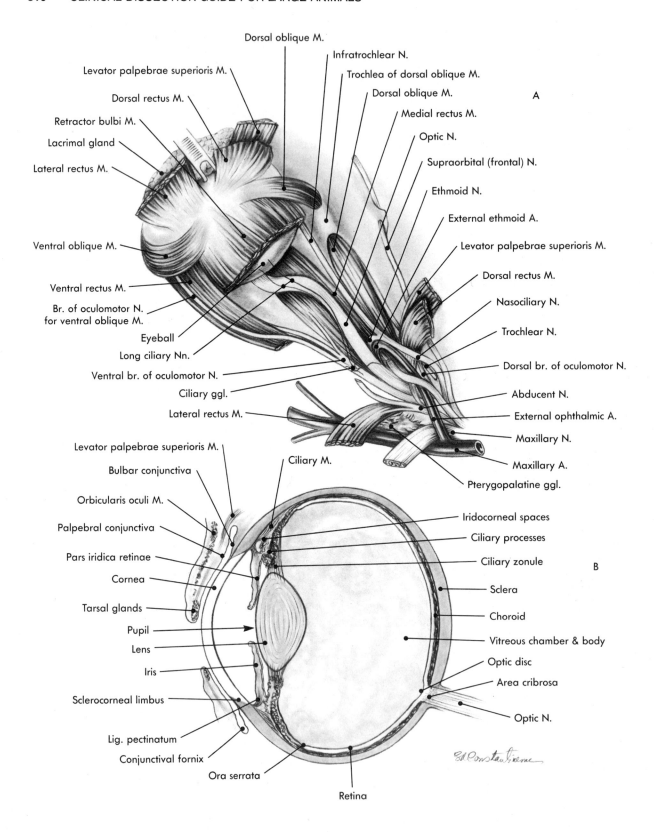

Dorsal oblique M.

Levator palpebrae superioris M.

Dorsal rectus M.

Retractor bulbi M.

Lacrimal gland

Lateral rectus M.

Ventral oblique M.

Ventral rectus M.

Br. of oculomotor N.
for ventral oblique M.

Eyeball

Long ciliary Nn.

Ventral br. of oculomotor N.

Ciliary ggl.

Lateral rectus M.

Infratrochlear N.

Trochlea of dorsal oblique M.

Dorsal oblique M.

Medial rectus M.

Optic N.

Supraorbital (frontal) N.

Ethmoid N.

External ethmoid A.

Levator palpebrae superioris M.

Dorsal rectus M.

Nasociliary N.

Trochlear N.

Dorsal br. of oculomotor N.

Abducent N.

External ophthalmic A.

Maxillary N.

Maxillary A.

Pterygopalatine ggl.

A

Levator palpebrae superioris M.

Bulbar conjunctiva

Orbicularis oculi M.

Palpebral conjunctiva

Pars iridica retinae

Cornea

Tarsal glands

Pupil

Lens

Iris

Sclerocorneal limbus

Lig. pectinatum

Conjunctival fornix

Ora serrata

Retina

Ciliary M.

Iridocorneal spaces

Ciliary processes

Ciliary zonule

Sclera

Choroid

Vitreous chamber & body

Optic disc

Area cribrosa

Optic N.

B

Fig. 7-13

A, Deep structures in orbital area. **B,** Vertical section of eye.

The following section on the dissection of the head shows the structures of the soft palate, pharynx, and larynx from the median or medial aspect (Fig. 7-14).

First examine the *nasal cavity*. Remove the nasal septum with a knife and examine its attachments. Between it and the nasal conchae, there is a space named the common meatus. Examine the communication of the common meatus with the dorsal, middle, and ventral meatus, which separate the dorsal and ventral conchae from one another and from the dorsal and ventral walls of the nasal cavity. Examine the dorsal and middle meatus. These stop at the middle concha, whereas the ventral meatus leads through the choana into the nasopharynx. Examine the three nasal conchae, their extensions, their divisions, and the details of their structure. With a knife, make a window in each concha to explore the cavities sculptured within them. Identify the conchal cells of the dorsal nasal concha, which are separated from the conchofrontal sinus by a bony wall. Identify the sinus of the ventral concha, middle nasal concha, and ethmoid conchae (see Fig. 7-9).

Examine the hard palate, which is composed of the palatine processes of the incisive and maxillary bones and the horizontal laminae of the palatine bones. Examine the *soft palate;* note its position, extension, and structure. It separates the oropharynx (ventrally) from the nasopharynx (dorsally). In the resting position, its caudal border comes in permanent contact with the rostral aspect of the epiglottis. This is a specific feature of the horse and is due to the presence of the strong hyoepiglottic lig. in this species only. Identify the palatoglossal arch connecting the soft palate with the root of the tongue, with which it borders the isthmus faucium. Identify the palatopharyngeal arch, which connects the caudal border of the soft palate to the lateral wall of the pharynx to form the intrapharyngeal opening (ostium).

Properly identify the nasopharynx between the choana, the caudal border of the soft palate, and the palatopharyngeal arch, including the pharyngeal opening of the auditory tube and the pharyngeal fornix or roof of the nasopharynx. Carefully incise the mucosa of the nasopharynx, taking as a landmark the pterygoid process of the sphenoid bone (which is palpable). Reflect the mucosa, and expose the auditory tube and the levator and tensor veli palatini Mm. (in mediolateral order). Expose the hamulus, and identify the tendon of the tensor veli palatini M., which inserts on the rostral half of the soft palate after surrounding the hamulus. Expose the insertion of the levator veli palatini M. on the caudal half of the soft palate. Dissect the palatinus M., which is a tiny paramedian structure; its role is to shorten the soft palate.

The first constrictor of the pharynx (the pterygopharyngeal and palatopharyngeal Mm.) is exposed on the lateral aspect and crosses the levator veli palatini M. Identify the muscles on the medial aspect (see Figs. 7-9 and 7-10).

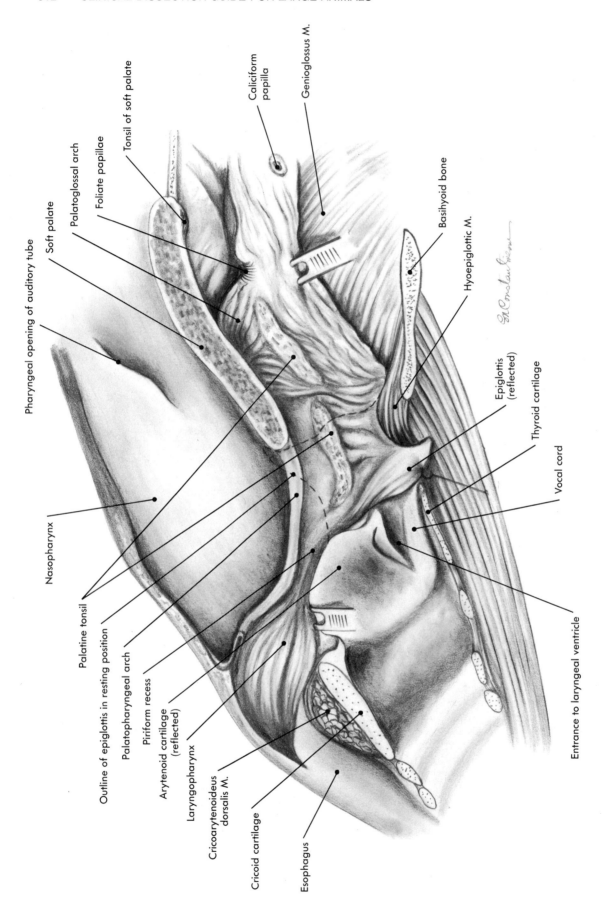

Caliciform papilla

Genioglossus M.

Tonsil of soft palate

Foliate papillae

Palatoglossal arch

Soft palate

Pharyngeal opening of auditory tube

Basihyoid bone

Hyoepiglottic M.

Epiglottis (reflected)

Thyroid cartilage

Vocal cord

Entrance to laryngeal ventricle

Nasopharynx

Palatine tonsil

Outline of epiglottis in resting position

Palatopharyngeal arch

Piriform recess

Arytenoid cartilage (reflected)

Laryngopharynx

Cricoarytenoideus dorsalis M.

Cricoid cartilage

Esophagus

Pharynx and larynx (medial aspect).

Fig. 7-14

To study the *oral cavity* in an organized manner, it is necessary first to distinguish the oral cavity proper from the vestibule. The oral cavity proper is located medial to and surrounded by the teeth; it includes the tongue, mucosa, and some of the salivary glands. The vestibule of the oral cavity is the space between the teeth and the lips and cheeks. In the resting position (with the mouth closed and the masticatory [occlusal] surface of the teeth against each other), communication is possible between the vestibule and the oral cavity proper, caudal to the last molars and through the diastema. Within the vestibule, some of the salivary ducts course and/or open.

First examine the structures located within the oral cavity proper. Examine the mucosa of the hard palate (with the palatine ridges and the palatin raphe) and the incisive papilla located caudal to the first pair of incisors. Examine the ventral aspect of the soft palate, the tonsil of the soft palate, the palatoglossal arch, and the palatine tonsil.

Notice that the palatine tonsil is located on the lateral wall of the oropharynx between the palatoglossal arch and the epiglottis.

Examine the tongue with the apex, body, and root. The apex lies on the corpus mandibulae, overlapping the sublingual caruncle and the orobasal organ. The mucosa of the ventral aspect of the apex in transition to the mucosa of the corpus mandibulae forms the frenulum linguae, an unpaired vertical fold of the mucosa.

Remember! The sublingual caruncle is a fold of the mucosa of the corpus mandibulae and covers and protects the opening of the mandibular duct. The orobasal organ is a paired rudiment of the rostral sublingual glands of reptiles and consists of two ducts that open caudal to the first lower incisors.

The palatoglossal arches, the soft palate, and the root of the tongue border the narrow communication between the oral cavity proper and the oropharynx called the isthmus faucium.

Remember! The oropharynx is also called the fauces.

Reflect the tongue medially (toward you). Still within the oral cavity proper, examine the lateral sublingual recess. Dissect and reflect the mucosa of that area, and expose the continuation of the lingual N. paralleling the styloglossus M.; the mandibular duct paralleling the hypoglossal N.; the sublingual polystomatic salivary glands; the geniohyoideus and mylohyoideus Mm.; and the rostral belly of the digastric M. Place the tongue back into the oral cavity proper, and examine the genioglossus M. with its fibers oriented in three directions: rostrally, dorsally, and caudally. Reflect the tendon of the genioglossus M. to expose part of the hyoglossus, geniohyoideus, and mylohyoideus Mm. and the lingual A.N., which run between the genioglossus and hyoglossus Mm. Make a longitudinal incision in the genioglossus M. to better expose the lingual A.

Reflect the tongue again, and examine the whole mucosa and the four kinds of papillae: filiform papillae (on the dorsal aspect); fungiform papillae (along the borders of the tongue); two vallate papillae (on the dorsal aspect and rostral to the root of the tongue); and two foliate papillae (on each side of the glossoepiglottic fold). The mucosa of the root of the tongue continues onto the rostral aspect of the epiglottis and forms a vertical median fold, called the glossoepiglottic fold. On the dorsal aspect of the root of the tongue, the lingual tonsil is visible.

Identify the median/sagittal section of the basihyoid bone and the lingual process. From the basihyoid toward the cervical area, identify the omohyoideus M. as the most ventral structure. Between the basihyoid and the rostral aspect of the epiglottis, identify two unpaired structures: the hyoepiglottic M. (rostrally) and the hyoepiglottic lig. (caudal and parallel to the hyoepiglottic M.). The glossoepiglottic fold is caused by the prominence of the hyoepiglottic M. (see Fig. 7-14).

Examine the *teeth*. The classification, evolution, dental formula, and estimation of age of the specimen are not described in this dissection guide. Only the morphology of the teeth is discussed.

There are three symmetrical incisors (upper and lower): I_1 or central, I_2 or intermediate, and I_3 or corner incisors. There are three symmetrical molars (both upper and lower).

First, examine the incisors. There are fundamental differences between permanent and deciduous incisors. Examine a deciduous incisor (Fig. 7-15, *A* to *C*), both intact and sectioned in the long axis. Each tooth has a vestibular surface, a lingual surface, two contact surfaces (except the corner incisors), and an occlusal surface. Each incisor is curved, with the concave surface toward the tongue. A rostral view of an incisor shows a broad crown, a narrow root, and an obvious transitional area, the incisor's neck.

Notice that the vestibular surface of the crown is named the clinical crown and the part of the root embedded in the dental alveolus is named the clinical root.

Examine the dental infundibulum, which is the invagination of enamel on the occlusal surface.

On a sectioned incisor, examine the external enamel surrounding the crown, neck, and root and the internal enamel lining the infundibulum. The internal and external enamels are continuous with one another until the tooth becomes worn. When the enamel of the occlusal surface that connects the external enamel with the internal enamel is worn off, the result is two visible areas of enamel: an elliptical or circular peripheral band and an inner one surrounding the infundibulum. They are separated by the dentine. The external enamel is protected by a thin layer of cement, which also partially lines the internal enamel within the infundibulum. Examine the dentine and the cement, and note the large dental cavity.

Next, examine a permanent incisor (Fig. 7-15, *D* to *F*), both intact and sectioned in the long axis. The permanent incisor is a curved prismatic (cylindrical) tooth without an apparent neck. A frontal view shows an even surface with the same width from the occlusal surface toward the root. Examine the occlusal surface, the infundibulum, and the two areas of enamel. A sectioned permanent incisor (Fig. 7-16) shows that the external enamel is much shorter than in a deciduous incisor; it does not extend distally, to the top of the root. The dental cavity is very narrow. When the dental cavity reaches the occlusal surface, it resembles a star (the dental star).

Both the crown and root of the canines are conical. Fully-developed canine teeth are regularly present in the male horse.

The premolars and molars are similar to one another. They are straight, prismatic teeth with multiple roots and a very discrete neck. The four surfaces (vestibular, lingual, and contact surfaces) border a square-shaped occlusal surface. The first premolar and the last molar teeth are different (having only one contact surface and an occlusal surface that is not regularly square shaped). The vestibular and lingual surfaces bear vertical ridges and grooves. The occlusal surface is provided with pronounced folds and ridges of enamel (both external and internal) (Fig. 7-15, *G* and *H*).

Return to the oropharynx and, through the intrapharyngeal opening (ostium), enter the laryngopharynx. The laryngopharynx is the common compartment of the pharynx for the respiratory and digestive apparatus. It is located caudal to the palatopharyngeal arch. The nasopharynx is the compartment of the pharynx located rostral to the palatopharyngeal arch and dorsal to the soft palate, whereas the oropharynx is located ventral to the soft palate. Consequently, two intrapharyngeal openings (communications) can be considered: one between the nasopharynx and laryngopharynx and another between the oropharynx and laryngopharynx.

Examine the aryepiglottic fold, connecting the epiglottis to the corniculate process of the arytenoid cartilage. The piriform recess, which is the narrow groove between the oropharynx and the esophagus, is bordered by the aryepiglottic fold and the palatopharyngeal arch.

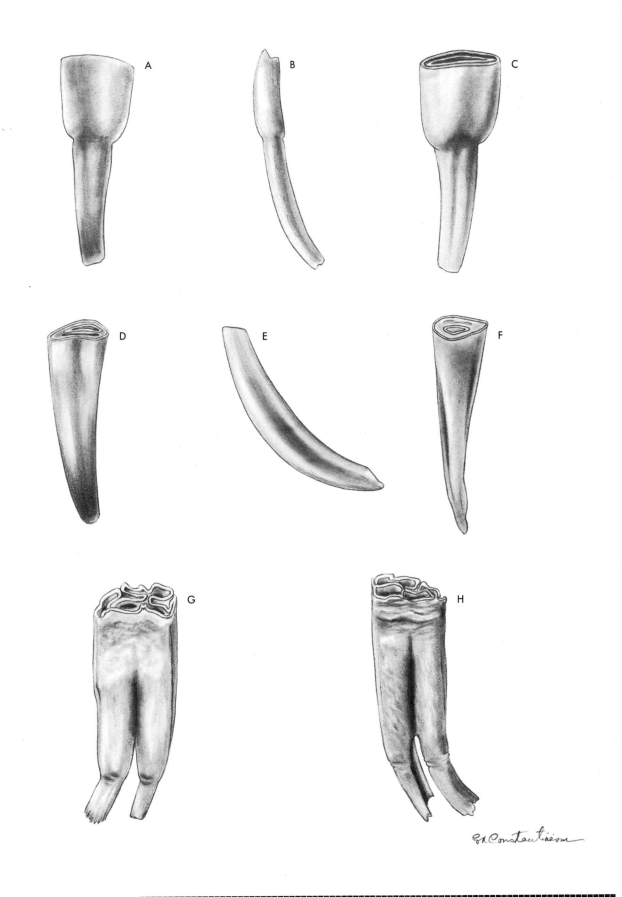

First right lower deciduous incisor: **A,** Labial aspect; **B,** Left side; **C,** Lingual aspect. First right lower permanent incisor: **D,** Labial aspect; **E,** Left side; **F,** Lingual aspect. **G,** Third right lower permanent premolar (vestibular aspect). **H,** Second left lower permanent molar (vestibular aspect).

Fig. 7-15

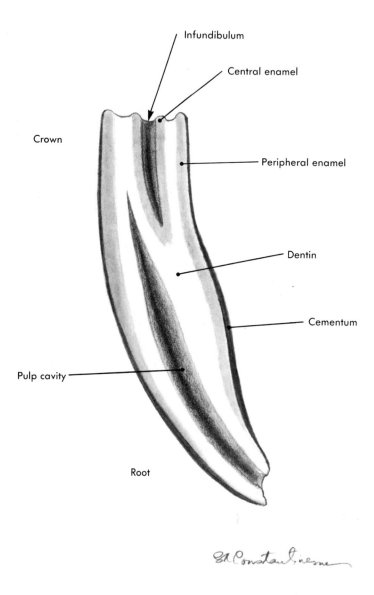

Fig. 7-16   Vertical section in permanent incisor.

Examine the following: the median section of the thyroid cartilage (the body) and the cricothyroid membrane; the median section of the cricoid cartilage (both the plate-lamina and the ring-arcus); and the first tracheal rings. Examine the median laryngeal recess and the two folds bordering the laryngeal ventricle. The rostral fold is the vestibular fold, whereas the caudal fold is the vocal fold. Two pairs of ligaments and muscles relate the arytenoid cartilages to the thyroid cartilage. The first is the vestibular lig. and the ventricular M., which is covered by the vestibular fold of the mucosa (and located rostral to the laryngeal ventricle); the second is the vocal lig. and muscle, covered by the vocal fold (located caudal to the laryngeal ventricle). Dissect all these structures (Fig. 7-17).

Notice that in the horse only, the vestibular lig. connects the arytenoid cartilage to the cuneiform process of the epiglottis (see Fig. 7-17).

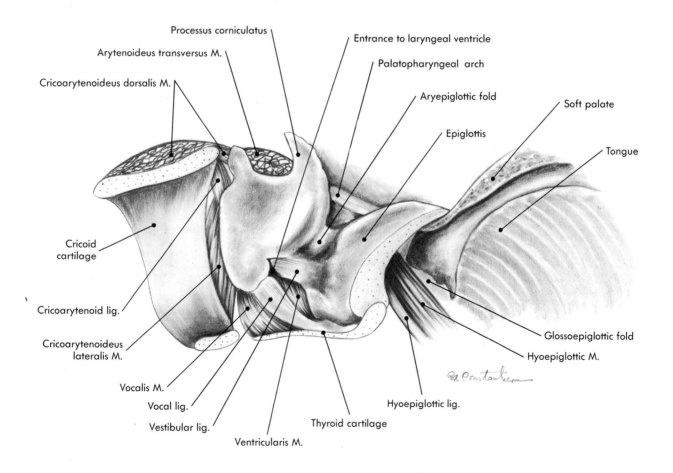

Larynx after removing mucosa (left medial aspect).   **Fig. 7-17**

Remove the hyoid apparatus with the tongue and larynx attached. Free the stylohyoid bone from the hyoid process of the temporal bone by transecting the tympanohyoid cartilage. Section the occipitohyoideus M. and the attachments of the rostral belly of the digastric, mylohyoideus, geniohyoideus, and genioglossus Mm. from the mandible. Section the attachment of the lingual mucosa to the oral cavity and oropharynx. Section the attachments of the dorsal walls of the esophagus and pharynx, transect the origin of the cranial laryngeal N., and free the larynx from any other attachments (Fig. 7-18, *B*). Remove the hyoid apparatus, tongue, and larynx; place them on a tray. Turn all these structures so that their lateral sides are up. Make a window within the wing (lamina) of the thyroid cartilage, and expose the cricoarytenoideus lateralis, ventricularis, and vocalis Mm. and the bottom of the laryngeal ventricle (Fig. 7-18, *A*). Dissect the cranial and caudal laryngeal Nn. on the inside of the larynx.

After removing the structures already dissected, free the hyoid bones and cartilages and the laryngeal cartilages from muscles, ligaments, membranes, and mucosa. Examine them separately.

Because the *paranasal sinuses* differ from species to species, it is necessary to briefly enumerate them. The horse is provided with the following paired paranasal sinuses: the sinus of the middle concha, the sinus of the ventral concha, the rostral maxillary sinus, the caudal maxillary sinus, the sphenopalatine sinus (the palatine sinus continuous with the sphenoid sinus), and the conchofrontal sinus (the dorsal conchal sinus continuous with the frontal sinus).

The approach to the paranasal sinuses depends on the specific sinus. In addition, some of them have clinical importance and are susceptible to investigation or surgical intervention.

Starting with those that are clinically significant, the rostral and caudal maxillary and the conchofrontal sinuses will be examined. Each of these three sinuses can be investigated by an external approach.

The maximum extension of the quadrilaterally shaped area of the combined rostral and caudal maxillary sinuses is outlined between the medial angle of the eye, the infraorbital foramen, the facial tubercle, and the facial crest. An oblique osseous septum separates the sinuses; it is located between the medial angle of the eye and (approximately) the middle of the facial crest. Sculpture these two sinuses with a chisel and hammer and explore their compartments and communications.

Caution! The extension of the maxillary sinuses depends on the age of the specimen. The older the specimen, the larger the sinuses. The capacity of the sinuses increases as the molars shorten.

Notice that the maxillary sinuses are lodged within the maxilla, lacrimal, zygomatic, and ethmoid bones.

Identify the infraorbital (osseous) canal, and examine the extension of the sinuses in the dissected specimen. In old specimens, the maxillary sinuses extend ventral to the facial crest, which is not relevant for the surgical approach.

Remember! In the area ventral to the facial crest, the masseter and buccinator Mm. and facial vessels and nerves are located.

Notice that in the horse both the rostral and caudal maxillary sinuses communicate with the middle nasal meatus through the nasomaxillary aperture and that the caudal maxillary sinus also communicates with the conchofrontal, middle conchal, and sphenopalatine sinuses.

Notice that the rostral maxillary sinus communicates with the middle meatus through the nasomaxillary aperture via the ventral conchal sinus.

Notice that the nasomaxillary aperture can be probed on your specimen. It is located within the middle nasal meatus toward the bottom of the nasal cavity, rostral to the middle nasal concha.

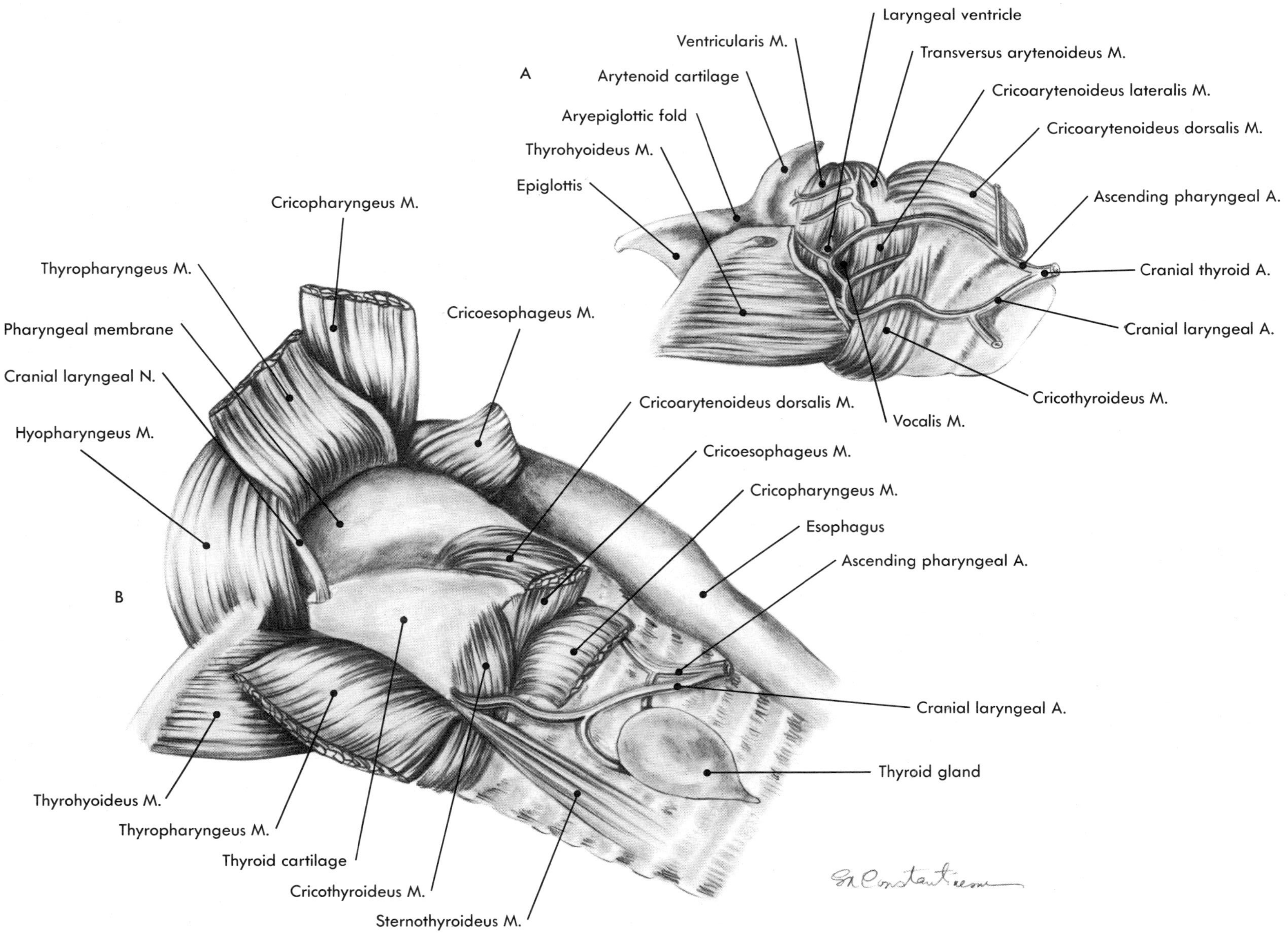

A, Arterial supply of larynx. **B,** Deep larynx (lateral aspect).

Fig. 7-18

A

Laryngeal ventricle

Transversus arytenoideus M.

Cricoarytenoideus lateralis M.

Cricoarytenoideus dorsalis M.

Ascending pharyngeal A.

Cranial thyroid A.

Cranial laryngeal A.

Cricothyroideus M.

Vocalis M.

Ventricularis M.

Arytenoid cartilage

Aryepiglottic fold

Thyrohyoideus M.

Epiglottis

Cricoarytenoideus dorsalis M.

Cricoesophageus M.

Cricopharyngeus M.

Esophagus

Ascending pharyngeal A.

Cranial laryngeal A.

Thyroid gland

Cricoesophageus M.

Cricopharyngeus M.

Thyropharyngeus M.

Pharyngeal membrane

Cranial laryngeal N.

Hyopharyngeus M.

B

Sternothyroideus M.

Cricothyroideus M.

Thyroid cartilage

Thyropharyngeus M.

Thyrohyoideus M.

The conchofrontal sinus occupies the area between the orbit and the interfrontal suture and extends rostrally into the area of the nasal bone. The extension of this sinus depends on the age of the specimen. With a chisel and hammer, sculpture this sinus; identify its two main components: the dorsal conchal sinus and the frontal sinus.

Notice that the horse is the only species in which these two sinuses communicate. Both sinuses (together) are considered a unique conchofrontal sinus, which communicates with the caudal maxillary sinus through the frontomaxillary aperture.

Explore the rostral, medial, and caudal recesses of the frontal sinus and the extentions of both the dorsal conchal and frontal sinuses.

At the border between the maxillary sinuses and the conchofrontal sinus is the nasolacrimal canal. When trephining, be very careful. Take into consideration the age of the specimen and the extension of the sinuses (Fig. 7-19).

The middle conchal sinus, which is enclosed in the middle nasal concha (from the ethmoid bone), and the sphenopalatine sinus cannot be approached externally. The sphenopalatine sinus is located within the presphenoid and palatine bones.

Turn the head so that the sectioned side is up. Examine the median/sagittal aspect of the *brain* and *spinal cord*.

Find the half of the head with the most intact meninges. Examine the leptomeninx (the pia mater and the arachnoid) and the pachymeninx (the dura mater). The dura mater, which separates the two hemispheres of the telencephalon and which is located in the longitudinal cerebral fissure, is the falx cerebri. The cerebrum is separated from the cerebellum by the tentorium cerebelli membranaceum. The hypophysis is protected and anchored in the hypophyseal fossa by the diaphragma sellae (Fig. 7-20, *A*).

Systematically identify the structures belonging to the telencephalon, diencephalon, mesencephalon, metencephalon, and myelencephalon (Fig. 7-20, *B*), as well as the arteries and veins (Fig. 7-20, *C*).

Identify the following structures of the *telencephalon:* the olfactory bulb, which sends the fila olfactoria (cranial nerve I) through the cribriform plate; the corpus callosum and rostral commissure (the white matter connecting the two hemispheres); and the fornix (fibers connecting the hippocampus to the mammillary body).

The rostral limit of the body of the corpus callosum is known as the genu corporis callosi; it bends caudally and ends as the rostrum corporis callosi, a beak-shaped structure connecting the genu corporis callosi to the rostral commissure. The caudal extent of the corpus callosum is the splenium corporis callosi. The fornix is a unique X-shaped structure; its rostral arms are called the columns of the fornix, and the caudal arms are called the crura of the fornix. The columns and the crura meet together to form the body. The corpus callosum represents the roof of both the first and second lateral ventricles; the fornix represents part of the floor of these ventricles and the roof of the third ventricle.

Notice that the septum pellucidum is a median, vertical wall of white matter between the corpus callosum (dorsally) and the fornix (ventrally) that separates the two lateral ventricles from each other.

Identify the following structures of the *diencephalon:* the interthalamic adhesion (a circular structure on the section), surrounded by the third ventricle; the interventricular foramen, allowing the communication between the lateral ventricles and the third ventricle; the epiphysis (pineal body), anchored by the habenula, which is located dorsal to the interthalamic adhesion; the hypophysis (pituitary gland), anchored by the infundibulum to the hypothalamus (the tuber cinereum [gray tubercle]); the optic chiasma and optic N. (cranial nerve II), located rostral to the hypophysis; and the mammillary body (located caudal to the hypophysis).

Identify and explore the recesses of the third ventricle. Examine the choroid plexus of the third ventricle.

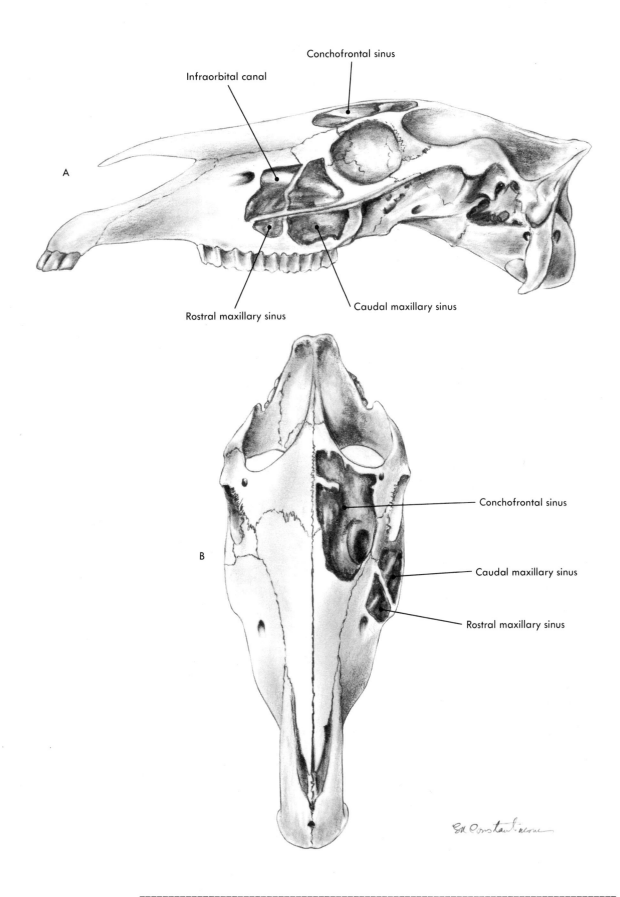

Infraorbital canal

Conchofrontal sinus

A

Rostral maxillary sinus

Caudal maxillary sinus

B

Conchofrontal sinus

Caudal maxillary sinus

Rostral maxillary sinus

A, Paranasal sinuses (lateral aspect). **B,** Paranasal sinuses (frontal aspect).

Fig. 7-19

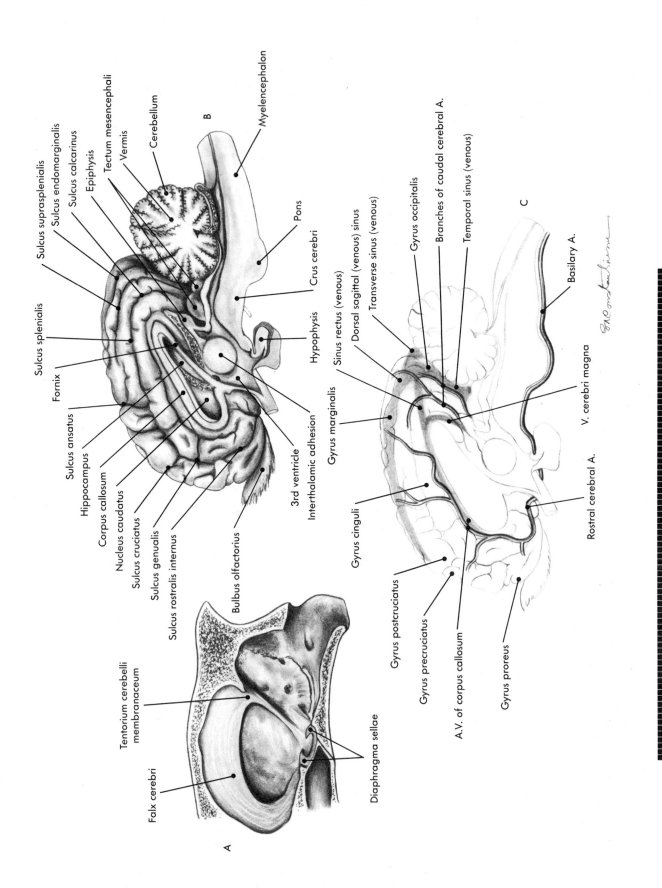

Myelencephalon

B

Cerebellum

Tectum mesencephali

Vermis

Epiphysis

Sulcus calcarinus

Sulcus endomarginalis

Sulcus suprasplenialis

Pons

Sulcus splenialis

Crus cerebri

Fornix

Hypophysis

Sulcus ansatus

Hippocampus

Corpus callosum

Nucleus caudatus

3rd ventricle

Sulcus cruciatus

Interthalamic adhesion

Sulcus genualis

Sulcus rostralis internus

Bulbus olfactorius

Gyrus marginalis

Sinus rectus (venous)

Dorsal sagittal (venous) sinus

Transverse sinus (venous)

Gyrus occipitalis

Branches of caudal cerebral A.

Temporal sinus (venous)

C

Basilary A.

Gyrus cinguli

V. cerebri magna

Rostral cerebral A.

Gyrus postcruciatus

Gyrus precruciatus

A.V. of corpus callosum

Gyrus proreus

Tentorium cerebelli membranaceum

Falx cerebri

Diaphragma sellae

A

Fig. 7-20   A, Encephalic dura mater. B, Encephalon (sagittal aspect). C, Arterial supply of encephalon (sagittal aspect).

Remove the septum pellucidum, and examine the lateral ventricle. Identify the caudate nucleus (rostrolaterally) and the hippocampus (caudomedially) on the floor of the lateral ventricle. They are separated by the choroid plexus of the lateral ventricle.

Notice that the caudate nucleus is the dorsal part of corpus striatum (the striate body), whereas the hippocampus is morphologically part of the olfactory brain.

Explore the lateral ventricle.

Notice that the hypophysis has two main parts: the adenohypophysis (the glandular part) rostrally and the neurohypophysis (the nervous part) caudally. The latter is attached to the hypothalamus by the infundibulum. The intermediate part of the hypophysis is sometimes visible.

On a median section, the *mesencephalon* consists of the tegmentum including the tectum (dorsally) and the cerebral crus (ventrally), which are separated by the mesencephalic aqueduct (the junction between the third and fourth ventricles). The tectum is represented by the rostral colliculus and caudal colliculus. The transitional area between the tectum and the epiphysis is the caudal commissure. From the ventral aspect of the cerebral crus originates the oculomotor N. (cranial nerve III).

The *metencephalon* is represented by the pons (ventrally) and the cerebellum (dorsally). The cerebellum also covers the medulla oblongata, which is continuous caudally with the pons. The fourth ventricle is surrounded by the pons and the medulla oblongata (ventrally, representing the floor) and the cerebellum (dorsally, on the roof). The roof of the fourth ventricle is a very thin membrane that parallels the ventral aspect of the cerebellum. The membrane is composed of the rostral and caudal medullary vela. The rostral medullary velum belongs to the metencephalon.

The median/sagittal section of the cerebellum shows the nine lobules of the vermis: the lingula, lobulus centralis, culmen, declive, folium vermis, tuber vermis, pyramis, uvula, and nodulus (starting from over the fourth ventricle and moving in a rostrolorsal, then clockwise, direction).

The *myelencephalon* is represented by the medulla oblongata and the caudal medullary velum. Examine the choroid plexus of the fourth ventricle. Examine the continuation of the fourth ventricle with the central canal of the spinal cord.

On the midsection, identify the veins (and the dural venous sinuses) and the arteries of the brain, such as the dorsal sagittal (venous) sinus within the falx cerebri; the sinus rectus and its branches (the V. of the corpus callosum, the V. cerebri magna, represented by the internal cerebral Vv., and so on); the transverse sinus and the temporal sinus; and the cavernous sinus, enclosing a segment of the internal carotid A. (which surrounds the hypophysis). Also identify the rostral cerebral A., the A. of the corpus callosum, and branches of the caudal cerebral A.

Examine the gyri and sulci (grooves) of the medial aspect of the hemisphere and the distribution of the white and gray matter in the cerebellum (see Fig. 7-20, *B*).

Carefully remove the brain and the spinal cord from the cranial cavity and the vertebral canal, respectively. With a fine scissors, free the hypophysis from the diaphragma sellae. Place the brain and the spinal cord on a tray with the ventral side up, and identify the structures in the same order they were examined on the midsection (Fig. 7-21, *A*).

However, before proceeding, observe that the cranial cavity is divided into two compartments (cerebral and cerebellar) by the tentorium cerebelli osseum.

On the ventral aspect of the *telencephalon,* identify the olfactory bulb; the olfactory peduncle; the lateral and medial olfactory tracts bordering the olfactory trigone; the piriform lobe; and the two medial and lateral olfactory grooves. The lateral olfactory groove separates the olfactory brain from the hemisphere.

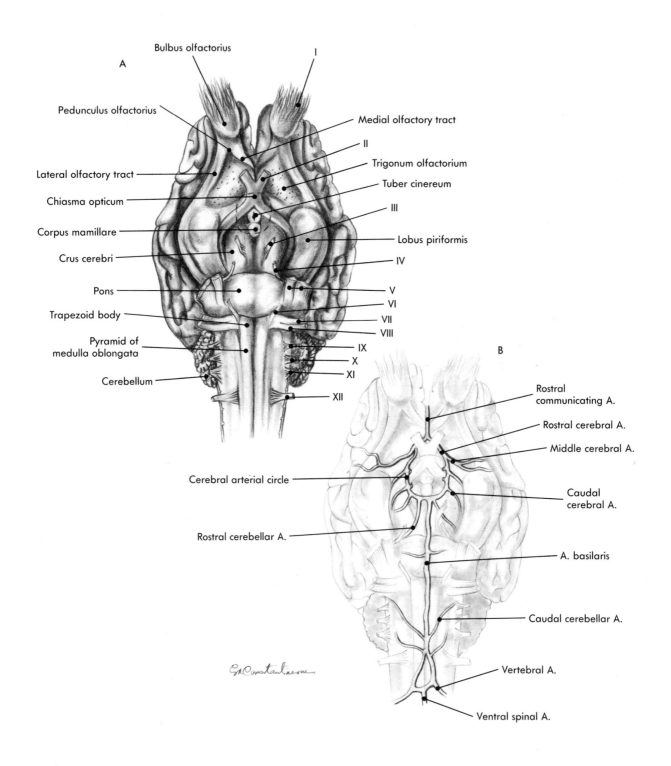

A

Bulbus olfactorius

I

Pedunculus olfactorius

Medial olfactory tract

II

Trigonum olfactorium

Lateral olfactory tract

Tuber cinereum

Chiasma opticum

III

Corpus mamillare

Lobus piriformis

Crus cerebri

IV

Pons

V

VI

Trapezoid body

VII

VIII

Pyramid of medulla oblongata

IX

X

XI

Cerebellum

XII

Rostral communicating A.

Rostral cerebral A.

B

Middle cerebral A.

Cerebral arterial circle

Caudal cerebral A.

Rostral cerebellar A.

A. basilaris

Caudal cerebellar A.

Vertebral A.

Ventral spinal A.

Fig. 7-21   **A,** Ventral structures of the brain. **B,** Arterial and venous supply of brain (ventral aspect).

The *diencephalon* is represented by the optic N., optic chiasma, and optic tract; the gray tubercle (tuber cinereum), with the infundibulum and the hypophysis; and the mammillary body.

The *mesencephalon* is represented on the ventral aspect only by the cerebral crus (separated from the symmetrical structure by the interpeduncular fossa), the oculomotor N., and the trochlear N. (cranial nerve IV), which is usually regarded as a dorsal structure, originating from tectum (the dorsal major component of the mesencephalon) and surrounding the mesencephalon in a ventral direction.

The ventral *metencephalon* exposes the pons with the basilar groove and the trigeminal N. (V) at the lateral extension. The lateral lobe of the cerebellum exceeds the limit of the medulla oblongata laterally.

The ventral aspect of the *medulla oblongata* bears the last seven (out of twelve) cerebral nerves. Identify the following structures: the pyramid (along the ventral median fissure), the trapezoid body (parallel with the caudal border of the pons), and the ventrolateral and dorsolateral sulci.

The abducent N. (VI) and hypoglossal N. (XII) leave the medulla oblongata from the ventrolateral sulcus, whereas the glossopharyngeal N. (IX), vagus N. (X), and accessory N. (XI) leave the central nervous system from the dorsolateral sulcus.

Remember! The accessory N. has two roots: cranial and spinal.

The intermediofacial N. (VII) and vestibulocochlear N. (VIII) leave the lateral extent of the trapezoid body. On the ventral aspect, the choroid plexus of the fourth ventricle is also visible between the lateral border of the medulla oblongata and the cerebellum.

Identify the following vessels: the ventral spinal A. (located in the ventral median fissure of the spinal cord) and the anastomosis with the vertebral A.; the A. basilaris; the caudal cerebellar A.; the rostral cerebellar A.; the caudal cerebral A.; the middle cerebral A.; the rostral cerebral A.; the two caudal and rostral communicating Aa.; and the cerebral arterial circle (Fig. 7-21, *B*).

Turn the brain so that the lateral side is up. Identify the olfactory brain and the hemisphere, which are the only components of the *telencephalon* seen from a lateral view. The olfactory brain is represented by the olfactory bulb, the olfactory peduncle, the olfactory tracts, the olfactory trigone, and the piriform lobe. On the lateral aspect of the hemisphere, examine the gyri, the sulci, and the fissures (Fig. 7-22, *A*).

The *diencephalon* is represented by the optic N. and optic chiasma; the hypophysis with the infundibulum and gray tubercle; and the mammillary body.

The *mesencephalon* shows the crus cerebri, the caudal colliculus, the oculomotor N., the trochlear N., and the rostral cerebellar peduncle.

The *metencephalon* is represented by the pons, the trigeminal N., the middle cerebellar peduncle, and the lateral lobe of the cerebellum.

The *myelencephalon* contains the trapezoid body with the intermediofacial and vestibulocochlear Nn.; the pyramid; the ventrolateral sulcus with the abducent and hypoglossal Nn.; the dorsolateral sulcus with the glossopharyngeal, vagus, and accessory Nn.; and the caudal cerebellar peduncle.

Before removing the cerebellum, examine the dorsal aspect of the hemisphere and identify the sulci and gyri; some of them are continuous on the lateral or medial sides of the hemisphere (Fig. 7-22, *B*).

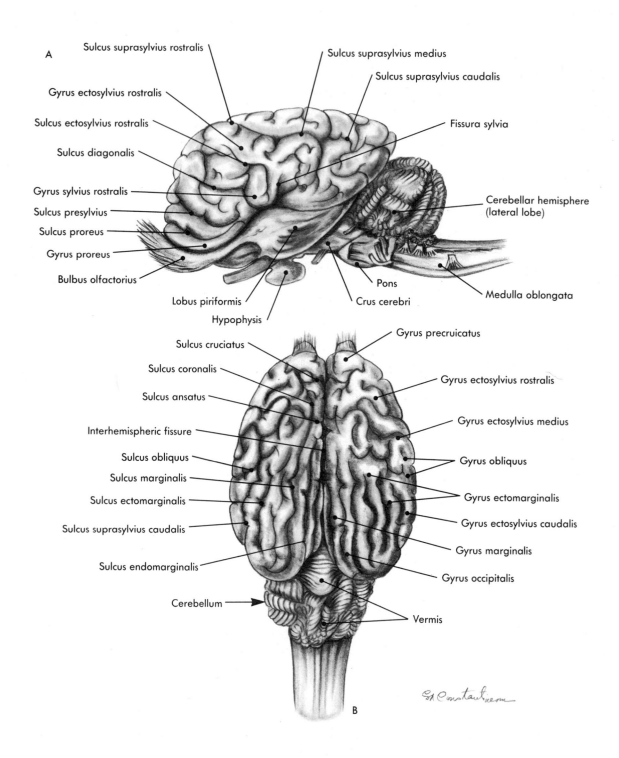

Fig. 7-22

A, Brain (lateral aspect). **B,** Brain (dorsal aspect).

Remove the cerebellum by transecting the three cerebellar peduncles: rostral (from the mesencephalon), middle (from the pons), and caudal (from the myelencephalon). The dorsal aspect of the myelencephalon, including the floor of the fourth ventricle, is exposed. To better expose the fourth and lateral ventricles, remove the cerebral hemisphere from the striate body, hippocampus, and diencephalon (Fig. 7-23, *A*).

The floor of the lateral ventricle was previously examined after the removal of the septum pellucidum.

On the dorsal aspect of the medulla oblongata, identify the median sulcus, which continues on the floor of the fourth ventricle, and the gracilis and cuneatus fascicles.

On the transverse section of the striate body (corpus striatum), identify the caudate and lentiform nucleus (Fig. 7-23, *B*).

Using another brain, make a horizontal section through the hemisphere at the level of the floor of the lateral ventricle and identify the caudate nucleus, the internal capsule, and the claustrum (structures of the striate body) (Fig. 7-23, *C*).

Make a second horizontal section (parallel with the previous) at the level of the rostral commissure, and identify the caudate and lentiform nucleus represented by the putamen and the pallidum, also structures of the striate body (Fig. 7-23, *D*).

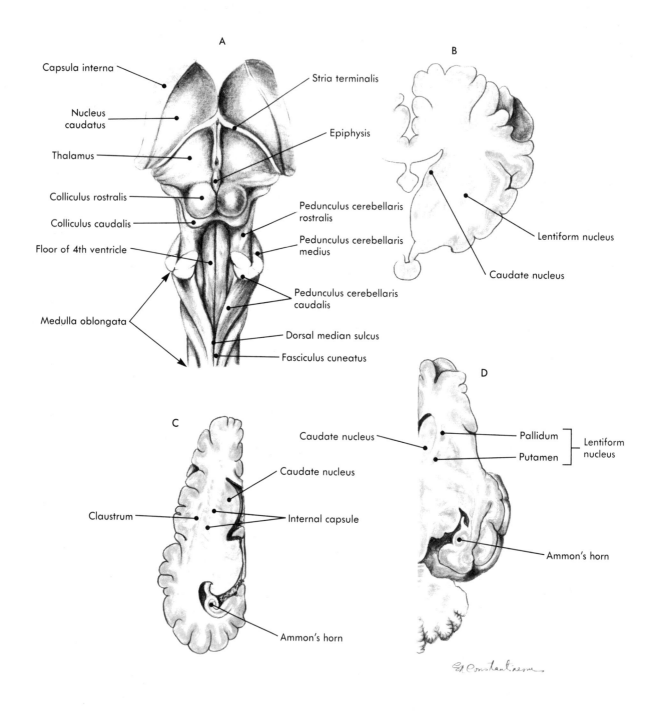

Fig. 7-23   **A,** Brain stem (dorsal aspect). **B,** Structures of striate body (transverse section). **C,** Structures of striate body (longitudinal section through floor of lateral ventricle). **D,** Structures of striate body (longitudinal section through rostral commissure).

Compare the skull of a large ruminant with the skull of a horse and then attempt to identify the same structures of the horse's head (described in the previous section) in the large ruminant. Although many structures are similar, some characteristics are specific to the large ruminants.

On the *skull* (in a lateral view), (Fig. 7-24), no superior incisors are present in ruminants (including the small ruminants); the infraorbital foramen is in a rostral position, located dorsal to the first premolar; the facial crest is irregular, (therefore the facial tubercle looks like an isolated structure); the temporal fossa is deep, narrow, and bordered dorsally by the temporal line; the parietal bone is reduced and lies on the bottom of the temporal fossa; a cornual process is present, even in dehornated specimens; the tympanic bulla is well developed and extends to the level of the distal end of the jugular process of the occipital bone; and the muscular process of the temporal bone is well developed. No canine teeth are present in ruminants!

Within the orbita, identify the three foramina of the pterygopalatine fossa, the well-developed and irregular pterygoid crest, and the following foramina (which are located only rostral to this crest in comparison with the horse). In dorsoventral order, those foramina are the orbital opening of the supraorbital canal; the ethmoidal foramen; the optic canal; the trochlear foramen; and the foramen orbitorotundum. (Only the latter three are located within the pterygoid fossa.)

Remember! The three foramina of the pterygopalatine fossa are the sphenopalatine, maxillary, and caudal palatine foramina (in dorsoventral order in ruminants).

In a frontal view (Fig. 7-25), the frontal bones exceed the rostral limits of the orbitae and occupy the rest of the skull up to and including the intercornual protuberance; the supraorbital foramen is continued by the supraorbital canal (inside of the frontal bone) and lies within the supraorbital sulcus.

In a ventral view (the base of the skull) several characteristics should be noted: again, there are no superior incisors; the major palatine foramina are very close to each other; the choanae are narrow; a well-developed lacrimal bulla is present inside each orbit; there is no foramen lacerum in ruminants (including the small ruminants); of the structures of the foramen lacerum in the horse, only the oval incisure is present in the ox (as the oval foramen); and a very narrow petrooccipital fissure (between the tympanic part of the temporal bone and the basioccipital) communicates caudally with the jugular foramen. Identify the tympanic bulla and the muscular process of the temporal bone.

Remember! No canine teeth are present in ruminants!

Compare the mandible of the ox with that of the horse.

Notice that the ruminants have eight incisors (only the inferior), instead of six superior and six inferior as in the horse.

On the *head* examine the planum nasolabiale, the rostral structure common to the upper lip and nostrils and specific to the large ruminants. It contains glands but not hair. The nasal cartilages are not palpable. Examine the relationship between the horns (or cornual processes) and the ears.

Compare the brain cavity of the ox with that of the horse.

Split the head of your specimen into two halves and skin and dissect them, following the same procedure as described in the dissection of the horse (Fig. 7-26).

Large ruminants
■■■■■■■■■■■■■■■■■■■

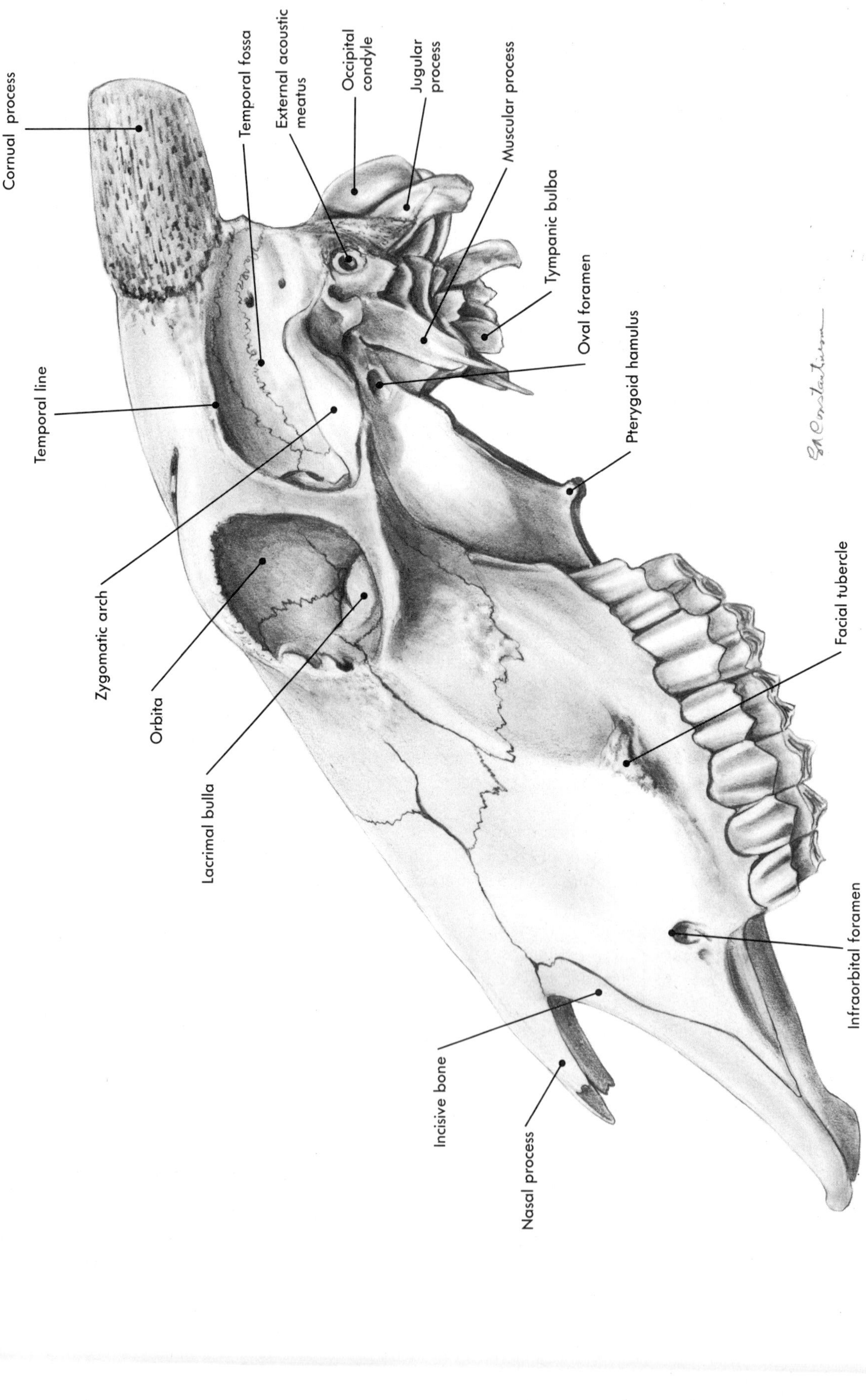

Fig. 7-24

Cornual process

Temporal fossa

External acoustic meatus

Occipital condyle

Jugular process

Muscular process

Temporal line

Tympanic bulba

Oval foramen

Pterygoid hamulus

Zygomatic arch

Orbita

Facial tubercle

Lacrimal bulla

Infraorbital foramen

Incisive bone

Nasal process

Skull (lateral aspect).

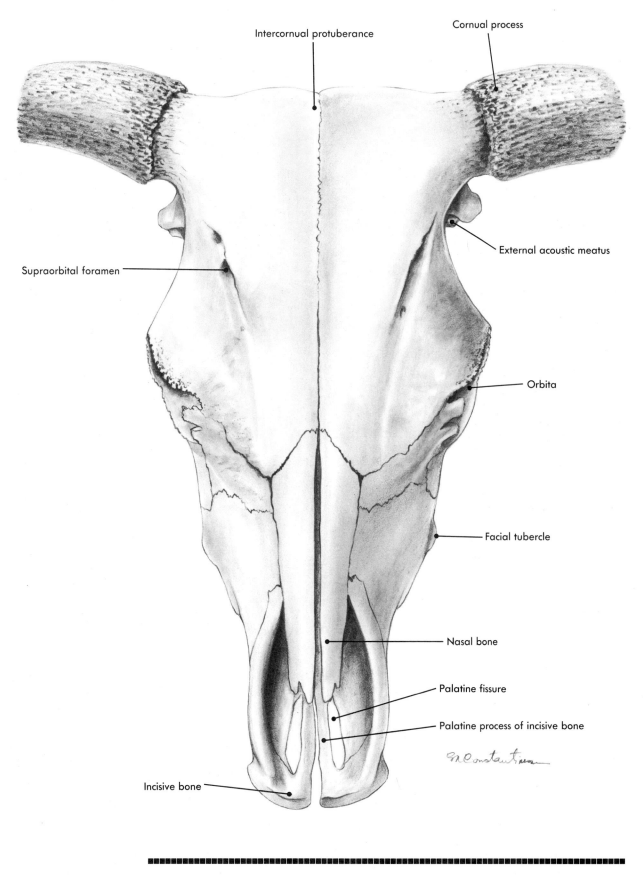

Intercornual protuberance

Cornual process

External acoustic meatus

Supraorbital foramen

Orbita

Facial tubercle

Nasal bone

Palatine fissure

Palatine process of incisive bone

Incisive bone

Skull (frontal aspect).

Fig. 7-25

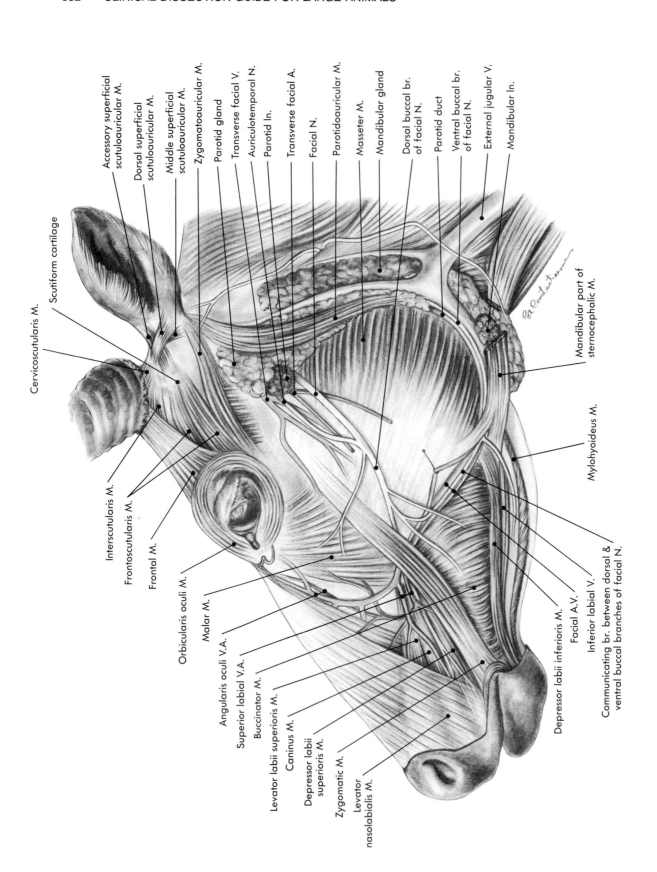

Accessory superficial scutuloauricular M.

Dorsal superficial scutuloauricular M.

Middle superficial scutuloauricular M.

Zygomatoauricular M.

Parotid gland

Transverse facial V.

Auriculotemporal N.

Parotid ln.

Transverse facial A.

Facial N.

Parotidoauricular M.

Masseter M.

Mandibular gland

Dorsal buccal br. of facial N.

Parotid duct

Ventral buccal br. of facial N.

External jugular V.

Mandibular ln.

Cervicoscutularis M.

Scutiform cartilage

Mandibular part of sternocephalic M.

Mylohyoideus M.

Interscutularis M.

Frontoscutularis M.

Frontal M.

Orbicularis oculi M.

Malar M.

Angularis oculi V.A.

Superior labial V.A.

Buccinator M.

Levator labii superioris M.

Caninus M.

Depressor labii superioris M.

Zygomatic M.

Levator nasolabialis M.

Depressor labii inferioris M.

Facial A.V.

Inferior labial V.

Communicating br. between dorsal & ventral buccal branches of facial N.

Superficial structures (lateral aspect).

Fig. 7-26

The cornual vessels are paralleled by the cornual br. (N.) (a branch of the lacrimal or zygomatic N.); these run together on the temporal line. Ventral to these structures, dissect the zygomaticotemporal br. of the lacrimal or zygomatic N. (Fig. 7-27, *A* and *B*).

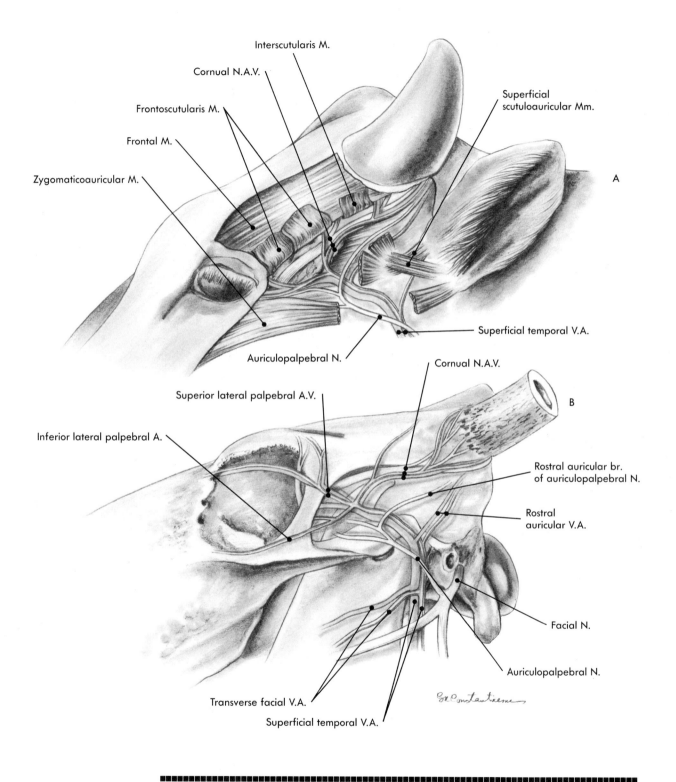

A, Structures between eye, ear, and horn. **B,** Vessels and nerves between eye, ear, and horn.

Fig. 7-27

Notice that, in the ruminants, the monostomatic sublingual gland is located in a rostral position in comparison with the polystomatic gland, which is located caudally (Fig. 7-28).

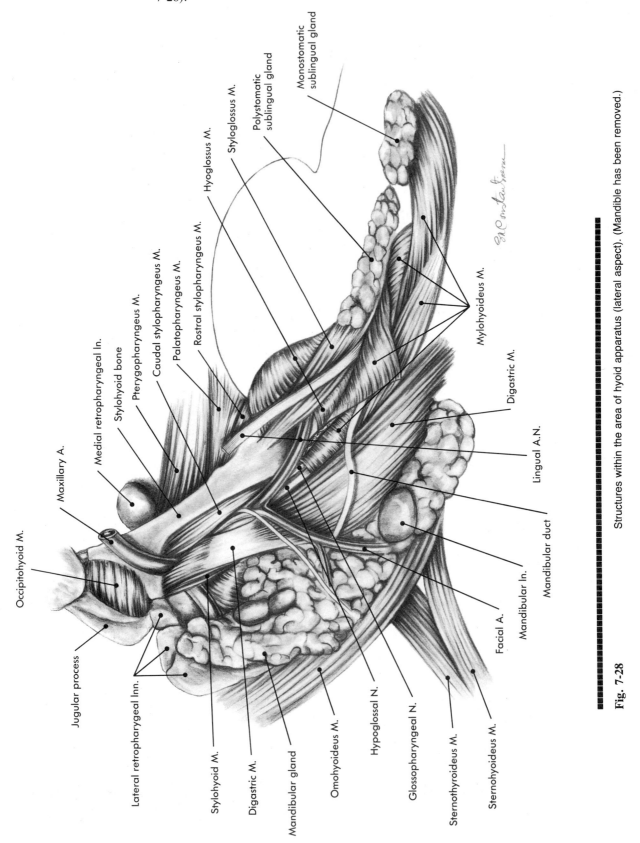

Fig. 7-28   Structures within the area of hyoid apparatus (lateral aspect). (Mandible has been removed.)

Finish the dissection of the deep structures of the head (Fig. 7-29).

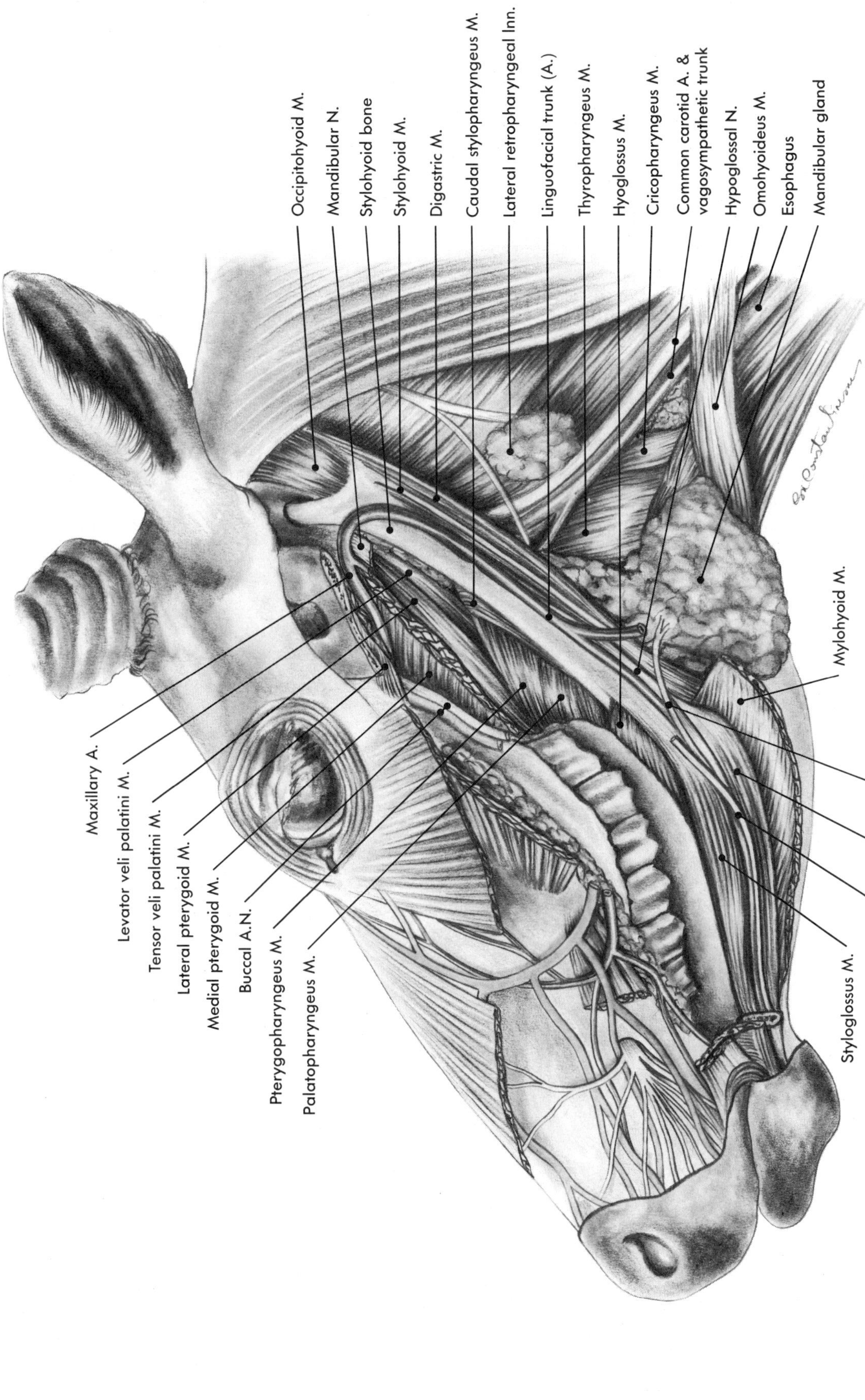

Occipitohyoid M.
Mandibular N.
Stylohyoid bone
Stylohyoid M.
Digastric M.
Caudal stylopharyngeus M.
Lateral retropharyngeal Inn.
Linguofacial trunk (A.)
Thyropharyngeus M.
Hyoglossus M.
Cricopharyngeus M.
Common carotid A. & vagosympathetic trunk
Hypoglossal N.
Omohyoideus M.
Esophagus
Mandibular gland

Maxillary A.
Levator veli palatini M.
Tensor veli palatini M.
Lateral pterygoid M.
Medial pterygoid M.
Buccal A.N.
Pterygopharyngeus M.
Palatopharyngeus M.

Mylohyoid M.

Styloglossus M.
Lingual N.
Geniohyoid M.
Mandibular duct

Deep structures of head (lateral aspect).

Fig. 7-29

If the half of the head you are working on was not split exactly on the midline and is the biggest half, you may easily identify the pharyngeal septum (located within the nasopharynx), the pharyngeal opening of the auditory tube, the pharyngeal tonsil, and the medial retropharyngeal lnn.

Examine the *nasal cavity*. Remove the nasal septum with a knife if necessary. Identify the nasal conchae and meatus.

Notice that, in the ox, there are three nasal conchae: the dorsal, middle, and ventral. Each of them encloses corresponding sinuses (dorsal, middle, and ventral conchal sinuses).

Remember! The same meatus described for the horse are present in the large ruminants: the dorsal, middle, ventral, and common meatus. The middle meatus is split caudally as a result of the presence of the developed middle nasal concha.

Remember! The oropharynx is also called the fauces, whereas the communication between the oral cavity proper and the oropharynx is called the isthmus faucium.

Examine the *oral cavity*. Identify the oral cavity proper and the vestibule. Examine the mucosa of the hard palate, the palatine ridges, and the raphe palati. In the ox, the ridges are serrated on the caudal border.

Notice that, in the ruminants (including the small ruminants), there is a dental pad, which replaces the superior incisors.

Identify the incisive papilla and the incisive ducts (of the vomeronasal organs). Examine the mucosal papillae of the lips and cheek; they are conical and cornified. Identify the orifice of the parotid duct, the sublingual caruncle with the openings of the mandibular and major sublingual ducts, and the orobasal organ.

Examine the shape of the tongue with the pointed apex, the torus linguae (the large prominence of the body), and the fossa linguae (Fig. 7-30).

Notice that the torus and fossa linguae are specific for the ruminants.

Examine the heavily cornified filiform papillae, the conical and lenticular papillae, the fungiform papillae, and the vallate papillae. There are 8 to 17 vallate papillae in the ox.

Notice that, as a rule, there are no foliate papillae in the ruminants (including the small ruminants). However, a pair of foliate papillae may be found in the ox.

Reflect the tongue medially and pull it out from the oral cavity, causing the glossoepiglottic fold and the palatoglossal arches to tense. Identify the lingual tonsil, which is located on the dorsolateral aspect of the root of the tongue. Identify the frenulum linguae. Dissect and reflect the mucosa of the lateral aspect of the tongue and expose the lingual N. which is parallel to the styloglossus M.; the mandibular duct with the hypoglossal N.; the sublingual glands (the monostomatic gland located rostral to the polystomatic gland); and the geniohyoideus, mylohyoideus, and digastric Mm. (the rostral belly of the latter).

Notice that the muscular fibers of the mylohyoideus M. are oriented in different directions.

Dissect the muscles and vessels of the tongue as shown in Fig. 7-31.

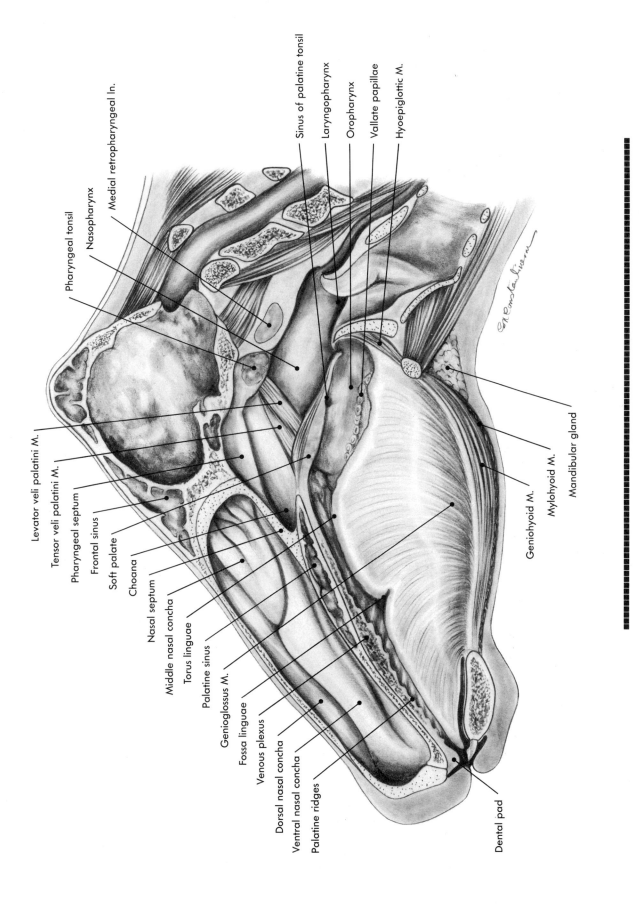

Pharyngeal tonsil

Nasopharynx

Medial retropharyngeal ln.

Sinus of palatine tonsil

Laryngopharynx

Oropharynx

Vallate papillae

Hyoepiglottic M.

Levator veli palatini M.

Tensor veli palatini M.

Pharyngeal septum

Frontal sinus

Soft palate

Choana

Nasal septum

Middle nasal concha

Torus linguae

Palatine sinus

Genioglossus M.

Fossa linguae

Venous plexus

Dorsal nasal concha

Ventral nasal concha

Palatine ridges

Dental pad

Geniohyoid M.

Mylohyoid M.

Mandibular gland

Fig. 7-30

Head (median aspect).

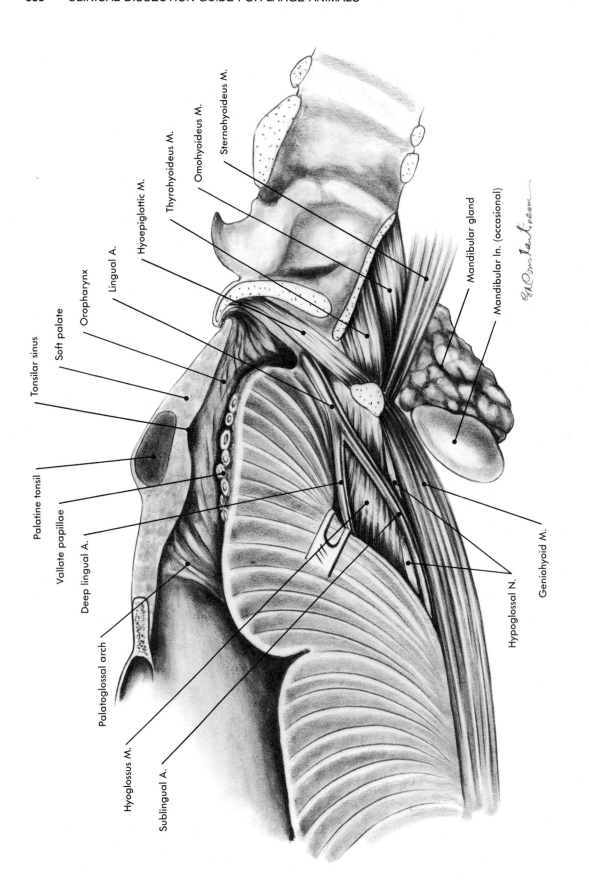

Fig. 7-31   Tongue, oropharynx, and larynx (median aspect).

Next, focus on the oropharynx and the laryngopharynx, which communicate with each other via an intrapharyngeal opening. Identify the aryepiglottic fold, which connects the epiglottis to the corniculate process of the corresponding arytenoid cartilage; also identify the piriform recess, which is the narrow groove between the oropharynx and the esophagus and is bordered by the aryepiglottic fold and the palatopharyngeal arch.

Examine the median section of the body of the thyroid cartilage and of both the plate (lamina) and ring (arch) of the cricoid cartilage of the larynx. Identify the vocal process of the arytenoid cartilage (the distocaudal angle of the cartilage) and the vocal fold, connecting the arytenoid to the caudal extent of the thyroid cartilage.

Remember! The vocal fold corresponds to the vocal lig.

Rostral to the vocal fold identify the lateral laryngeal fossa.

Notice that the ox is the only large animal with a lateral laryngeal fossa. There is no vestibular fold in the ruminants (including the small ruminants).

Caution! No lateral laryngeal ventricle is present in the ruminants (including the small ruminants).

Examine the *teeth*. Incisors, premolars, and molars are encountered in the ruminants (including the small ruminants); there are no canine teeth. An additional incisor is present on each side (the ruminants have eight incisors, instead of six as in all the other large animals). From the developmental point of view, the lateral incisors may be considered canine teeth. According to their number, they can be referred to as either I1 to 4 or central, first intermediate, second intermediate, and corner incisors. In incisors, differences exist between the deciduous and the permanent teeth. However, an incisor tooth generally has a shovel-shaped crown that is separated from the thin rounded or squared root by a distinct neck. The deciduous incisors are smaller than the permanent teeth and are set in divergent positions. The specific shape and size of the roots and their large alveoli allow for a slight movement of the teeth.

The free edge of the incisors, corresponding to the level at which the vestibular and lingual surfaces meet, is sharp on both deciduous and permanent teeth. The crown is covered with enamel. No infundibulum is observed in ruminants. The enamel is worn from the free edge in a caudal direction, exposing the dentine on the occlusal surface. The dental cavity and the root canal are very large in the deciduous incisors. In the middle of the occlusal surface, a small dental star appears in specimens that are between five and twelve years of age. In these specimens, secondary dentine obliterates the dental cavity.

The premolars and molars resemble those of the horse. However, the vestibular ridges of the superior cheek teeth are more prominent and the grooves separating them are deeper that those of the horse. The lingual surface of these teeth is convex (unique for premolars and separated into two halves by a vertical groove in molars). Both vestibular and lingual surfaces of the inferior cheek teeth are convex and are provided with vertical grooves. The occlusal surface of all the cheek teeth is very irregular and shows dentine, enamel, and cementum (Fig. 7-32, *A-G*).

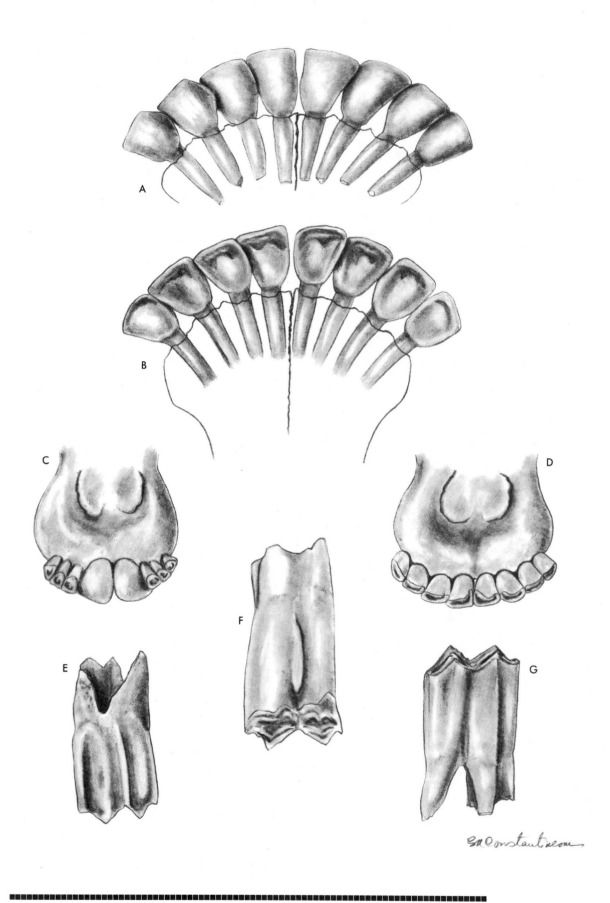

Fig. 7-32   **A,** Labial aspect of incisor teeth (5 years old). **B,** Lingual aspect of incisor teeth (5 years old). **C,** Lingual aspect of incisor teeth (2 years old). **D,** Lingual aspect of incisor teeth (6 years old). **E,** Vestibular aspect of superior third molar, (permanent). **F,** Lingual aspect of superior third molar (permanent). **G,** Vestibular aspect of inferior third molar (permanent).

Examine the *paranasal sinuses*. In the large ruminants, the following sinuses are present: the frontal, maxillary, lacrimal, palatine, and sphenoid sinuses along with the dorsal, middle, and ventral conchal sinuses.

On the midsagittal surface of the head of your specimen, identify the internal and external plates of the frontal bone. A median septum separates the two symmetrical bones. The frontal sinus is sculptured within the frontal, parietal, interparietal, temporal, and occipital bones. With a hammer and chisel, remove the external plate of the frontal bone to expose the compartments of the frontal sinus (also called frontal sinuses). A caudal frontal sinus and three rostral frontal sinuses are present. The caudal frontal sinus is separated from the rostral frontal sinuses by a transverse septum, which joins the middle of the orbita to the median septum. The caudal frontal sinus is divided into diverticula. Identify the following diverticula: cornual, nuchal, and postorbital diverticula. Rostral to the transverse septum, identify the rostral frontal sinuses: the medial, intermediate, and lateral rostral frontal sinuses.

The other sinuses have no diverticula (Fig. 7-33, *A* and *B*).

Turn the head so that the midsagittal side is up, and examine the *spinal cord* and the *brain*.

Identify the gray and white matter and the central canal of the spinal cord. Separate and identify the pia mater from the arachnoid and the dura mater. Probe the subarachnoid, subdural, and epidural spaces (the latter only in the vertebral canal).

Remember! The pia mater and the arachnoid, together, are called the leptomeninx. The dura mater is called the pachimeninx.

Identify the meninges of the brain, especially the falx cerebri, tentorium cerebelli membranaceum, and diaphragma sellae.

Systematically identify the same structures of the brain as were described in the section on horse and compare them (Figs. 7-34, *A-C*, and 7-35, *A* and *B*).

A few differences are necessary to be mentioned: in ruminants the cavernous sinus (venous) which surrounds the hypophysis does not enclose the internal carotid A.; in ruminants there is an arterial network protected within the cavernous sinuses called the rete mirabile epidurale.

Notice that the rete mirabile epidurale is characteristic for the ruminants (including the small ruminants) and pig. In the ox there are two retia mirable epidurale: one rostral and one caudal.

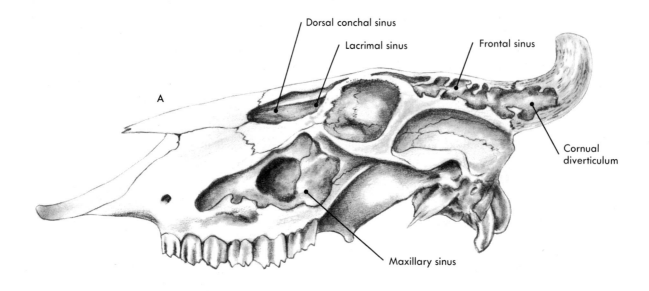

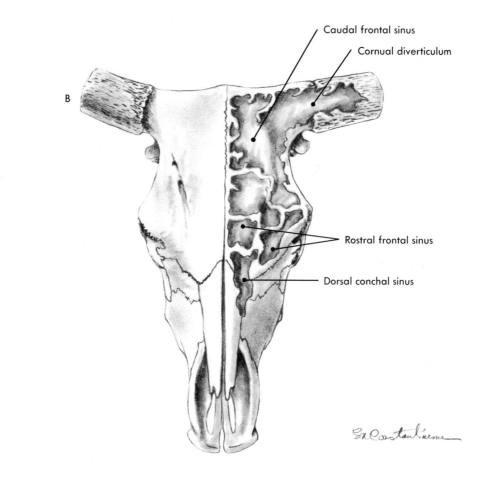

Fig. 7-33   **A,** Paranasal sinuses (lateral aspect). **B,** Paranasal sinuses (frontal aspect).

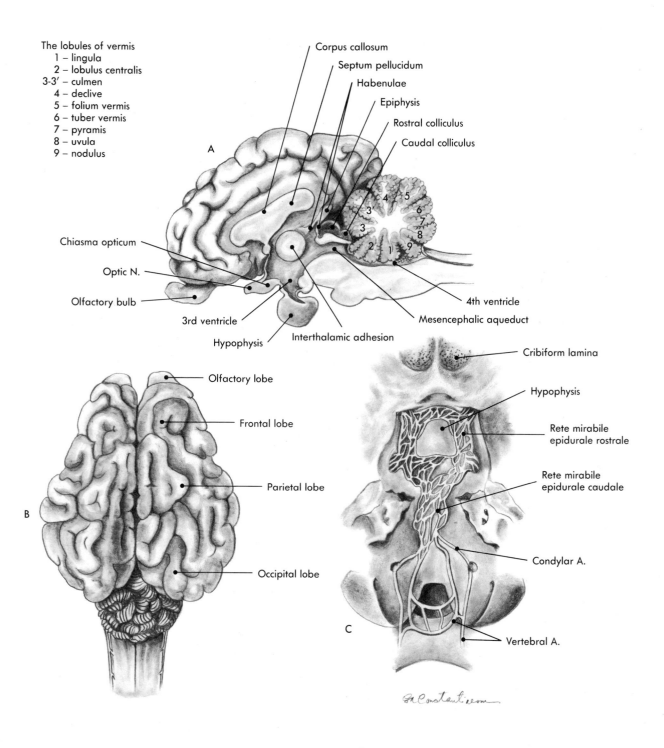

The lobules of vermis
1 – lingula
2 – lobulus centralis
3-3' – culmen
4 – declive
5 – folium vermis
6 – tuber vermis
7 – pyramis
8 – uvula
9 – nodulus

A

Corpus callosum
Septum pellucidum
Habenulae
Epiphysis
Rostral colliculus
Caudal colliculus

Chiasma opticum
Optic N.
Olfactory bulb
3rd ventricle
Hypophysis
Interthalamic adhesion
Mesencephalic aqueduct
4th ventricle

B

Olfactory lobe
Frontal lobe
Parietal lobe
Occipital lobe

C

Cribiform lamina
Hypophysis
Rete mirabile epidurale rostrale
Rete mirabile epidurale caudale
Condylar A.
Vertebral A.

A, Brain (median aspect). **B,** Brain (dorsal aspect). **C,** Arteries of the floor of the brain cavity.

Fig. 7-34

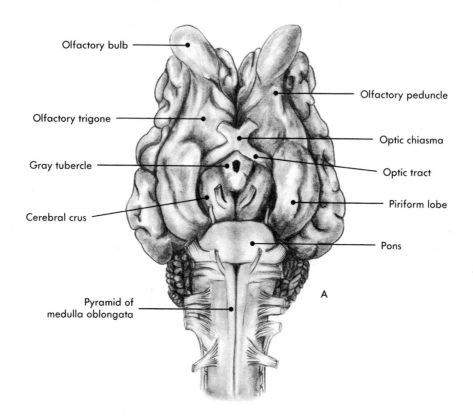

Olfactory bulb

Olfactory trigone

Gray tubercle

Cerebral crus

Pyramid of
medulla oblongata

Olfactory peduncle

Optic chiasma

Optic tract

Piriform lobe

Pons

A

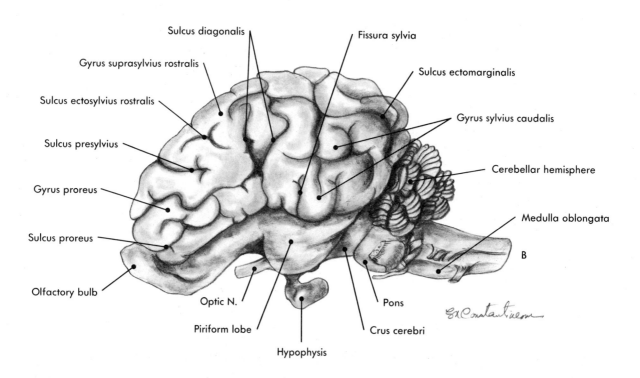

Sulcus diagonalis

Gyrus suprasylvius rostralis

Sulcus ectosylvius rostralis

Sulcus presylvius

Gyrus proreus

Sulcus proreus

Olfactory bulb

Optic N.

Piriform lobe

Hypophysis

Fissura sylvia

Sulcus ectomarginalis

Gyrus sylvius caudalis

Cerebellar hemisphere

Medulla oblongata

B

Pons

Crus cerebri

Fig. 7-35

Brain. **A,** Ventral aspect. **B,** Lateral aspect.

Turn the head so that the lateral side is up, and inspect the *eye*. Dissect the fibrous structures, muscles, vessels, and nerves in a similar manner with the horse (Figs. 7-36, *A* and *B* and 7-37, *A* and *B*).

Examine the eyeball and compare all its structures with those of the horse.

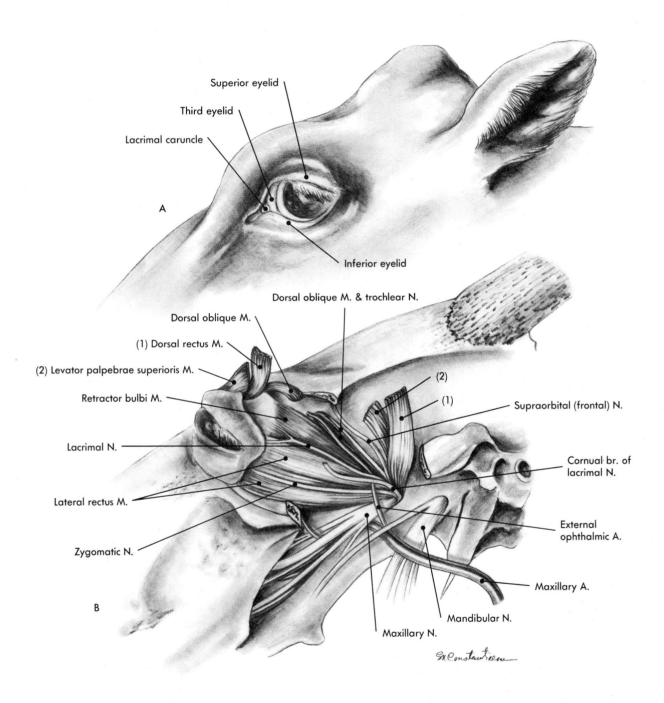

A, Eye (external aspect). **B,** Structures in orbital area.

Fig. 7-36

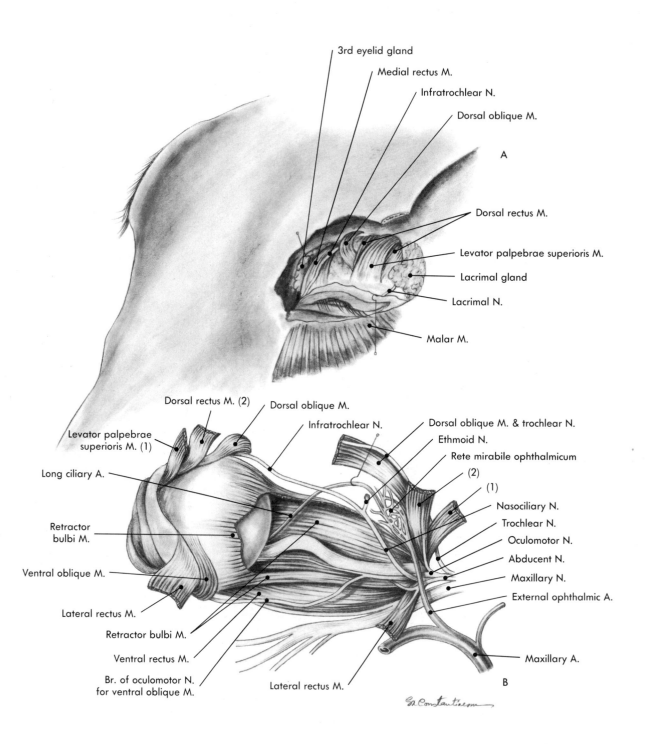

Fig. 7-37   **A,** Annex structures of eye (dorsal view). **B,** Deep structures in orbital area.

There are, of course, some significant differences, between the large and small ruminants and even between the sheep and the goat. Following the directions for the dissection for the large ruminants, the more important differences are pointed out for the small ruminants (Fig. 7-38 to 7-42, *A-D*).

Notice that the route of the parotid duct in sheep is specific among the large animals.

Notice that the transverse facial A. gives off the superior and the inferior labial Aa. The superior labial A. gives off the angularis oris A.

Notice that there is no linguofacial A. in the small ruminants (including the goat). The facial A. is absent; however, the lingual A. is present.

Notice that there is no sternomandibular M. in the sheep (only sternomastoideus M.).

Notice that there is no angularis oculi A. in the small ruminants (including the goat) but that the angularis oculi V. is present as a branch of the facial V.

Notice that the malaris A. branches off outside of the orbita the dorsal nasal and the lateral nasal Aa. The lateral nasal V. is a branch of the facial V., whereas the dorsal nasal V. is a branch of the angularis oculi V.

Notice that, in the small ruminants (including the goat), there are no superficial and deep inferior labial Aa.Vv., only inferior labial A.V.

Notice that, in the sheep, the parotid duct opens in the vestibule at the level of the first superior molar (the fourth upper cheek tooth).

Dissect both lateral and medial aspects of the head, the eye, the paranasal sinuses, and the brain and compare all these structures with those of the large ruminants.

Sheep
(Figs 7-38 to 7-42)

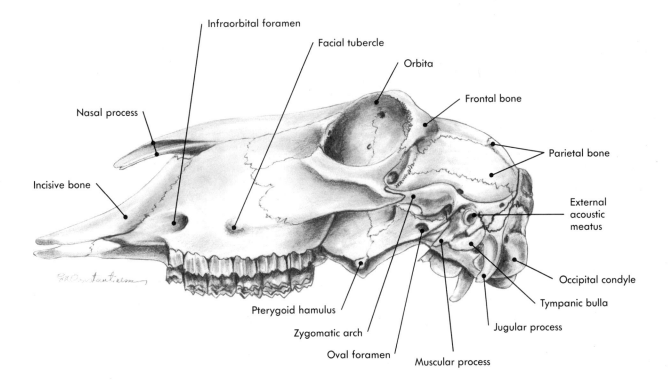

Skull (lateral aspect).

Fig. 7-38

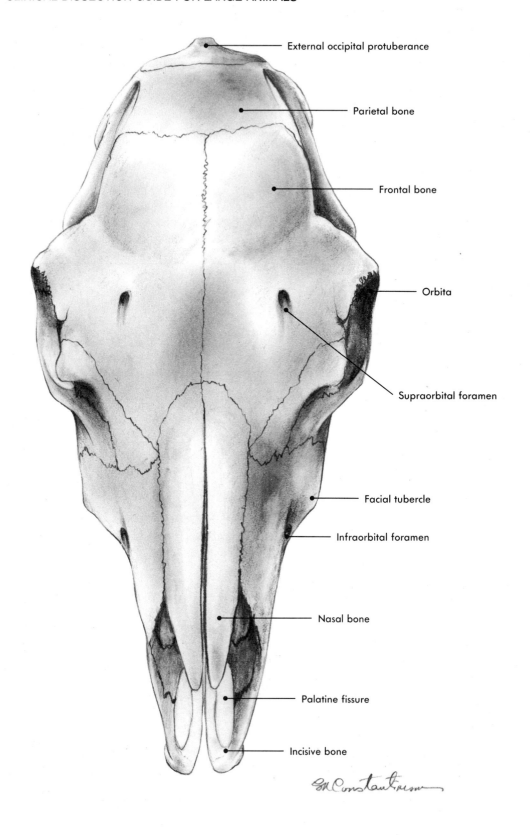

External occipital protuberance

Parietal bone

Frontal bone

Orbita

Supraorbital foramen

Facial tubercle

Infraorbital foramen

Nasal bone

Palatine fissure

Incisive bone

Fig. 7-39   Skull (frontal aspect).

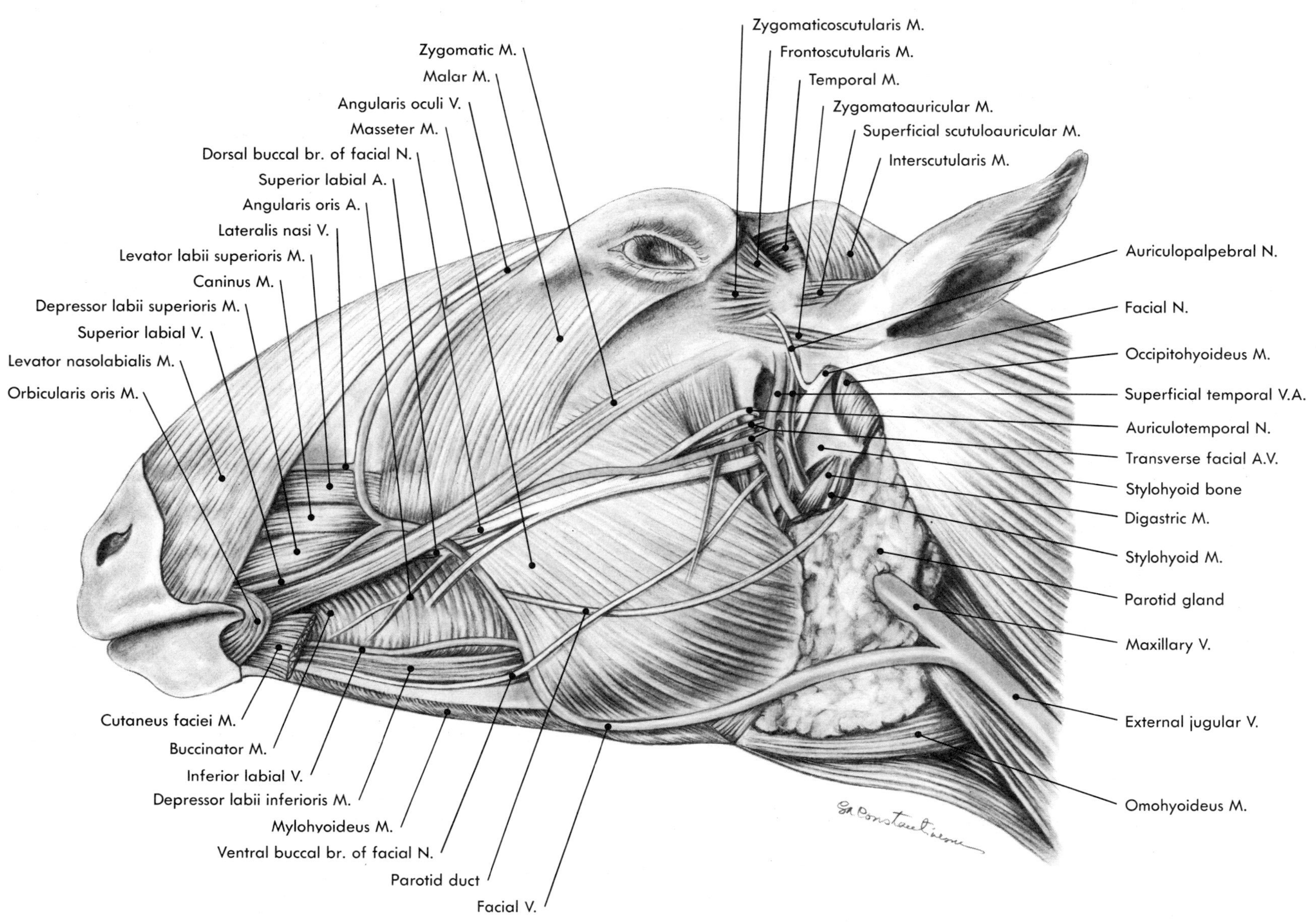

Zygomaticoscutularis M.
Frontoscutularis M.
Temporal M.
Zygomatoauricular M.
Superficial scutuloauricular M.
Interscutularis M.

Auriculopalpebral N.
Facial N.
Occipitohyoideus M.
Superficial temporal V.A.
Auriculotemporal N.
Transverse facial A.V.
Stylohyoid bone
Digastric M.
Stylohyoid M.
Parotid gland
Maxillary V.
External jugular V.
Omohyoideus M.

Zygomatic M.
Malar M.
Angularis oculi V.
Masseter M.
Dorsal buccal br. of facial N.
Superior labial A.
Angularis oris A.
Lateralis nasi V.
Levator labii superioris M.
Depressor labii superioris M.
Superior labial V.
Levator nasolabialis M.
Orbicularis oris M.
Caninus M.

Cutaneus faciei M.
Buccinator M.
Inferior labial V.
Depressor labii inferioris M.
Mylohyoideus M.
Ventral buccal br. of facial N.
Parotid duct
Facial V.

Fig. 7-40

Superficial structures of head (lateral aspect).

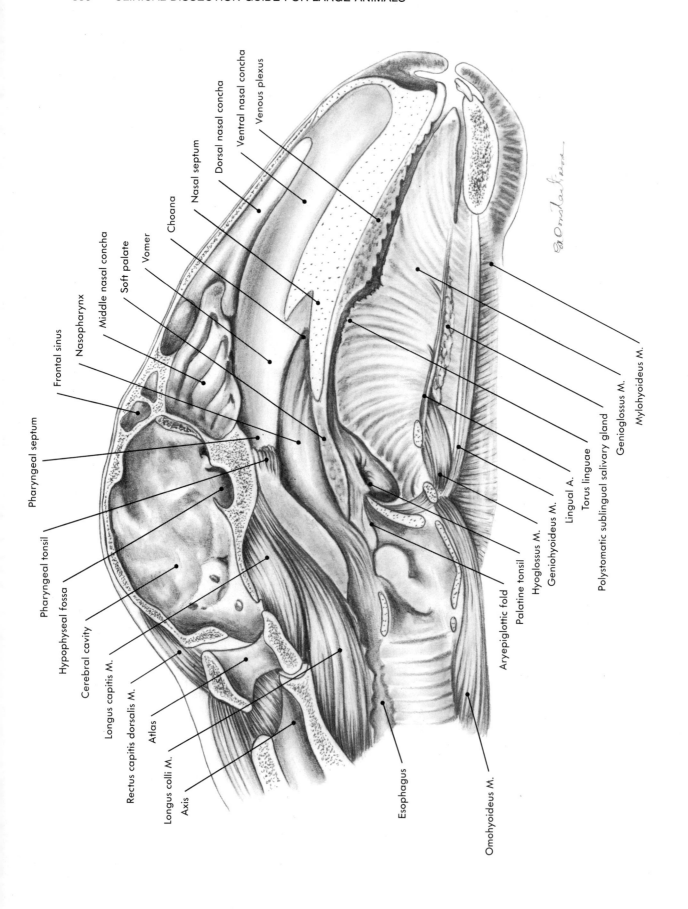

Pharyngeal septum

Frontal sinus

Nasopharynx

Middle nasal concha

Soft palate

Vomer

Choana

Nasal septum

Dorsal nasal concha

Ventral nasal concha

Venous plexus

Pharyngeal tonsil

Hypophyseal fossa

Cerebral cavity

Longus capitis M.

Rectus capitis dorsalis M.

Atlas

Longus colli M.

Axis

Esophagus

Omohyoideus M.

Aryepiglottic fold

Palatine tonsil

Hyoglossus M.

Geniohyoideus M.

Lingual A.

Torus linguae

Geniohyoideus M.

Polystomatic sublingual salivary gland

Genioglossus M.

Mylohyoideus M.

Head (median aspect).

Fig. 7-41

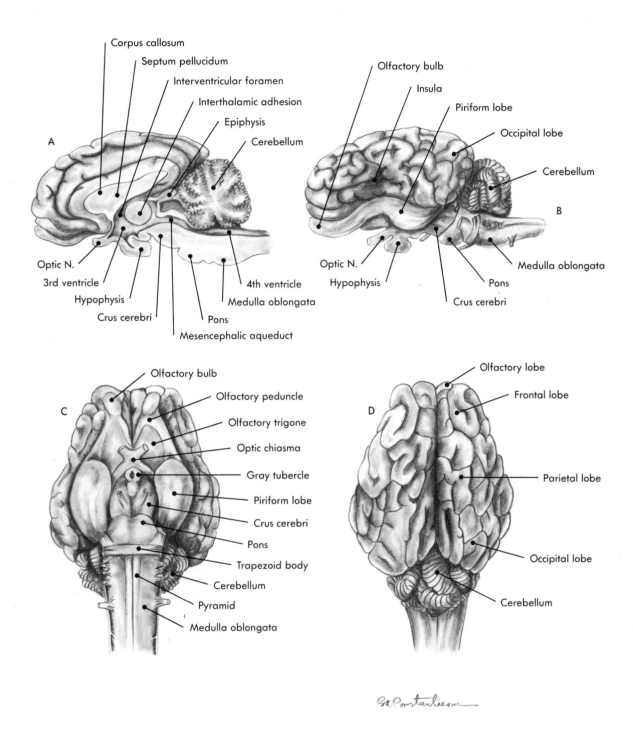

A, Brain (median aspect). **B,** Brain (lateral aspect). **C,** Brain (ventral aspect). **D,** Cerebral lobes (dorsal view).

Fig. 7-42

Goat
■■■■■■■■
(Figs. 7-43
to 7-47)

The significant characteristics of the goat are illustrated in Figs. 7-43 to 7-47.

Carefully identify and dissect the following structures: the sternozygomaticus and depressor of the lower lid (retractor anguli oculi lateralis) Mm.; the parotid duct whose route and opening are similar to those of the horse; the dorsal and ventral buccal branches of the facial N. which are similar to those of the sheep; the beard; the horn glands; and the cervical appendix (tassel).

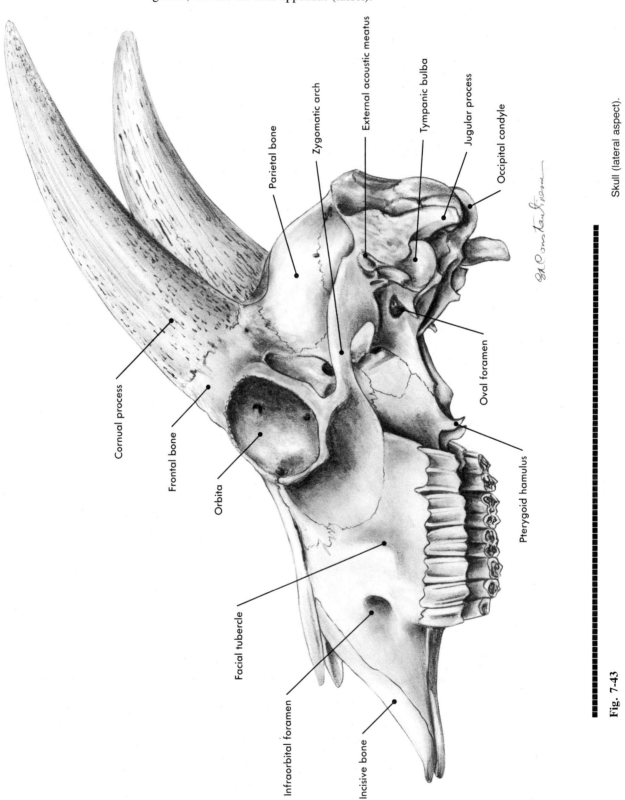

Skull (lateral aspect).

Fig. 7-43

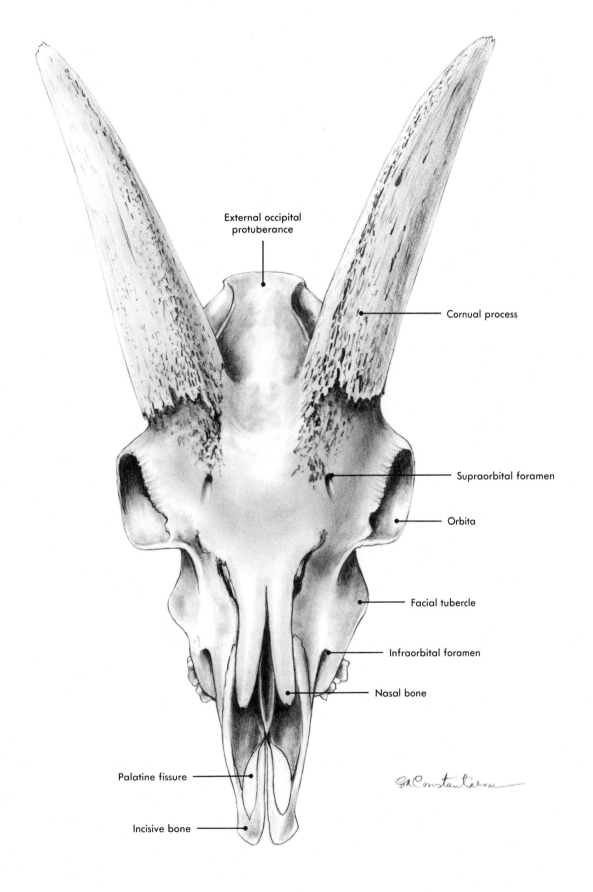

External occipital
protuberance

Cornual process

Supraorbital foramen

Orbita

Facial tubercle

Infraorbital foramen

Nasal bone

Palatine fissure

Incisive bone

Skull (frontal aspect).

Fig. 7-44

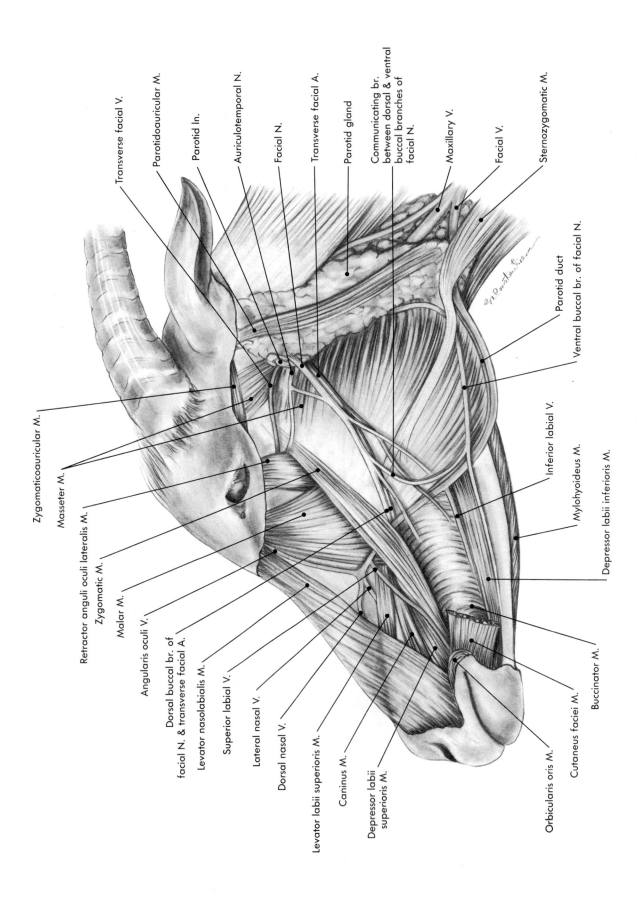

Transverse facial V.

Parotidoauricular M.

Parotid ln.

Auriculotemporal N.

Facial N.

Transverse facial A.

Parotid gland

Communicating br. between dorsal & ventral buccal branches of facial N.

Maxillary V.

Facial V.

Sternozygomatic M.

Zygomaticoauricular M.

Masseter M.

Retractor anguli oculi lateralis M.

Zygomatic M.

Malar M.

Angularis oculi V.

Dorsal buccal br. of facial N. & transverse facial A.

Levator nasolabialis M.

Superior labial V.

Lateral nasal V.

Dorsal nasal V.

Levator labii superioris M.

Caninus M.

Depressor labii superioris M.

Orbicularis oris M.

Cutaneus faciei M.

Buccinator M.

Depressor labii inferioris M.

Mylohyoideus M.

Inferior labial V.

Ventral buccal br. of facial N.

Parotid duct

Fig. 7-45   Superficial structures of head (lateral aspect).

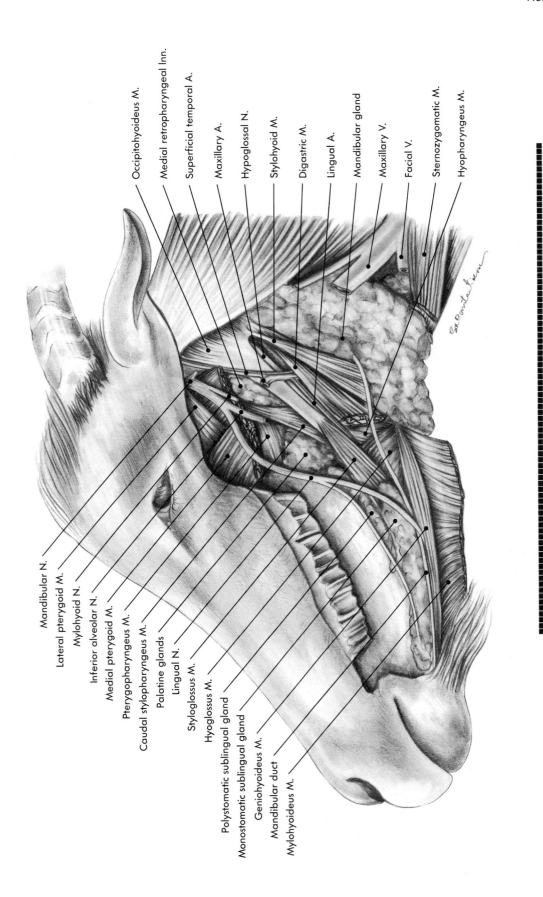

Mandibular N.
Lateral pterygoid M.
Mylohyoid N.
Inferior alveolar N.
Medial pterygoid M.
Pterygopharyngeus M.
Caudal stylopharyngeus M.
Palatine glands
Lingual N.
Styloglossus M.
Hyoglossus M.
Polystomatic sublingual gland
Monostomatic sublingual gland
Geniohyoideus M.
Mandibular duct
Mylohyoideus M.

Occipitohyoideus M.
Medial retropharyngeal Inn.
Superficial temporal A.
Maxillary A.
Hypoglossal N.
Stylohyoid M.
Digastric M.
Lingual A.
Mandibular gland
Maxillary V.
Facial V.
Sternozygomatic M.
Hyopharyngeus M.

Fig. 7-46

Deep structures of head (lateral aspect). (Mandible has been removed.)

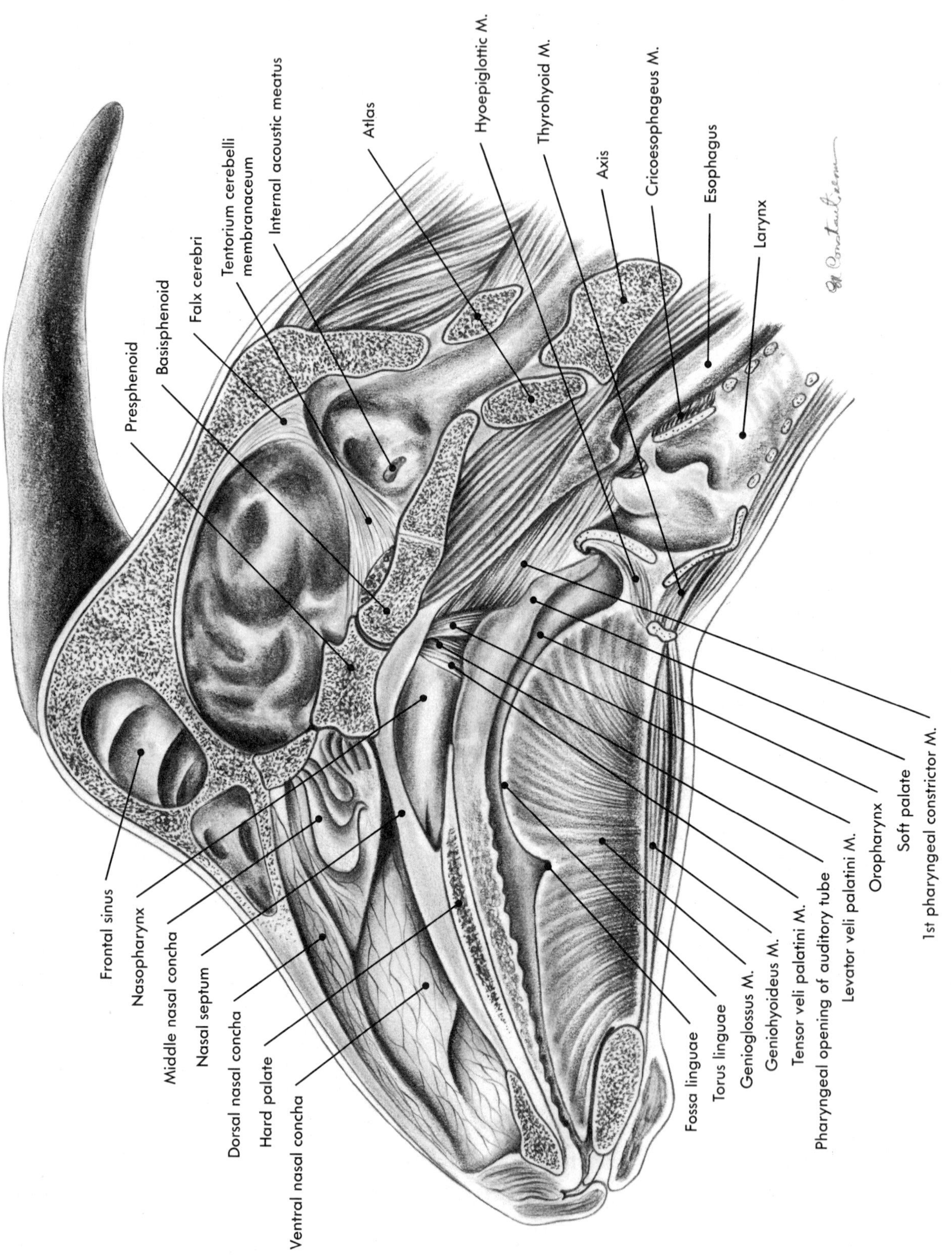

Fig. 7-47   Head (median aspect).

Identify the landmarks and structures superficial and deep on both lateral and medial aspects of the head following the same procedures as in the previous species.

Some specific structures are to be mentioned on skull, such as the rostral bone and the tricuspid process formed by the fusion of the hamulus, pterygoid process of the sphenoid bone and the pyramidal process of the palatine bone.

Notice that the pig is the only domestic mammal with four premolars and three molars on both the maxilla and mandible.

On the lateral aspect of the head, identify the specific planum rostrale (the snout) with its very short philtrum (in the ventral area) and two circular nostrils; this disc-shaped structure represents the apex of the nose fused with the upper lip. In the pig, the sublingual polystomatic gland is located rostral to the sublingual monostomatic gland.

Caution! Between the caudal extent of the hard palate (ventrally) and the vomer (dorsally) there is a long pharyngeal septum. Also examine the very long choanae just identified on the skull.

Notice that the pig has no palatine tonsil.

Notice that the only species with a pharyngeal diverticulum is the pig.

Notice that the only species with a median laryngeal recess are the horse and the pig.

Examine the teeth, the eye, the paranasal sinuses, and the brain of the pig.

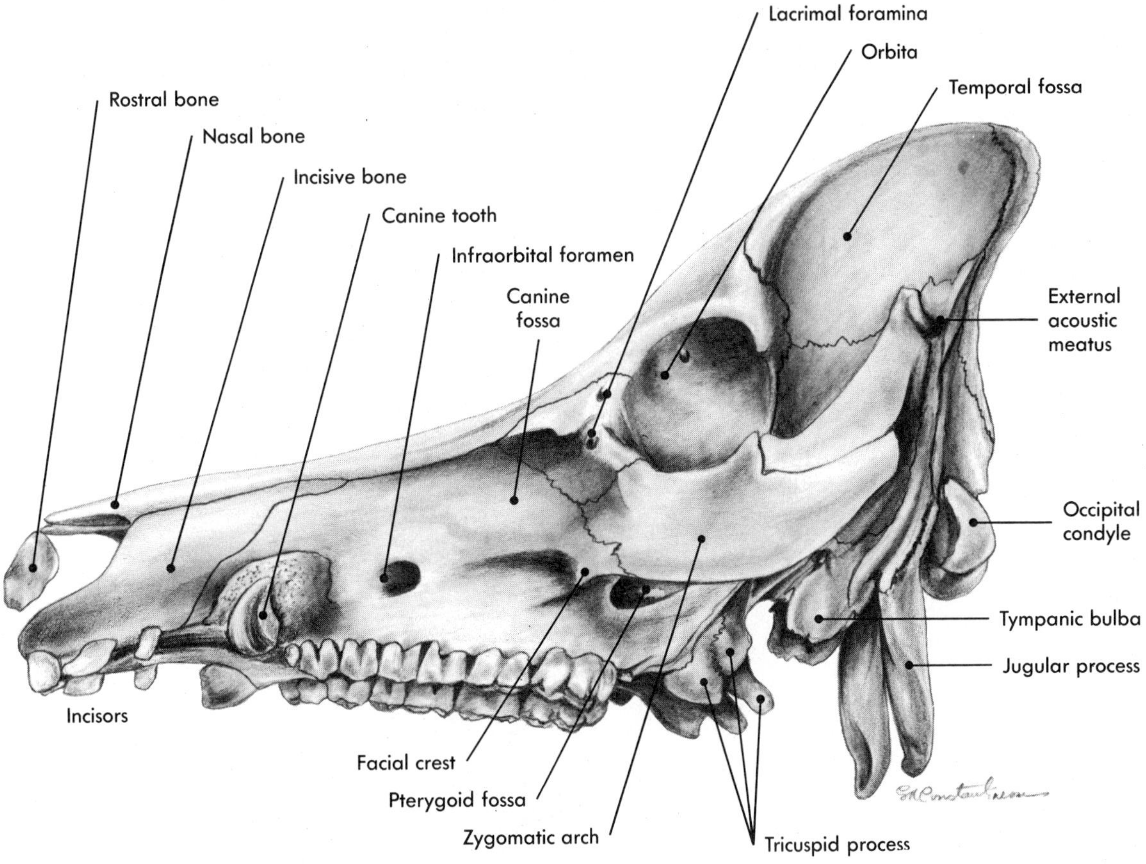

Pig
■■■■■
(Figs. 7-48 to 7-54)

Skull (lateral aspect)

Fig. 7-48

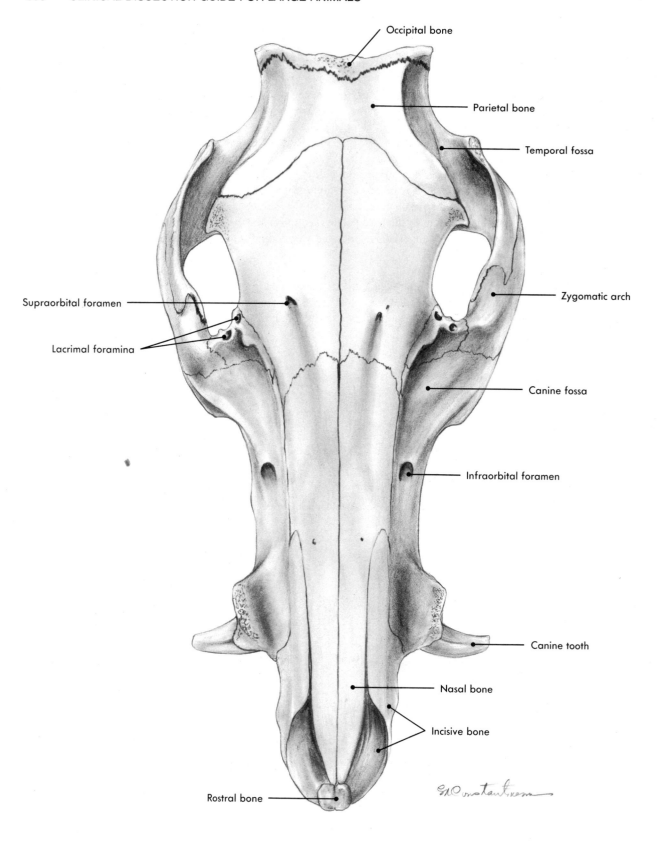

Occipital bone

Parietal bone

Temporal fossa

Supraorbital foramen

Zygomatic arch

Lacrimal foramina

Canine fossa

Infraorbital foramen

Canine tooth

Nasal bone

Incisive bone

Rostral bone

Fig. 7-49

Skull (frontal aspect).

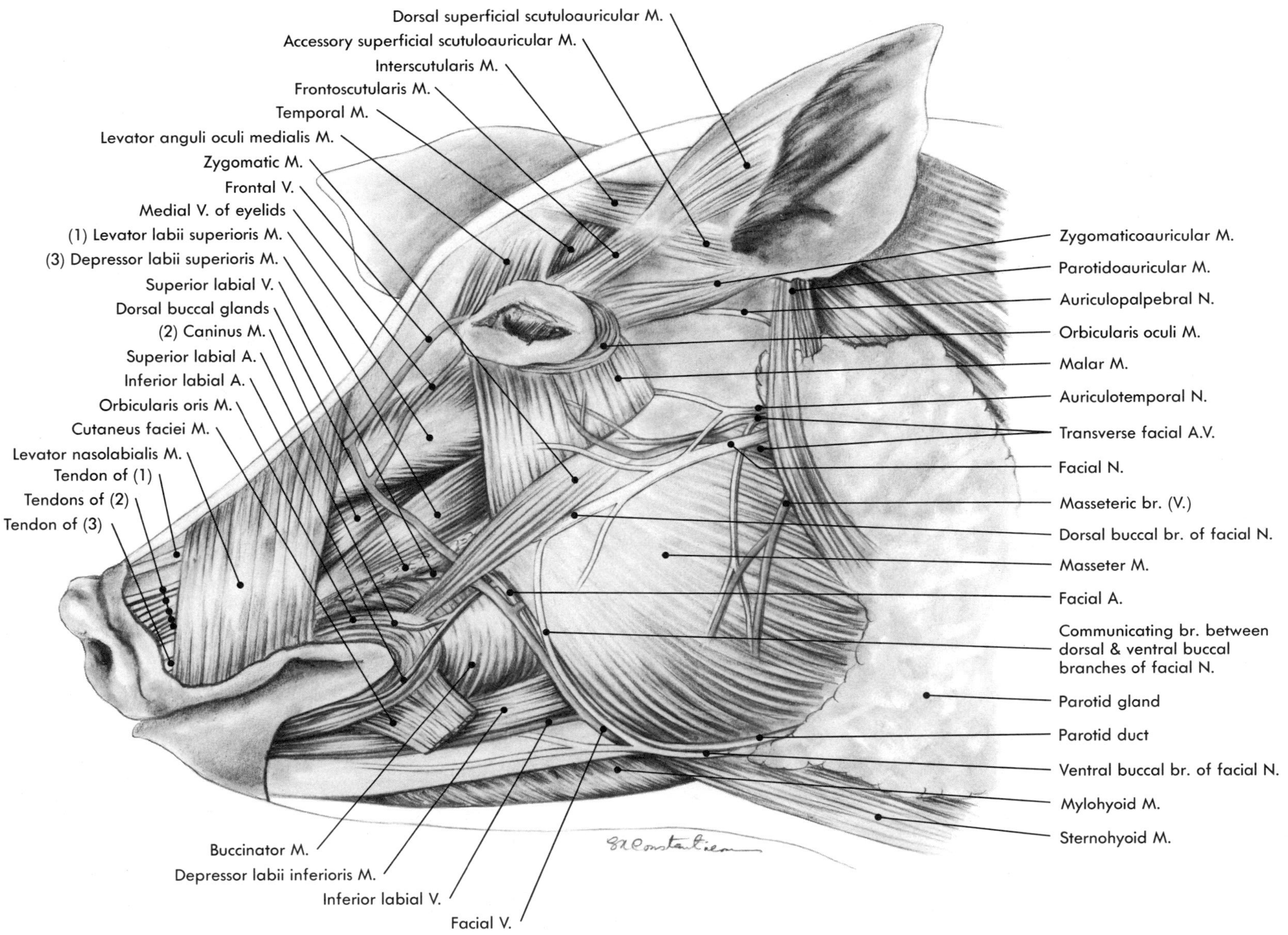

Dorsal superficial scutuloauricular M.
Accessory superficial scutuloauricular M.
Interscutularis M.
Frontoscutularis M.
Temporal M.
Levator anguli oculi medialis M.
Zygomatic M.
Frontal V.
Medial V. of eyelids
(1) Levator labii superioris M.
(3) Depressor labii superioris M.
Superior labial V.
Dorsal buccal glands
(2) Caninus M.
Superior labial A.
Inferior labial A.
Orbicularis oris M.
Cutaneus faciei M.
Levator nasolabialis M.
Tendon of (1)
Tendon of (2)
Tendon of (3)

Zygomaticoauricular M.
Parotidoauricular M.
Auriculopalpebral N.
Orbicularis oculi M.
Malar M.
Auriculotemporal N.
Transverse facial A.V.
Facial N.
Masseteric br. (V.)
Dorsal buccal br. of facial N.
Masseter M.
Facial A.
Communicating br. between dorsal & ventral buccal branches of facial N.
Parotid gland
Parotid duct
Ventral buccal br. of facial N.
Mylohyoid M.
Sternohyoid M.

Buccinator M.
Depressor labii inferioris M.
Inferior labial V.
Facial V.

Fig. 7-50

Superficial structures of head (lateral aspect).

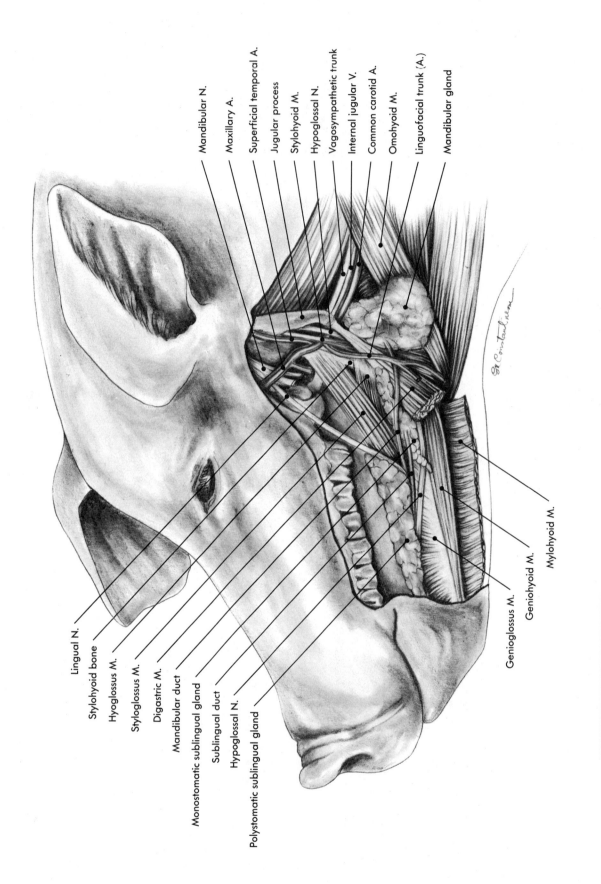

Lingual N.

Stylohyoid bone

Hyoglossus M.

Styloglossus M.

Digastric M.

Mandibular duct

Monostomatic sublingual gland

Sublingual duct

Hypoglossal N.

Polystomatic sublingual gland

Mandibular N.

Maxillary A.

Superficial temporal A.

Jugular process

Stylohyoid M.

Hypoglossal N.

Vagosympathetic trunk

Internal jugular V.

Common carotid A.

Omohyoid M.

Linguofacial trunk (A.)

Mandibular gland

Genioglossus M.

Geniohyoid M.

Mylohyoid M.

Fig. 7-51

Deep structures of head (lateral aspect). (Mandible has been removed.)

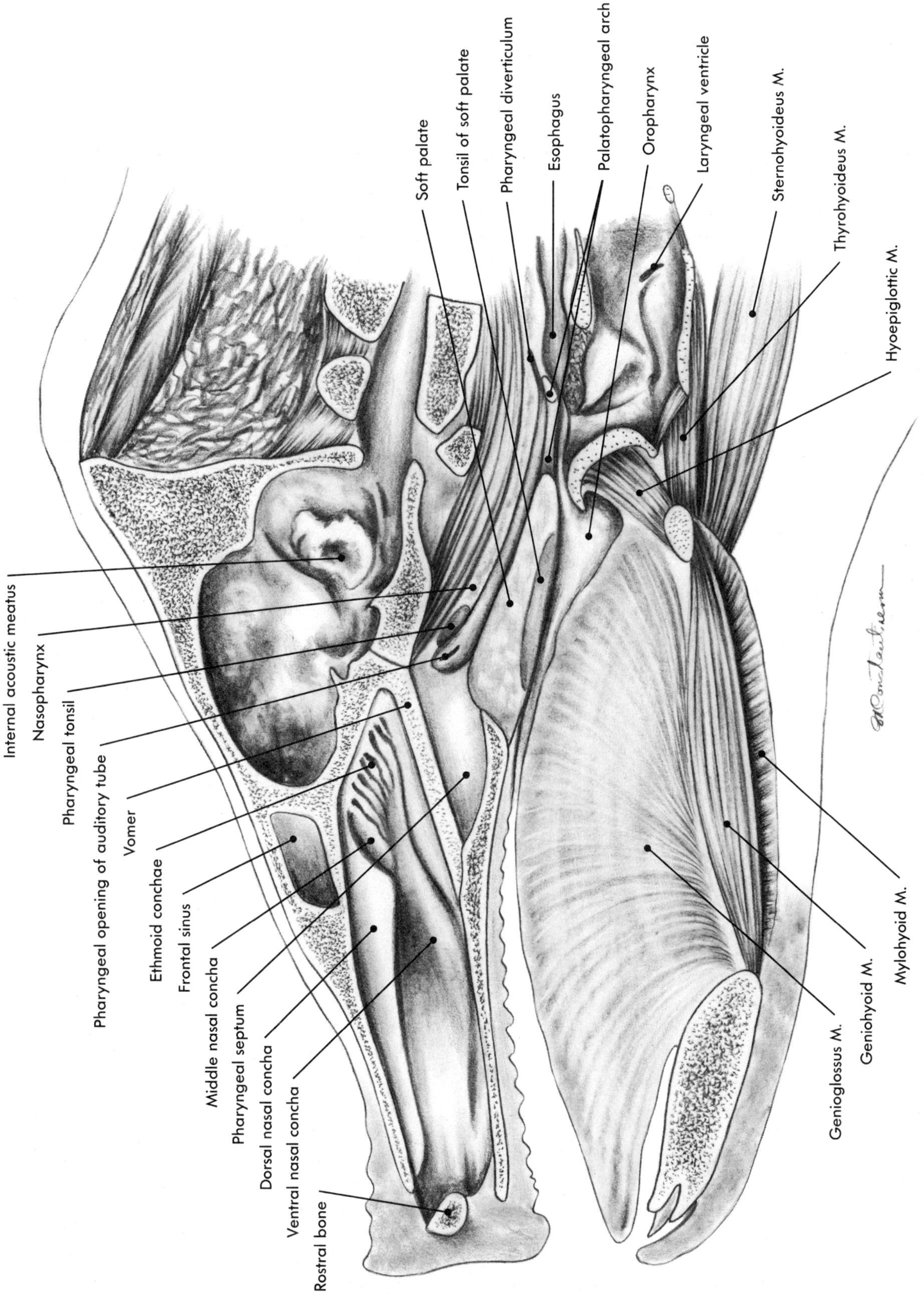

Internal acoustic meatus
Nasopharynx
Pharyngeal tonsil
Pharyngeal opening of auditory tube
Vomer
Ethmoid conchae
Frontal sinus
Middle nasal concha
Pharyngeal septum
Dorsal nasal concha
Ventral nasal concha
Rostral bone

Soft palate
Tonsil of soft palate
Pharyngeal diverticulum
Esophagus
Palatopharyngeal arch
Oropharynx
Laryngeal ventricle
Sternohyoideus M.
Thyrohyoideus M.
Hyoepiglottic M.

Genioglossus M.
Geniohyoid M.
Mylohyoid M.

Head (median aspect).

Fig. 7-52

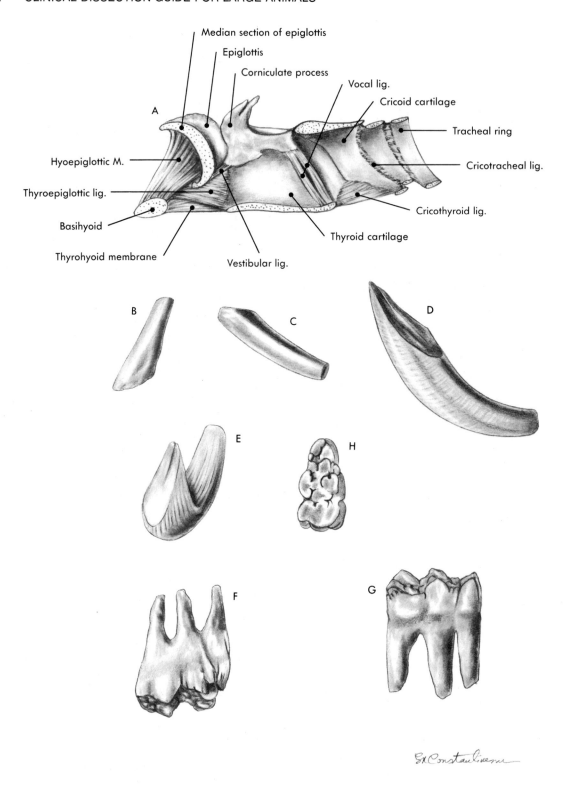

Fig. 7-53 A, Median section through larynx. B, Second left upper incisor (labial aspect). C, Second left lower incisor (labial aspect). D, Left lower canine tooth (caudolateral view). E, Left upper canine tooth (rostrolateral view). F, Vestibular aspect of last left upper molar. G, Lingual aspect of last left lower molar. H, Occlusal surface of last left lower molar.

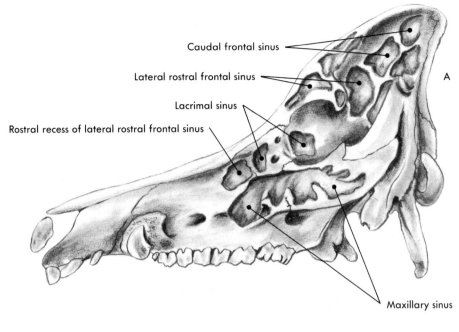

Caudal frontal sinus

Lateral rostral frontal sinus

Lacrimal sinus

Rostral recess of lateral rostral frontal sinus

A

Maxillary sinus

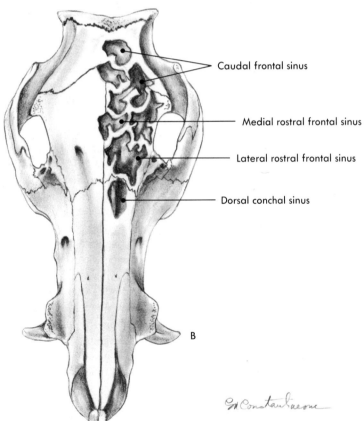

Caudal frontal sinus

Medial rostral frontal sinus

Lateral rostral frontal sinus

Dorsal conchal sinus

B

A, Paranasal sinuses (lateral aspect). **B,** Paranasal sinuses (frontal aspect).

Fig. 7-54

Chapter 8

Parotid Region

Horse

Reflect the skin carefully, because it is very adherent to the underlying anatomical structures as a result of the sparse loose connective tissue.

At the first anatomic level (most superficial level) of dissection (Fig. 8-1), the loose connective tissue includes terminal branches of the ventral branches of the second and third cervical spinal Nn. and branches of the caudal auricular V., which lie on the cutaneous colli and parotidoauricularis Mm., respectively.

Another reason to reflect the skin carefully is to protect the structures included in the second level of dissection. These include the parotid fascia; the parotidoauricularis M.; the parotid salivary gland; the masseteric branch (A.) and the ventral masseteric V.; the external jugular V.; the maxillary and linguofacial Vv. (covered at the ventral third of the region by the cutaneus colli M. and the transverse N. of the neck); the omohyoideus, sternocephalicus (sternomandibularis), and cleidomastoideus Mm.; the origin of the parotid duct; and the great auricular N.

The rostral border of the parotid gland should be reflected carefully, starting from the temporomandibular joint.

Caution! The parotid ln. (lnn.) is (are) surrounded by the glandular tissue of the parotid gland (refer to the landmarks and approach for the exposure of the parotid lnn. on p. 429).

In the third anatomic level, the following vessels and nerves pass through the parotid gland: the maxillary V., the masseteric branch (A.), the ventral masseteric V., the caudal auricular A.V., and the ramus colli of the facial N.

In the fourth anatomic level, the subparotid route of the facial N. with all its branches is exposed. The branches are as follows: dorsally and rostrocaudally—the auriculopalpebral N., internal auricular branch, and caudal auricular N.; and ventrally and in the same direction—the ramus colli (the cervical branch of facial N.); the ansa (or loop, which surrounds the caudal auricular A.), parotid branches, and the digastric branch for the digastric (including the occipitomandibular part) and the occipitohyoideus Mm.; and the stylohyoid branch for the stylohyoideus M.

Caution! The ramus colli does not have a constant route. Only after the dissection of this nerve in the cervical area can one dissect the parotidoauricularis M. The nerve becomes subcutaneous by piercing both the parotid gland and the parotidoauricularis M., by piercing only the parotid gland (running between the gland and the muscle), or by emerging deep to the parotid gland. In each of these cases, it parallels the maxillary and external jugular Vv. One can reflect the parotidoauricularis M. by dissecting it in a ventrodorsal direction up to the base of the ear.

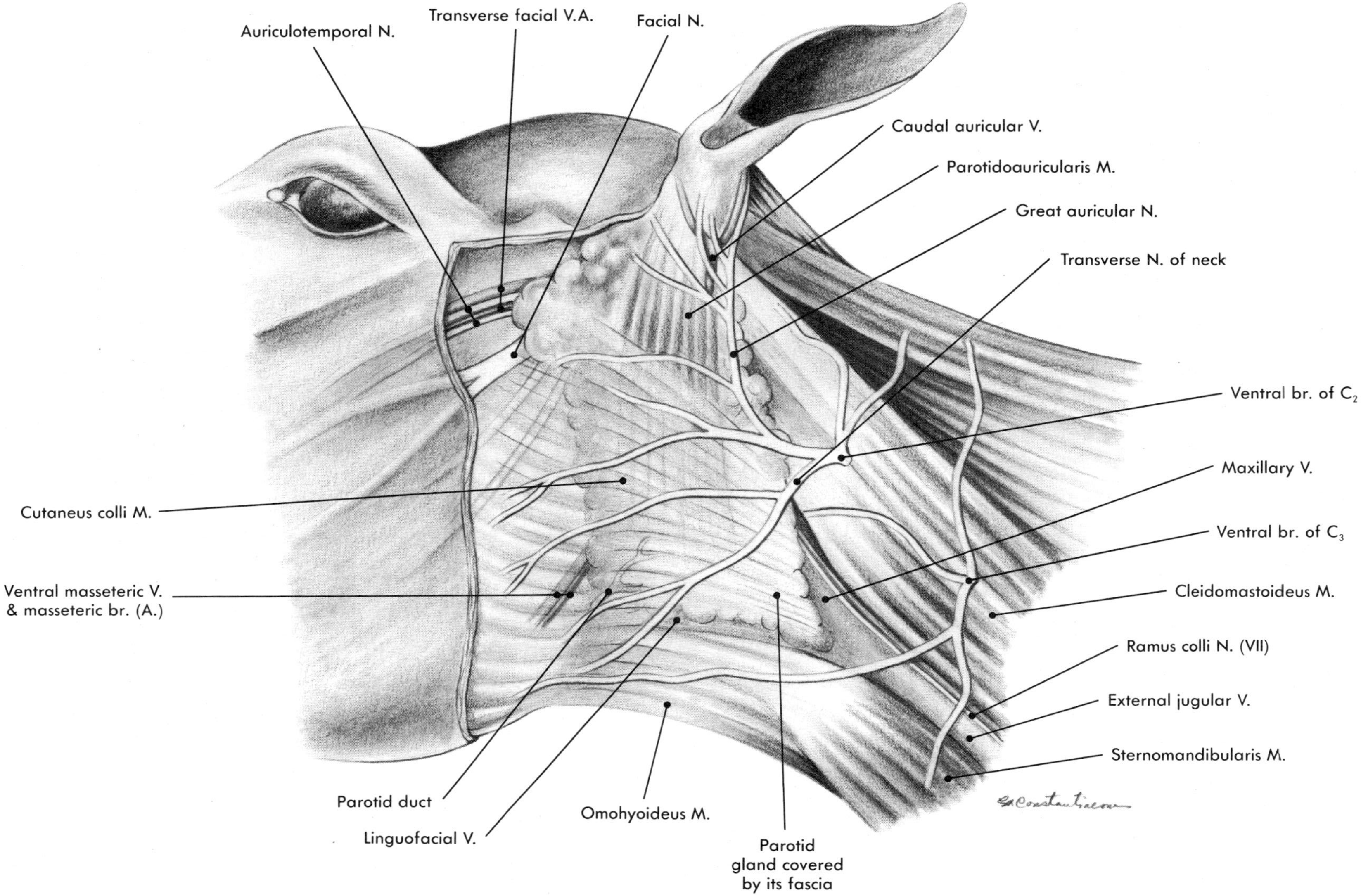

Auriculotemporal N.

Transverse facial V.A.

Facial N.

Caudal auricular V.

Parotidoauricularis M.

Great auricular N.

Transverse N. of neck

Ventral br. of C₂

Maxillary V.

Ventral br. of C₃

Cleidomastoideus M.

Ramus colli N. (VII)

External jugular V.

Sternomandibularis M.

Parotid gland covered by its fascia

Omohyoideus M.

Linguofacial V.

Parotid duct

Ventral masseteric V. & masseteric br. (A.)

Cutaneus colli M.

Fig. 8-1

Superficial structures.

Another group of structures, in the fifth anatomic level, consists of the auriculotemporal N. (a branch of the mandibular from the trigeminal), which joins the facial N. and parallels the transverse facial A.V.; the superficial temporal A.V. with the rostral auricular A.V.; the maxillary A.V.; the caudal auricular A.V. with all their branches; and the origin of the masseteric branch (A.) and ventral masseteric V.

Caution! To avoid transecting the ventral branch of the accessory N. at the deep aspect of the parotid gland the subparotid aponeurosis should not be incised. This is the structure between the sternomandibularis and cleidomastoideus tendons, which separates the parotid and mandibular salivary glands. The lateral retropharyngeal and cranial deep cervical lnn., as well as the mandibular salivary gland, will be exposed (Fig. 8-2).

At the sixth anatomic level in the area of the dorsal retromandibular fossa, the following structures are exposed by careful dissection: the external carotid A.; the maxillary and superficial temporal Aa. (the latter is situated between the guttural pouch and the parotid gland); the stylohyoideus M.; the stylohyoid bone; the occipitohyoideus M.; the aponeurosis of the cleidomastoideus M.; the rostral auricular A.; the transverse facial A.; the superficial temporal A.; the linguofacial arterial trunk; the masseteric branch (A.); and the glossopharyngeal and hypoglossal Nn. (Fig. 8-3).

In the area of the ventral retromandibular fossa, the first structure exposed is the mandibular salivary gland, which is located deep to the subparotid aponeurosis. The mandibular insertions of the sternomandibularis M. and the occipitomandibular part of the digastric M. are exposed.

Notice that the sternomandibularis M. crosses the digastric M. and actually overlaps the occipitomandibular part of that muscle.

Vessels and nerves lie on the ventral aspect of the guttural pouch (Fig. 8-4), except for the ventrolateral part of it, which corresponds to the Viborg's triangle. The mandibular salivary gland should be carefully and manually removed to protect the vessels and nerves.

It is possible to dissect the ventral branches of the first and second cervical spinal Nn. Deep to the roots of the external jugular V., the omohyoideus M. is exposed. At its cranial border, expose by blind dissection the common carotid A. with its three terminations (the occipital, internal carotid, and external carotid Aa.). Next, dissect the lateral retropharyngeal and cranial deep cervical lnn.; the cervical sympathetic trunk; the pharyngeal branch and the branch for the glomus caroticum and carotid sinus from the cranial cervical ggl.; the occipital V.; and the subparotid branches of the four last pairs of cranial nerves. The glossopharyngeal N. gives off the pharyngeal branch, the lingual branch, the branch for the glomus caroticum and carotid sinus, and the communicating branches. It runs rostroventrally and parallel to the stylohyoid bone (rostrally) and the linguofacial arterial trunk (caudally). The vagus N., which runs caudally, dorsal and parallel to the common carotid A., issues a pharyngeal branch, the cranial laryngeal N. and communicating branches. The accessory N. splits into the internal and external branches. The external branch runs caudally (dorsal to the vagus) and divides into dorsal and ventral branches close to the entrance of the occipital A. into the alar foramen. The hypoglossal N. parallels the glossopharyngeal N. caudally and is located between the linguofacial arterial trunk and the stylohyoideus M.

The deep muscles in this area are the occipitohyoideus, stylohyoideus, and rectus capitis ventralis Mm.

If the specimen is fresh, keep the head in extension and pull the sternomandibularis M., external jugular V., and common carotid A. caudally.

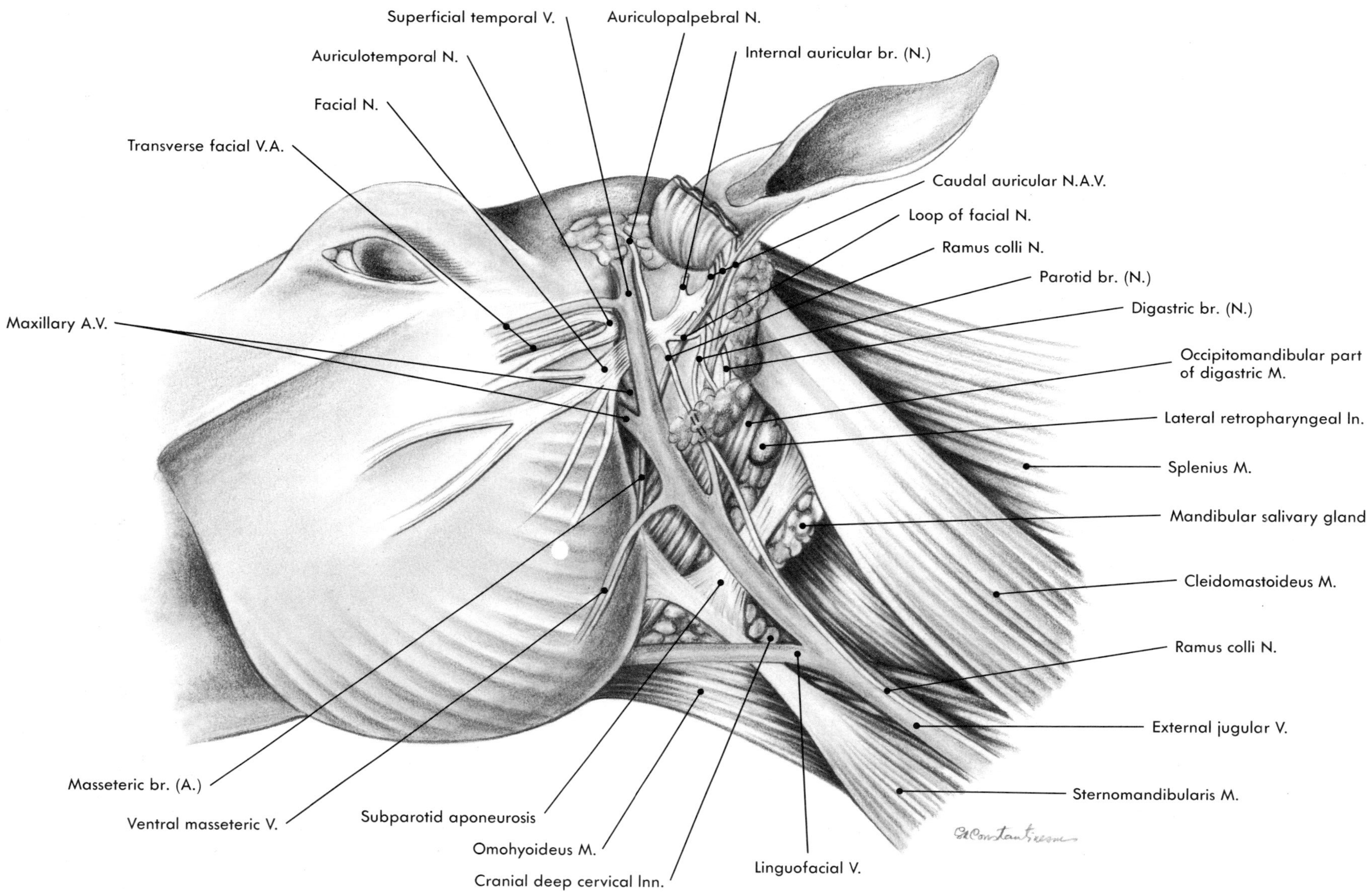

Superficial temporal V.

Auriculotemporal N.

Transverse facial V.A.

Facial N.

Maxillary A.V.

Auriculopalpebral N.

Internal auricular br. (N.)

Caudal auricular N.A.V.

Loop of facial N.

Ramus colli N.

Parotid br. (N.)

Digastric br. (N.)

Occipitomandibular part of digastric M.

Lateral retropharyngeal ln.

Splenius M.

Mandibular salivary gland

Cleidomastoideus M.

Ramus colli N.

External jugular V.

Sternomandibularis M.

Linguofacial V.

Cranial deep cervical lnn.

Omohyoideus M.

Subparotid aponeurosis

Ventral masseteric V.

Masseteric br. (A.)

Fig. 8-2

Structures located deep in parotid gland.

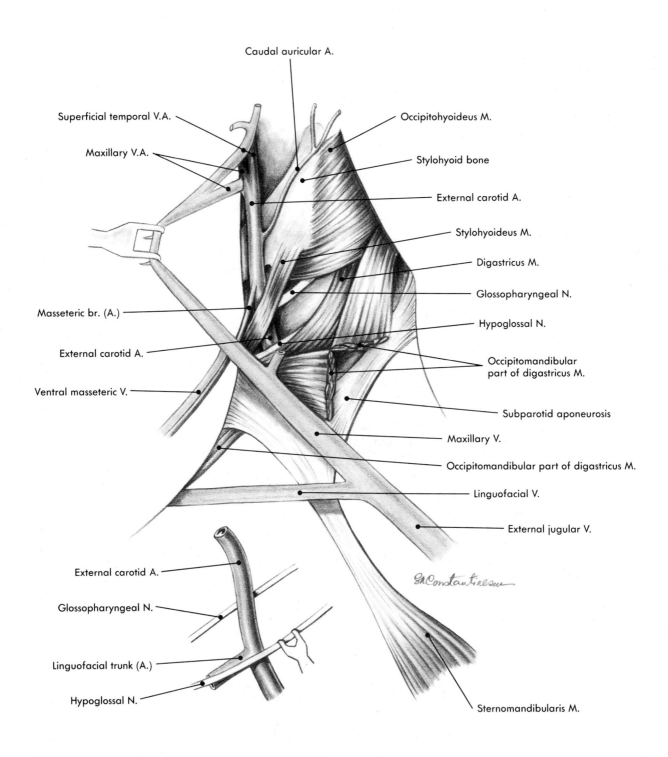

Caudal auricular A.

Superficial temporal V.A.

Maxillary V.A.

Occipitohyoideus M.

Stylohyoid bone

External carotid A.

Stylohyoideus M.

Digastricus M.

Glossopharyngeal N.

Masseteric br. (A.)

Hypoglossal N.

External carotid A.

Occipitomandibular part of digastricus M.

Ventral masseteric V.

Subparotid aponeurosis

Maxillary V.

Occipitomandibular part of digastricus M.

Linguofacial V.

External jugular V.

External carotid A.

Glossopharyngeal N.

Linguofacial trunk (A.)

Hypoglossal N.

Sternomandibularis M.

Fig. 8-3

Deep structures within parotid region.

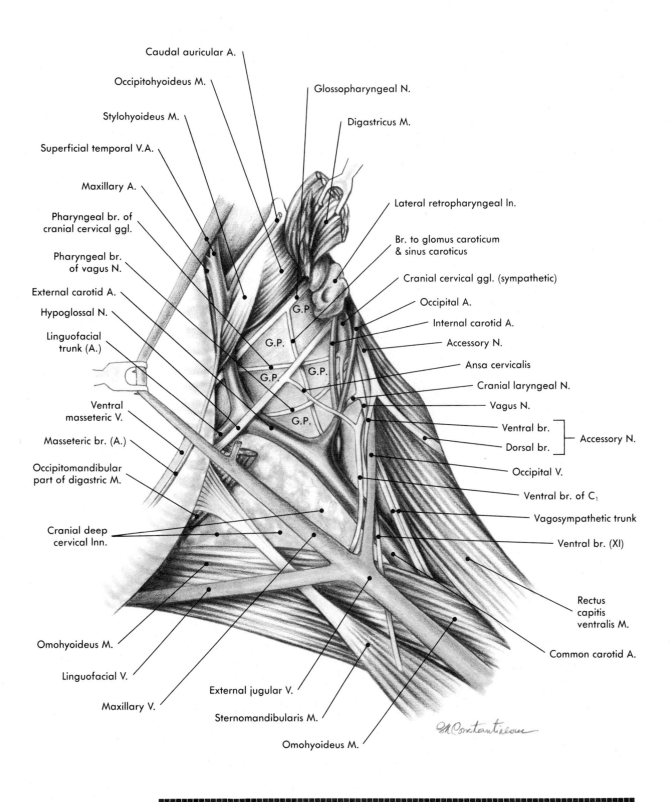

Caudal auricular A.

Occipitohyoideus M.

Stylohyoideus M.

Superficial temporal V.A.

Maxillary A.

Pharyngeal br. of cranial cervical ggl.

Pharyngeal br. of vagus N.

External carotid A.

Hypoglossal N.

Linguofacial trunk (A.)

Ventral masseteric V.

Masseteric br. (A.)

Occipitomandibular part of digastric M.

Cranial deep cervical lnn.

Omohyoideus M.

Linguofacial V.

Maxillary V.

External jugular V.

Sternomandibularis M.

Omohyoideus M.

Glossopharyngeal N.

Digastricus M.

Lateral retropharyngeal ln.

Br. to glomus caroticum & sinus caroticus

Cranial cervical ggl. (sympathetic)

Occipital A.

Internal carotid A.

Accessory N.

Ansa cervicalis

Cranial laryngeal N.

Vagus N.

Ventral br.

Dorsal br.

Accessory N.

Occipital V.

Ventral br. of C₁

Vagosympathetic trunk

Ventral br. (XI)

Rectus capitis ventralis M.

Common carotid A.

G.P.

Deep structures related to guttural pouch (G.P.).

Fig. 8-4

Ox
▪▪▪▪
(Figs. 8-5
to 8-7)

Follow the same directions as for the previous species, and carefully dissect the specific structures of the large ruminants.

Caution! The ventral buccal br. of the facial N. is partially embedded in glandular tissue. The dissection of this nerve may begin at the ventral end of the parotid gland.

Caution! In the ox, the course of the external carotid A. is very often not as close to the stylohyoid bone as in other species because of a fibrous band that is attached to the temporal bone, ventral to the external acoustic meatus and to the caudal border of the mandible. This band functions similar to a hinge, suspending the artery.

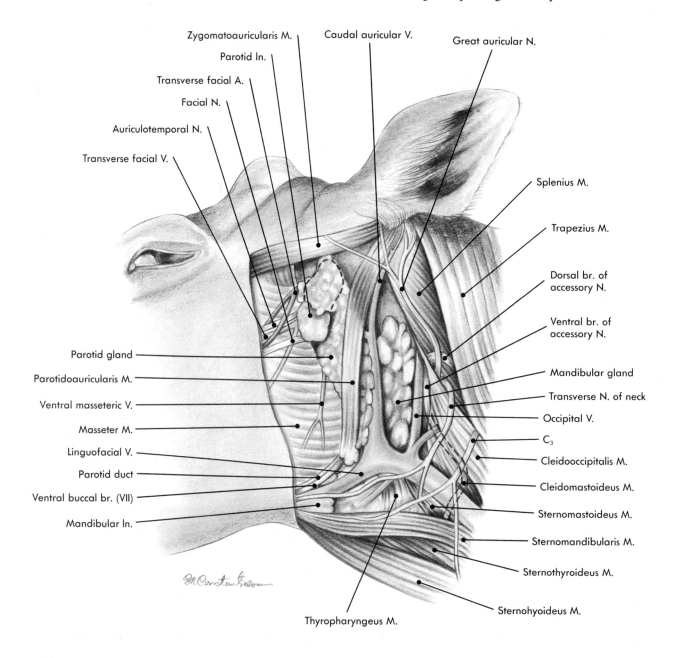

Zygomatoauricularis M.
Parotid ln.
Transverse facial A.
Facial N.
Auriculotemporal N.
Transverse facial V.
Caudal auricular V.
Great auricular N.
Splenius M.
Trapezius M.
Dorsal br. of accessory N.
Ventral br. of accessory N.
Mandibular gland
Transverse N. of neck
Occipital V.
C₃
Cleidooccipitalis M.
Cleidomastoideus M.
Sternomastoideus M.
Sternomandibularis M.
Sternothyroideus M.
Sternohyoideus M.
Thyropharyngeus M.
Mandibular ln.
Ventral buccal br. (VII)
Parotid duct
Linguofacial V.
Masseter M.
Ventral masseteric V.
Parotidoauricularis M.
Parotid gland

Fig. 8-5   Superficial structures.

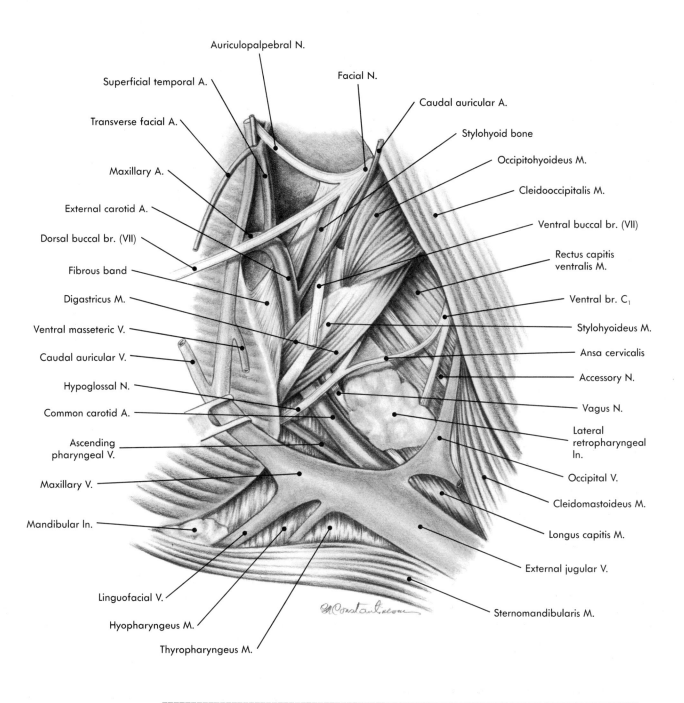

Auriculopalpebral N.

Superficial temporal A.

Transverse facial A.

Maxillary A.

External carotid A.

Dorsal buccal br. (VII)

Fibrous band

Digastricus M.

Ventral masseteric V.

Caudal auricular V.

Hypoglossal N.

Common carotid A.

Ascending pharyngeal V.

Maxillary V.

Mandibular ln.

Linguofacial V.

Hyopharyngeus M.

Thyropharyngeus M.

Facial N.

Caudal auricular A.

Stylohyoid bone

Occipitohyoideus M.

Cleidooccipitalis M.

Ventral buccal br. (VII)

Rectus capitis ventralis M.

Ventral br. C₁

Stylohyoideus M.

Ansa cervicalis

Accessory N.

Vagus N.

Lateral retropharyngeal ln.

Occipital V.

Cleidomastoideus M.

Longus capitis M.

External jugular V.

Sternomandibularis M.

Deep adjacent structures to facial N. and external carotid A.

Fig. 8-6

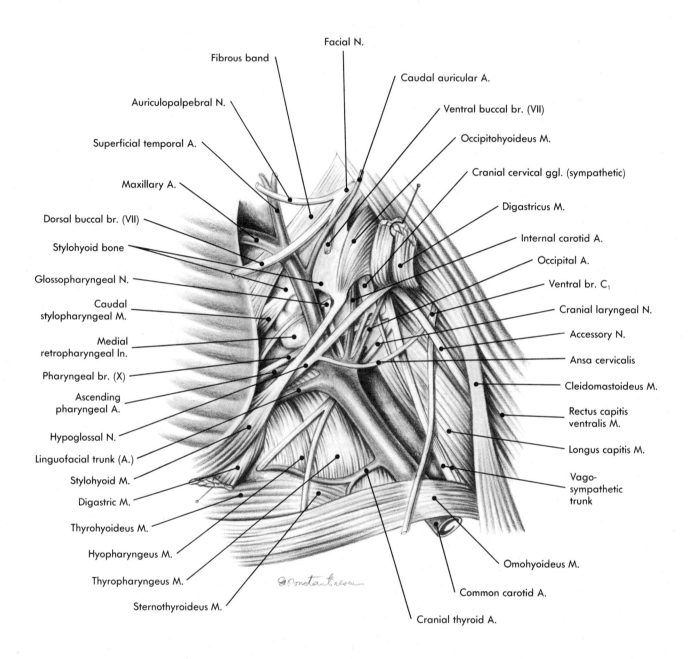

Fibrous band

Facial N.

Auriculopalpebral N.

Caudal auricular A.

Ventral buccal br. (VII)

Superficial temporal A.

Occipitohyoideus M.

Maxillary A.

Cranial cervical ggl. (sympathetic)

Dorsal buccal br. (VII)

Digastricus M.

Stylohyoid bone

Internal carotid A.

Glossopharyngeal N.

Occipital A.

Caudal stylopharyngeal M.

Ventral br. C₁

Medial retropharyngeal ln.

Cranial laryngeal N.

Pharyngeal br. (X)

Accessory N.

Ascending pharyngeal A.

Ansa cervicalis

Hypoglossal N.

Cleidomastoideus M.

Linguofacial trunk (A.)

Rectus capitis ventralis M.

Stylohyoid M.

Longus capitis M.

Digastric M.

Vago-sympathetic trunk

Thyrohyoideus M.

Hyopharyngeus M.

Omohyoideus M.

Thyropharyngeus M.

Common carotid A.

Sternothyroideus M.

Cranial thyroid A.

Fig. 8-7

Deepest structures of area.

Carefully follow the directions of the horse parotid region and adapt them to sheep.

Caution! Similar to the horse, a fibrous band unites the tendon of the sternomastoideus M. to the aponeurosis of the cleidooccipitalis M. and overlaps the occipital V., accessory N., and caudal border of the lateral retropharyngeal ln.

Sheep
■■■■■■■■■
(Figs. 8-8
to 8-9)

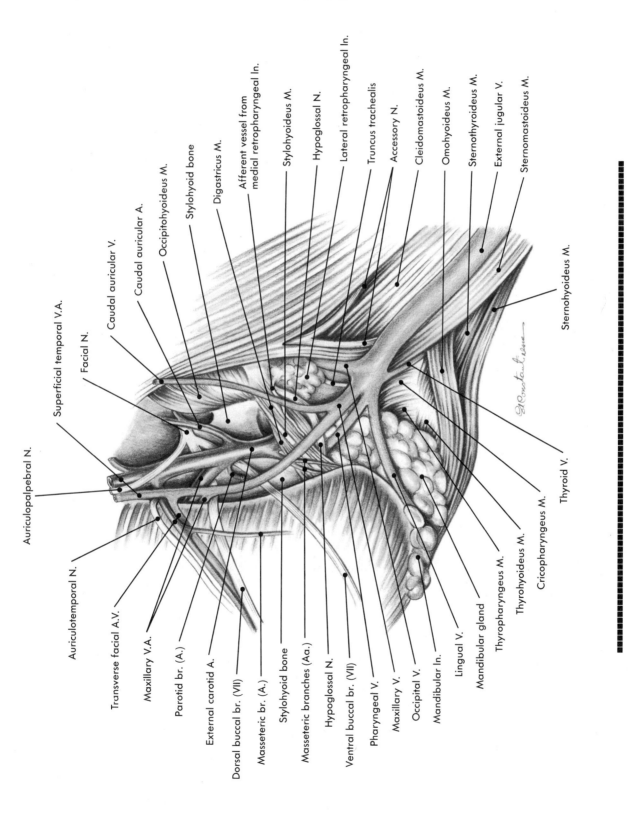

Fig. 8-8

Deep structures after parotid gland has been removed.

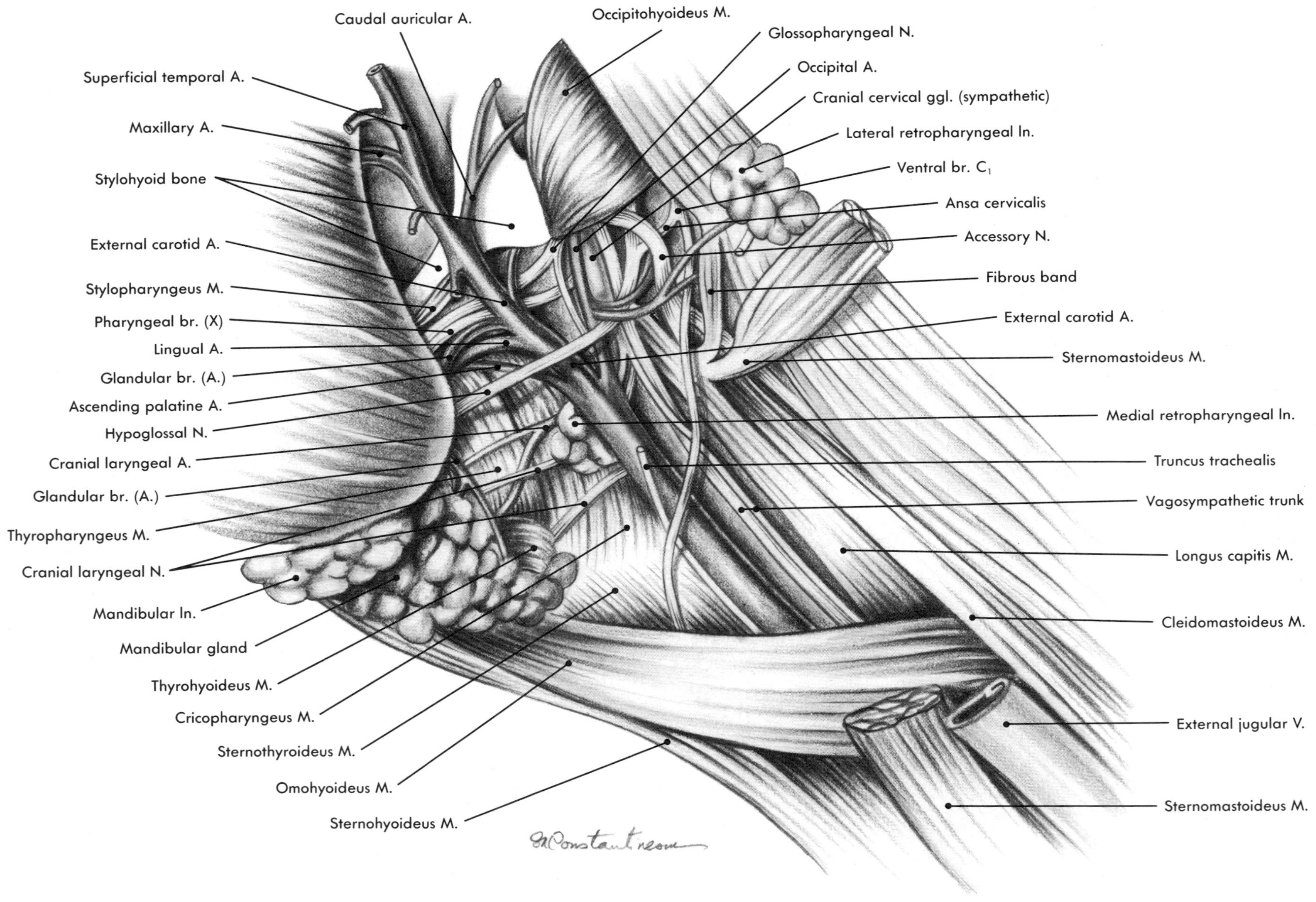

Superficial temporal A.

Maxillary A.

Stylohyoid bone

External carotid A.

Stylopharyngeus M.

Pharyngeal br. (X)

Lingual A.

Glandular br. (A.)

Ascending palatine A.

Hypoglossal N.

Cranial laryngeal A.

Glandular br. (A.)

Thyropharyngeus M.

Cranial laryngeal N.

Mandibular ln.

Mandibular gland

Thyrohyoideus M.

Cricopharyngeus M.

Sternothyroideus M.

Omohyoideus M.

Sternohyoideus M.

Caudal auricular A.

Occipitohyoideus M.

Glossopharyngeal N.

Occipital A.

Cranial cervical ggl. (sympathetic)

Lateral retropharyngeal ln.

Ventral br. C₁

Ansa cervicalis

Accessory N.

Fibrous band

External carotid A.

Sternomastoideus M.

Medial retropharyngeal ln.

Truncus trachealis

Vagosympathetic trunk

Longus capitis M.

Cleidomastoideus M.

External jugular V.

Sternomastoideus M.

Deepest structures of area.

Fig. 8-9

Take the horse as a model for the dissection of the parotid region in goat.

Goat
■■■■■■■■
(Figs. 8-10
to 8-12)

Caution! Ventral to the digastric M., a muscle for the cervical appendix emerges and is located between the digastric M. and the hypoglossal N. This muscle lies ventral and parallel with the caudal border of the mandibular salivary gland, between it and the cervical thymus.

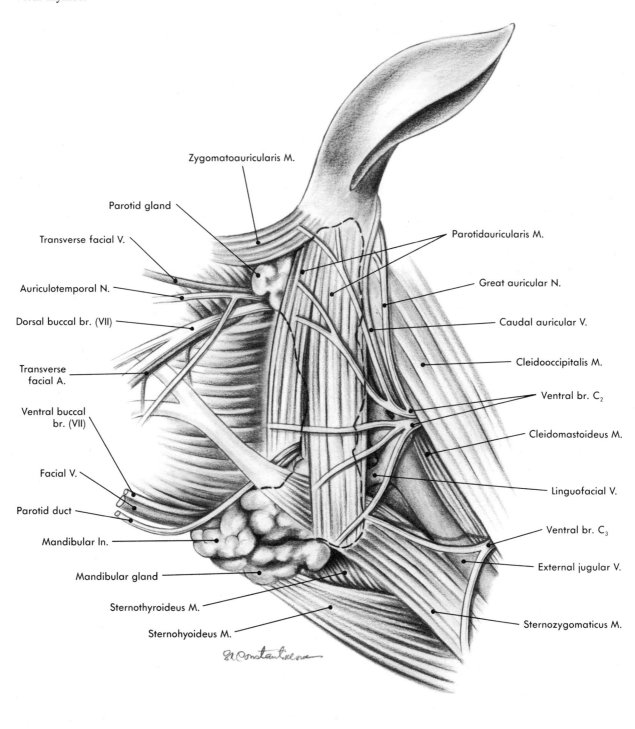

Zygomatoauricularis M.

Parotid gland

Transverse facial V.

Auriculotemporal N.

Dorsal buccal br. (VII)

Transverse facial A.

Ventral buccal br. (VII)

Facial V.

Parotid duct

Mandibular ln.

Mandibular gland

Sternothyroideus M.

Sternohyoideus M.

Parotidauricularis M.

Great auricular N.

Caudal auricular V.

Cleidooccipitalis M.

Ventral br. C$_2$

Cleidomastoideus M.

Linguofacial V.

Ventral br. C$_3$

External jugular V.

Sternozygomaticus M.

Superficial structures.

Fig. 8-10

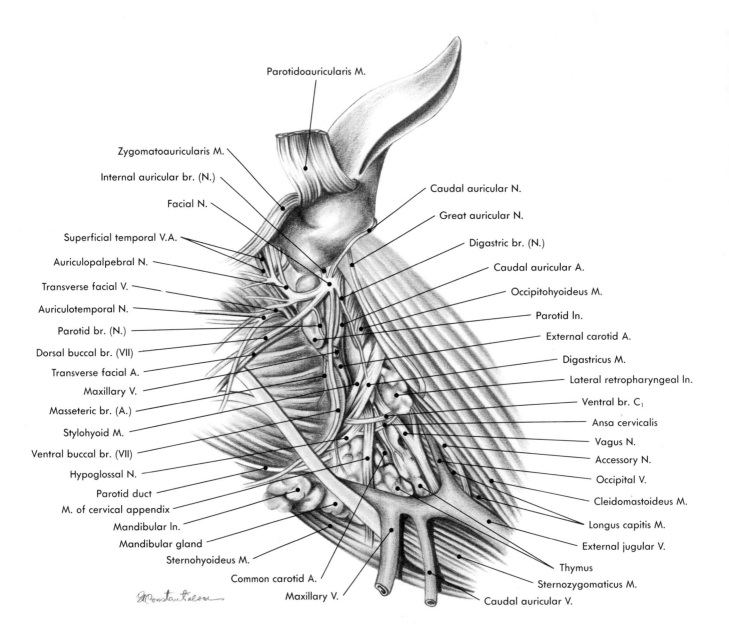

Fig. 8-11   Deep structures of area. (Parotid gland has been removed.)

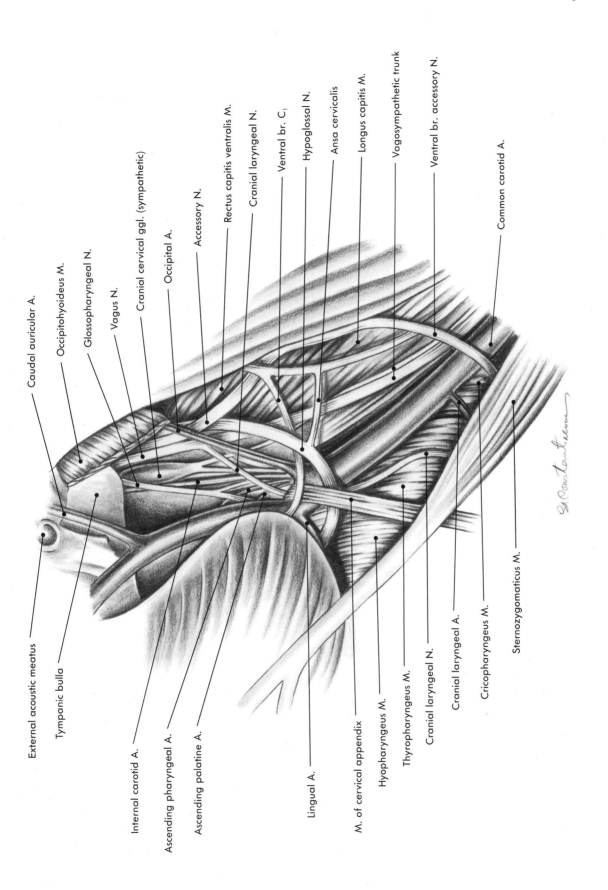

Caudal auricular A.

Occipitohyoideus M.

Glossopharyngeal N.

Vagus N.

Cranial cervical ggl. (sympathetic)

Occipital A.

Accessory N.

Rectus capitis ventralis M.

Cranial laryngeal N.

Ventral br. C₁

Hypoglossal N.

Ansa cervicalis

Longus capitis M.

Vagosympathetic trunk

Ventral br. accessory N.

Common carotid A.

External acoustic meatus

Tympanic bulla

Internal carotid A.

Ascending pharyngeal A.

Ascending palatine A.

Lingual A.

M. of cervical appendix

Hyopharyngeus M.

Thyropharyngeus M.

Cranial laryngeal N.

Cranial laryngeal A.

Cricopharyngeus M.

Sternozygomaticus M.

Fig. 8-12

Deepest structures of area.

Pig
▪▪▪▪▪
(Figs. 8-13
to 8-15)

The dissection of this region in pig is a little difficult due to the large amount of fat, the narrow transition between head and neck, and the short neck of the pig.

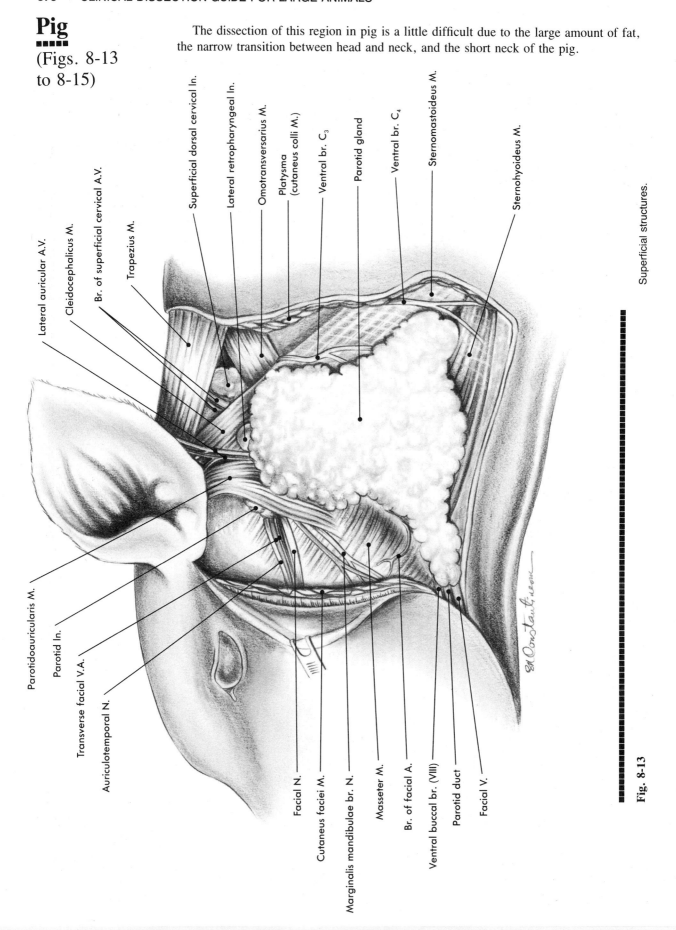

Superficial structures.

Fig. 8-13

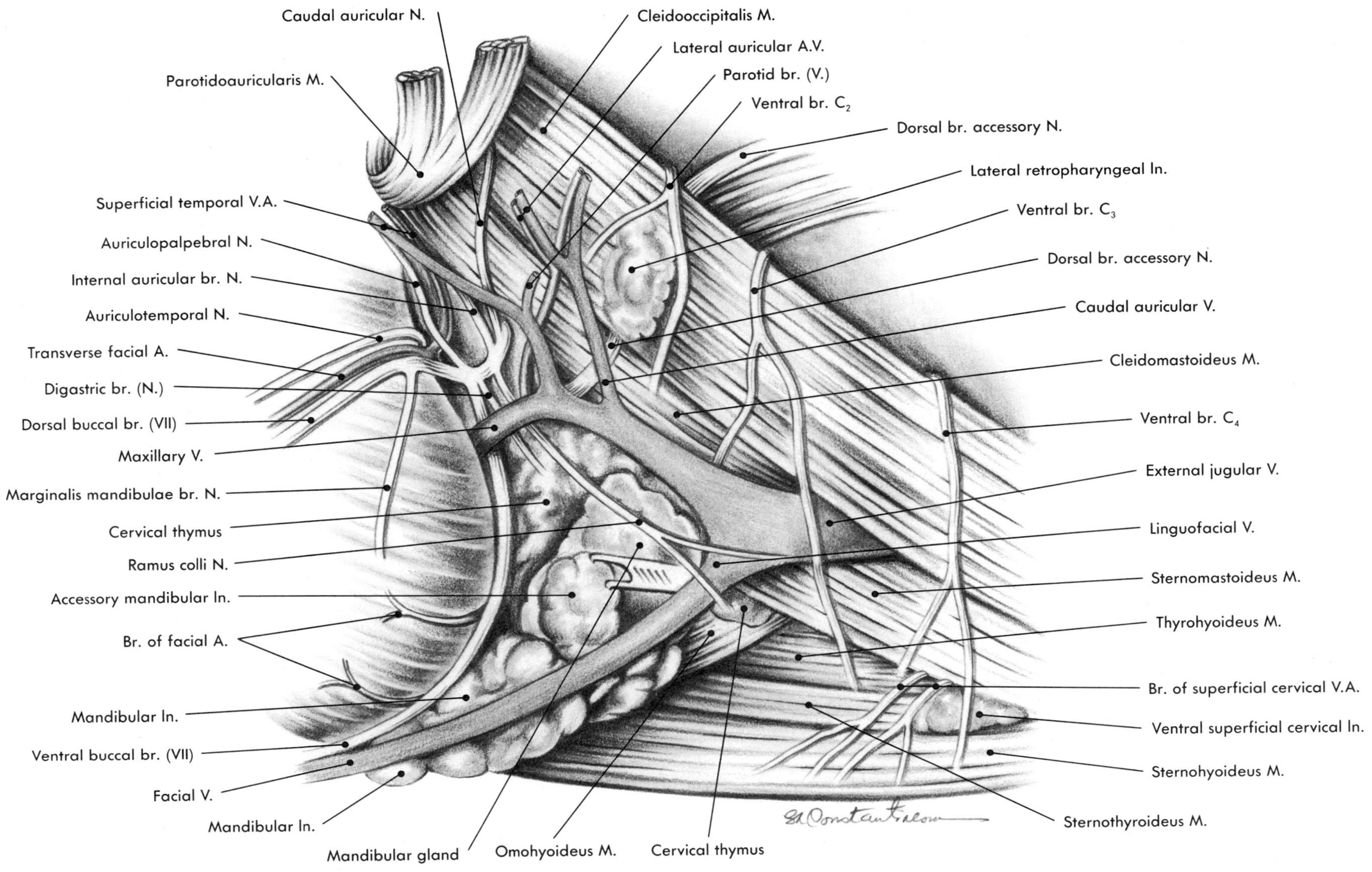

Caudal auricular N.

Parotidoauricularis M.

Cleidooccipitalis M.

Lateral auricular A.V.

Parotid br. (V.)

Ventral br. C₂

Dorsal br. accessory N.

Lateral retropharyngeal ln.

Ventral br. C₃

Dorsal br. accessory N.

Caudal auricular V.

Cleidomastoideus M.

Ventral br. C₄

External jugular V.

Linguofacial V.

Sternomastoideus M.

Thyrohyoideus M.

Br. of superficial cervical V.A.

Ventral superficial cervical ln.

Sternohyoideus M.

Sternothyroideus M.

Superficial temporal V.A.

Auriculopalpebral N.

Internal auricular br. N.

Auriculotemporal N.

Transverse facial A.

Digastric br. (N.)

Dorsal buccal br. (VII)

Maxillary V.

Marginalis mandibulae br. N.

Cervical thymus

Ramus colli N.

Accessory mandibular ln.

Br. of facial A.

Mandibular ln.

Ventral buccal br. (VII)

Facial V.

Mandibular ln.

Mandibular gland

Omohyoideus M.

Cervical thymus

Fig. 8-14

Deep structures. (Parotid gland has been removed.)

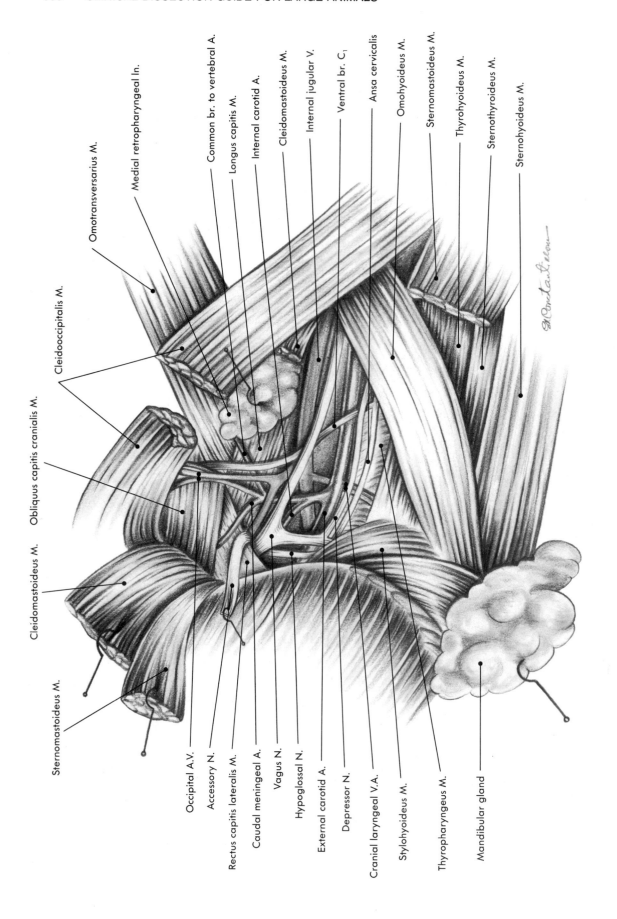

Omotransversarius M.

Medial retropharyngeal ln.

Common br. to vertebral A.

Longus capitis M.

Internal carotid A.

Cleidomastoideus M.

Internal jugular V.

Ventral br. C₁

Ansa cervicalis

Omohyoideus M.

Sternomastoideus M.

Thyrohyoideus M.

Sternothyroideus M.

Sternohyoideus M.

Cleidooccipitalis M.

Obliquus capitis cranialis M.

Cleidomastoideus M.

Sternomastoideus M.

Occipital A.V.

Accessory N.

Rectus capitis lateralis M.

Caudal meningeal A.

Vagus N.

Hypoglossal N.

External carotid A.

Depressor N.

Cranial laryngeal V.A.

Stylohyoideus M.

Thyropharyngeus M.

Mandibular gland

Deepest structures of area.

Fig. 8-15

Chapter 9

Landmarks and approaches of main anatomic structures of the horse susceptible to clinical intervention

In this chapter, the knowledge accumulated in the anatomy laboratory on cadavers is transposed to the living specimens. *"Hic locus est ubi mors gaudet succurrere vitae"*— "Here is the place where the death is glad to be of use to the life." This eloquent sentence is written on the wall of one of the lecture rooms of the School of Medicine in Vienna, Austria.

Keeping the meaning of this axiom in mind, some of the structures previously dissected are discussed again with regard to their landmarks and clinical approaches. The same sequential description of body regions throughout this book (neck, thorax, abdomen, and so on) is followed here. Structures of primary clinical significance are emphasized, and their topographic positioning and relation to pertinent structures encountered during clinical approaches is illustrated.

I. External jugular V., common carotid A., vagosympathetic trunk (N.), and recurrent laryngeal N. (Fig. 9-1, *A*)
  A. *Landmarks:* Jugular groove (bordered by the cleidomastoideus M. [dorsally] and the sternocephalicus M. [ventrally])
  B. *Approach:*
    1. The external jugular V. in the middle third of the groove, deep to the cutaneus colli M. and the superficial lamina of cervical fascia, separated by the omohyoideus M. from the common carotid A.
    2. The rest of the structures in the caudal third of the groove, deep to the superficial lamina of cervical fascia and within the carotid sheath
II. Superficial cervical A., cephalic V. (the brachial segment) (Fig. 9-1, *B*)
  A. *Landmarks:* Lateral pectoral groove (bordered by the cleidobrachialis M. [laterally] and the descending pectoral M. [medially])
  B. *Approach:* Through the skin and cutaneus colli M.
III. Superficial cervical lnn. (Fig. 9-1, *A* and *B*)
  A. *Landmarks:* Shoulder joint; subclavius M.; jugular groove; lateral pectoral groove; and omotransversarius and cleidomastoideus Mm.
  B. *Approach:* Midway between the point 10 cm dorsal to shoulder joint and the merger of jugular groove and lateral pectoral groove, deep to omotransversarius and cleidomastoideus Mm., and at the cranial border of subclavius M.

NECK
(Figs. 9-1 and 9-2)

IV. Trachea and esophagus (Fig. 9-1, *B*)

A. *Landmarks:* The two symmetrical sternocephalicus Mm.

B. *Approach:* Through the ventral midline of the neck and between the two sternocephalicus Mm. The sternothyroidei and sternohyoidei Mm. accompany the ventral aspect of the trachea. The trachea is superficial and easily palpable in the cranial two-thirds of the neck.

Caution! The external jugular V. lies superficially in the jugular groove, whereas the common carotid A., vagosympathetic trunk (N.), and recurrent laryngeal N. lie deep in the groove, along the lateral aspect of trachea. The esophagus accompanies the trachea dorsally and slightly toward the left side.

V. Vertebral bodies of C_3 to C_5 (Fig. 9-1, *B*)

A. *Landmarks:* Transverse processes of C_3, C_4, and C_5; the two symmetrical sternocephalicus Mm.

B. *Approach:* Through the ventral midline of the neck, between the two sternocephalicus Mm., and by reflecting the trachea, esophagus, vessels, and nerves at the corresponding level C_3, C_4, or C_5.

VI. Atlantooccipital space (Fig. 9-2, *A*)

A. *Landmarks:* Wing of atlas; nuchal crest; and funiculus nuchae.

B. *Approach:* Between the three landmarks.

VII. Cranial nuchal (atlantal) bursa (Fig. 9-2, *A*)

A. *Landmarks:* Wing of atlas and funiculus nuchae.

B. *Approach:* Midway between the cranial and caudal extents of the wing of atlas, between funiculus nuchae and the dorsal arch of atlas.

VIII. Caudal nuchal bursa (Fig. 9-2, *A*)

A. *Landmarks:* Transverse process of axis and funiculus nuchae.

B. *Approach:* At a point 5 cm dorsocranial to the transverse process of axis and between the funiculus nuchae and the spinous process of axis.

IX. Larynx (Fig. 9-2, *A*)

A. *Landmarks:* Ramus mandibulae; tendon of sternocephalicus M.; and linguofacial V.

B. *Approach:* With the head in extension, through the skin and omohyoideus M., the larynx is palpable slightly caudal to the mandible and cranial to the tendon of sternocephalicus M. and is obliquely crossed by the linguofacial V. or is between the two rami mandibulae. The corniculate processes of the arytenoid cartilages are prominent in palpation. The cricoarytenoidei dorsalis Mm. can be palpated dorsolaterally.

X. Cricothyroid lig. (Fig. 9-2, *B*)

A. *Landmarks:* Laryngeal prominence (body of the thyroid cartilage); caudal thyroid notch; and cricoid arch.

B. *Approach:* Between the three landmarks on the ventral midline of larynx.

XI. Laryngeal ventricle (Fig. 9-2, *B*)

A. *Landmarks:* Caudal thyroid notch; thyroid laminae (wings); and cricoid arch.

B. *Approach:* Between the three landmarks, on the medial aspect of the thyroid lamina, and within the larynx (for surgical approach). The laryngeal ventricles cannot be palpated externally but can readily be viewed endoscopically.

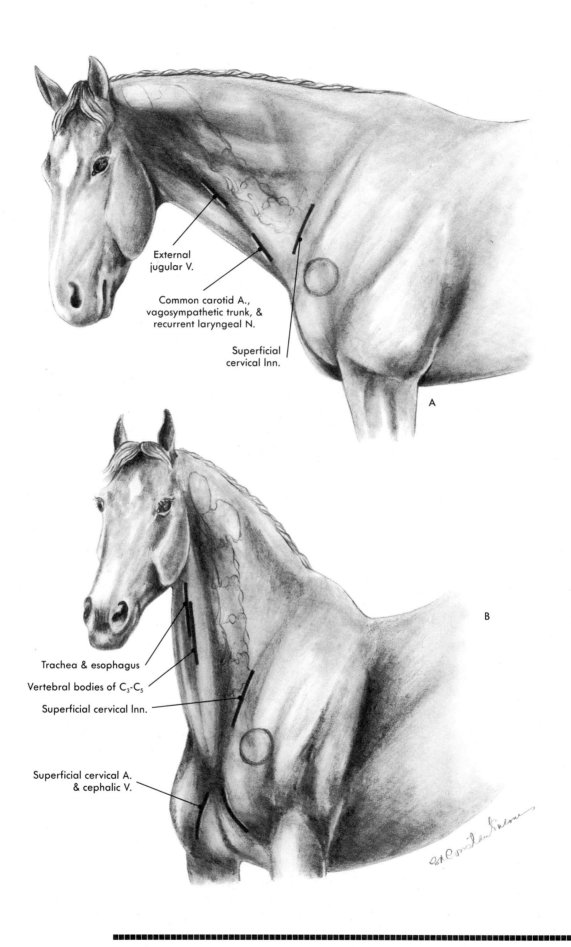

External
jugular V.

Common carotid A.,
vagosympathetic trunk, &
recurrent laryngeal N.

Superficial
cervical Inn.

A

B

Trachea & esophagus

Vertebral bodies of C₃-C₅

Superficial cervical Inn.

Superficial cervical A.
& cephalic V.

A, Structures in jugular groove and superficial cervical Inn. **B,** Jugular and lateral pectoral grooves
and superficial cervical Inn.

Fig. 9-1

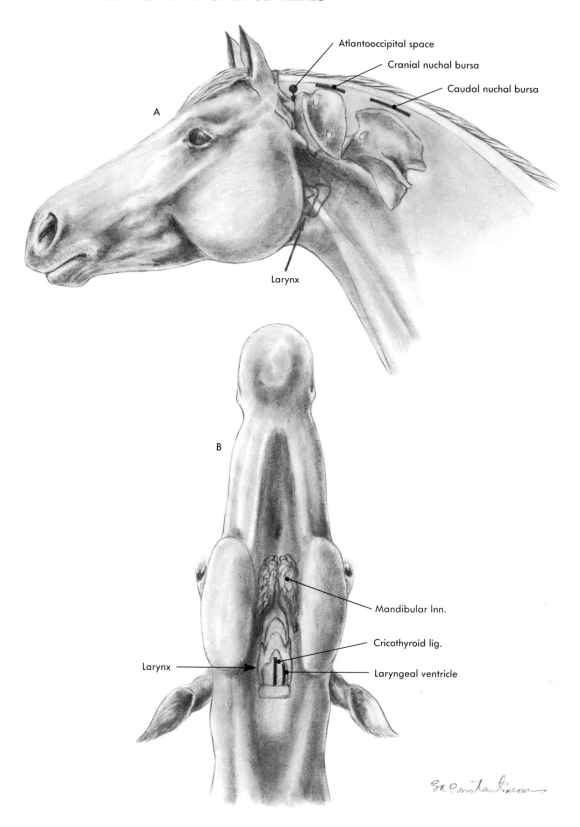

Fig. 9-2   **A,** Structures related to first two cervical vertebrae and larynx (lateral aspect). **B,** Laryngeal structures (ventral view).

I. Superficial thoracic V. (Fig. 9-3, *A*)
   A. *Landmarks:* Olecranon and dorsal border of deep pectoral M.
   B. *Approach:* From the olecranon and along the dorsal border of deep pectoral M., which is covered by cutaneus trunci M.
   Caution! The vein is paralleled by the lateral thoracic N.

II. Supraspinous bursa (Fig. 9-3, *A*)
   A. *Landmarks:* Supraspinous lig. and the highest point of withers, which corresponds to the spinous processes of T_2 to T_4.
   B. *Approach:* Between the two landmarks.

III. Intercostal Aa.Vv.Nn. (Fig. 9-3, *A*)
   A. *Landmarks:* Caudal border of ribs VI (or V) to XVII
   B. *Approach:* At the caudal border of a rib, between the internal intercostal M. and the endothoracic fascia (along the entire costal area in a dorsoventral direction) and in the middle of an intercostal space, between the two intercostal Mm. (in the proximal half of costal area). To reach these structures, pass through the skin, cutaneus trunci M., and other structures such as latissimus dorsi, serratus ventralis thoracis, serratus dorsalis, and obliquus abdominis externus Mm.

IV. Line of diaphragmatic pleural reflection (Fig. 9-3, *B*)
   A. *Landmarks:* The ribs and costal cartilages and the projection of the dorsal wall of the thoracic cavity between tuber coxae and tuberosity of scapular spine.
   B. *Approach:* From the sixth costochondral joint in a convex line to the intersection of the rib XVII with the projection of the dorsal wall of the thoracic cavity.

V. Area of auscultation and percussion of the lungs (Fig. 9-3, *B*)
   A. *Landmarks:* The projection of the dorsal wall of the thoracic cavity; the ribs and costal cartilages; and the caudal border of triceps brachii M.
   B. *Approach:* From a point 10 cm dorsal to the sixth costochondral joint in a slight ventrocaudally oriented convex line to the intersection of the sixteenth intercostal space with the projection of the dorsal wall of the thoracic cavity, then along the projection of the dorsal wall and the caudal border of the triceps brachii M.

VI. Auscultation of the heart (Fig. 9-4, *A*)
   A. *Landmarks:* Intercostal spaces III to V; a horizontal line passing through the shoulder joint; the caudal border of triceps brachii M.; and the olecranon.
   B. *Approach:*
      1. Pulmonary valve in the left third intercostal space (a hand ventral to the horizontal line passing through the shoulder joint).
      2. Aortic valve on or ventral to the horizontal line passing through the shoulder joint (in the left fourth intercostal space).
      3. Left atrioventricular valve in the fifth intercostal space and one and one-half hand (12 to 15 cm) dorsal to the olecranon or one hand ventral to the horizontal line passing through the shoulder joint.
      4. Right atrioventricular valve in the fourth intercostal space and one hand dorsal to the olecranon; on the left side, in the third intercostal space and one hand dorsal to the olecranon.
      Caution! Push the triceps brachii M. as far cranially as possible to widely expose the area of auscultation.

VII. Percussion of the heart (Fig. 9-4, *A*)
   A. *Landmarks:* The second to sixth intercostal spaces; the line passing through the shoulder joint; the caudal border of triceps brachii M.; the olecranon; and the sternum.

THORAX
(Figs. 9-3 and 9-4)

B. *Approach:*

1. The area of relative dullness extends between the second and sixth intercostal spaces, one hand dorsal to the line passing through the shoulder joint, and the sternum.

2. The area of absolute dullness extends on the left side between the third and fifth intercostal spaces, one hand dorsal to the olecranon in the fourth space and 2 to 3 cm dorsal to the olecranon in the fifth space, and the sternum. On the right side, the area of absolute dullness extends between the third and fourth intercostal spaces, 2 to 3 cm dorsal to the olecranon in the fourth space, and the sternum.

VIII. Cervicothoracic (stellate) ggl. (Fig. 9-4, *B*)

A. *Landmarks:* Jugular groove, last cervical vertebrae, and subclavius M.

B. *Approach:* A long gauge needle is inserted in the jugular groove at the intersection with the subclavius M. and through the skin and cutaneus colli M. in a horizontal position; the needle is pointed toward the thoracic cavity, between the cervical vertebrae and trachea, and along the medial aspect of rib I, after perforating the scalenus M.

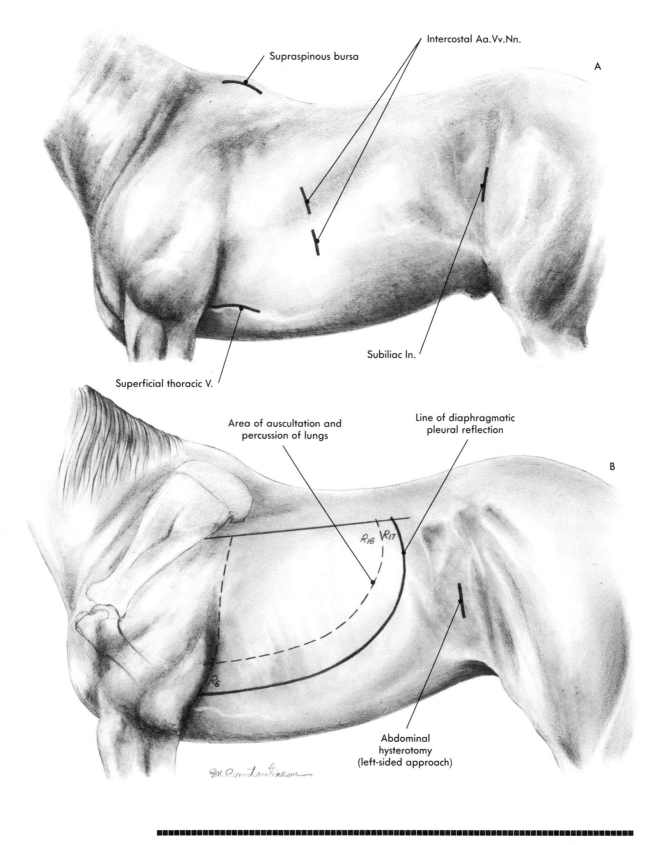

A, Structures on lateral aspect of body. **B,** Auscultation and percussion of lungs and surgical approach in left flank.

Fig. 9-3

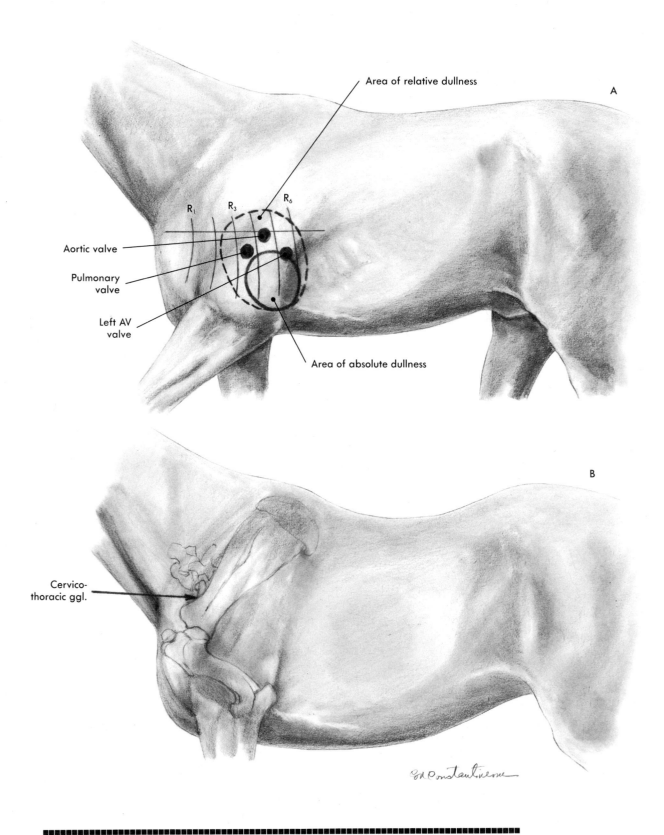

Fig. 9-4   **A,** Auscultation and percussion of heart. **B,** Approach to stellate ggl.

I. Subiliac ln. (Fig. 9-3, *A*)

    A. *Landmarks:* Tuber coxae; patella; and cranial border of tensor fasciae latae M.

    B. *Approach:* Midway between the tuber coxae and patella, on the cranial border of the tensor fasciae latae M.

II. Abdominal hysterotomy and hysterectomy (left-sided approach) (Fig. 9-3, *A*)

    A,B. *Landmarks and approach:* Same as in the exploration of the cecum or cecotomy, but on the left side.

III. External pudendal A. (Fig. 9-5, *A*)

    A. *Landmarks:* Linea alba; prepubic tendon; and superficial inguinal ring.

    B. *Approach:* The caudomedial commissure of the superficial inguinal ring, between the prepubic tendon and linea alba.

    Caution! In the male, the artery runs cranially and crosses the medial aspect of the spermatic cord, whereas, in the female, it runs 3 to 4 cm lateral to the linea alba. In both the male and female, the superficial inguinal (inguinofemoral) lnn. lie on this artery.

IV. Celiotomy (ventrotomy or ventral laparotomy) (Fig. 9-5, *A*)

    A. *Landmarks:* Umbilicus; linea alba; prepuce or mammary gland; and xiphoid cartilage.

    B. *Approach:* Length and placement of the incision line depends on the surgical procedure; for example, from the umbilicus toward the xiphoid cartilage on the linea alba in a cesarean section or from the umbilicus toward the prepuce or mammary gland on the linea alba in other interventions, including approaches to the urinary bladder.

V. Laparotomy (right-sided approach) (Fig. 9-5, *B*)

    α Cecocentesis

        A. *Landmarks:* Right paralumbar fossa (fossa of the flank).

        B. *Approach:* In the middle of the right paralumbar fossa, midway between the last rib and the tensor fasciae latae M., and one hand ventral to the transverse processes of the lumbar vertebrae.

    β Exploration of the cecum or cecotomy

        A. *Landmarks:* Slope of the right flank and cranial border of the tensor fasciae latae M.

        B. *Approach:* Vertical incision in the slope of flank, one hand cranial to and parallel with the cranial border of tensor fasciae latae M.

VI. Left celiac ganglion and plexus block (Fig. 9-6, *A*)

    A. *Landmarks:* The last intercostal space and the dorsal midline.

    B. *Approach:* At a point 10 cm paramedian in the last intercostal space, through the epaxial muscles, and to the roof of the abdominal (not peritoneal) cavity.

    Caution! The celiac ggl. and plexus are located outside of the peritoneal cavity, toward the roof of the abdominal cavity, and at the cranial pole of the kidney. The approach of the right celiac ggl. and plexus, located in the fifteenth to sixteenth intercostal spaces is difficult because of the close relationship between the kidney and the caudate lobe of the liver.

VII. Projection of viscera on the right abdominal wall (see p. 92, Fig. 3-3, *A*).

VIII. Projection of viscera on the left abdominal wall (see p. 92, Fig. 3-3, *B*).

IX. Projection of viscera on the ventral abdominal wall (see p. 94, Fig. 3-4).

ABDOMEN
(Figs. 9-5 to 9-9)

X. Rectal examination of abdominopelvic viscera and other structures (transrectal palpation, exploration of abdominopelvic viscera per rectum, or rectal exploration)

A. *Landmarks* (in caudocranial order) (Fig. 9-6, *B*):

1. The hand with the palmar aspect oriented dorsally.
   a. On the midline.
      i. The pelvic surface of sacrum (with the ventral sacral foramina).
      ii. The promontory (the most prominent structure of the sacrum located ventrocranially).
      iii. The bodies of the lumbar vertebrae, covered ventrally by the aorta (left) and caudal vena cava (right).
   b. On both lateral sides.
      i. The sacrosciatic ligament.
      ii. The body of ilium with the internal obturator M.
      iii. The flank.
      iv. The last rib.

2. The hand with the palmar aspect oriented ventrally.
   a. On the midline.
      i. The internal obturator M.
      ii. The cranial border of the pubis.

B. Palpable structures within the *pelvic cavity* (on the floor of the pelvic cavity in a caudocranial order);

1. In the male (Fig. 9-7, *A*).
   a. The flexure of urethra.
   b. The accessory genital glands.
      i. The bulbourethral glands.
      ii. The prostate gland (difficult to identify in horse).
      iii. The seminal vesicles.
   c. The pelvic part of urethra.
   d. The genital fold with the corresponding deferent ducts.
   e. The urinary bladder (the apex is sometimes dropped into the abdominal cavity depending on the plenitude of the bladder).

2. In the female (Fig. 9-7, *B*).
   a. The vestibule and vagina.
   b. The cervix.
   c. The body of the uterus and its left and right horns (in older, multiparous mares and pregnant mares, it is usually within the abdominal cavity; in young, nongravid females, these structures may still be within the pelvic cavity).
   d. The urinary bladder (see above).

C. Palpable structures within the *abdominal cavity*.

1. In the male.
   a. Both left and right ductus deferens continue to course from the pelvic cavity to the pelvic inlet where they enter the vaginal rings craniolateroventrally.

2. In the female.
   a. The uterine horns are convex ventrally and course laterally and dorsally toward the roof of the abdominal cavity.
   b. The uterine tubes are easily identified, only if enlarged.
   c. The two ovaries are located dorsal, lateral, and slightly cranial to the respective tips of the uterine horns. They lie in a caudoventral position to the corresponding kidneys.

3. In both the male and female (caudocranially) (Fig. 9-7, *C* and *D* to 9-9, *A*).
   a. On the left side.
      i. The descending colon (dorsal in the paralumbar fossa and with sacculations and bands).
      ii. The jejunum (medial and cranial, smooth and without sacculations and bands).
      iii. The left kidney and the left renal A. (only operators with a long reach can palpate the entire left kidney; the caudal pole of the left kidney is frequently palpable).
      iv. The spleen (the caudal border of the spleen can usually be palpated against the left abdominal wall, lateral to the left colons along the last rib).
      v. The pelvic flexure of the ascending colon (in the pelvic inlet ventrally but moveable; may be located in general area of pelvic inlet).
      vi. The left dorsal and left ventral colons (continue cranially from the pelvic flexure), just medial to the spleen and left body wall.
      vii. The intercolic fold (between the left dorsal and left ventral colons).
      viii. The left ureter on the left side of abdominal aorta (difficult, but not impossible to palpate, unless abnormal).
   b. On the right side.
      i. The base of cecum and the proximal part of its body, with the corresponding sacculations and bands.
      ii. The transverse colon (on the medial aspect of the base of cecum), often too far cranially to palpate.
      iii. The right dorsal colon.
      iv. The right ureter on the right side of caudal vena cava (difficult, but not impossible, to palpate, unless abnormal).
D. Palpation of the arteries and lymph nodes in the pelvic and abdominal cavities (in caudocranial order).
   1. Caudal gluteal A. with the sacral branches.
   2. Obturator A. (on the body of the ilium).
   3. Lateral vesical ligg. with the umbilical Aa.
   4. Internal iliac A. with the sacral lnn. (lateral side of L_6 and S_1).
   5. External iliac A. with the medial iliac lnn. (cranial to the internal iliac A.).
   6. The quadrifurcation of the aorta (ventral to L_5).
   7. Caudal mesenteric A. (can frequently be palpated).
   8. Cranial mesenteric A. with the mesenteric lnn. (can occasionally be palpated in association with the root of the cranial mesentery as it branches from the abdominal aorta. It can be located too far cranially to be palpated).
XI. Paravertebral thoracolumbar approach (Fig. 9-9, *B*).
   A. *Landmarks:* Last rib and transverse processes of L_1 to L_3.
   B. *Approach:* At a point 10 to 15 cm from the midline (depending on the size of the specimen), perpendicular and parallel to the caudal border of last rib, and through the skin, thoracolumbar fascia, serratus dorsalis caudalis, and retractor costae Mm.; between L_1 and L_2 and between L_2 and L_3 (when palpation of the extremities of transverse processes is possible), 2 to 3 cm deep in a vertical position, and through the skin, thoracolumbar fascia, external and internal abdominal oblique Mm. and up to transversus abdominis M.

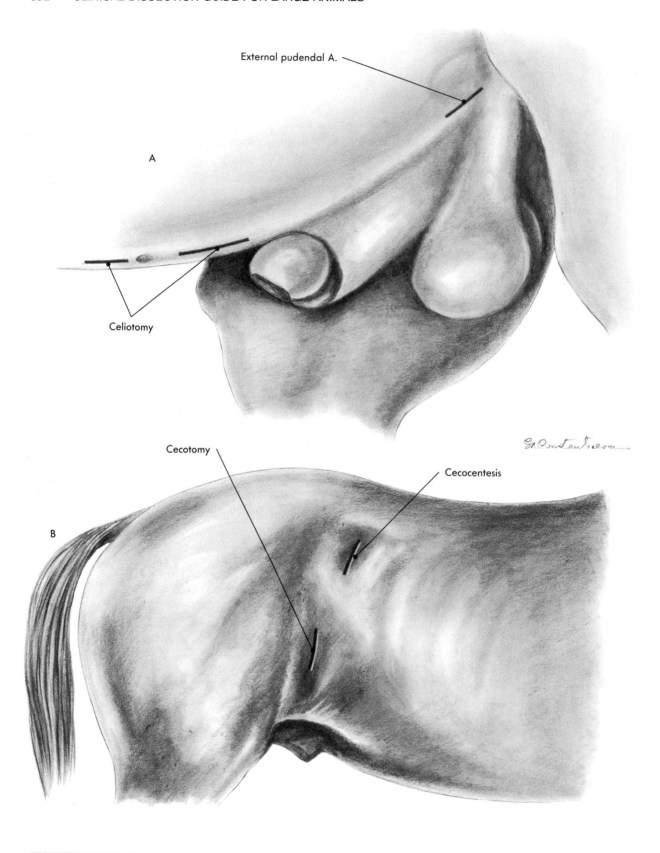

Fig. 9-5   **A,** Surgical approaches to genital and abdominal structures. **B,** Approaches to cecum.

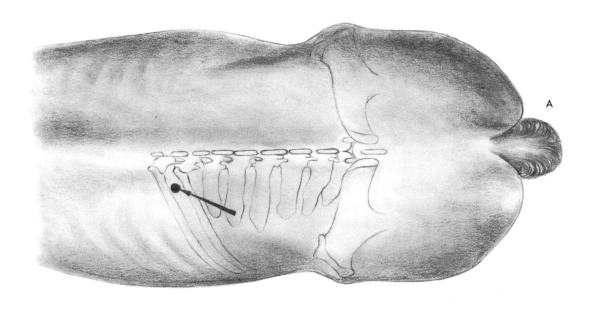

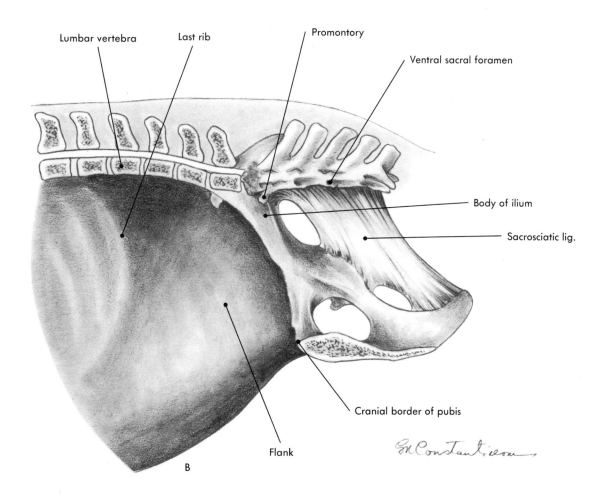

Lumbar vertebra   Last rib   Promontory   Ventral sacral foramen

Body of ilium

Sacrosciatic lig.

Cranial border of pubis

Flank

B

A, Left celiac ggl. and plexus block. *B*, Landmarks for rectal exploration.

Fig. 9-6

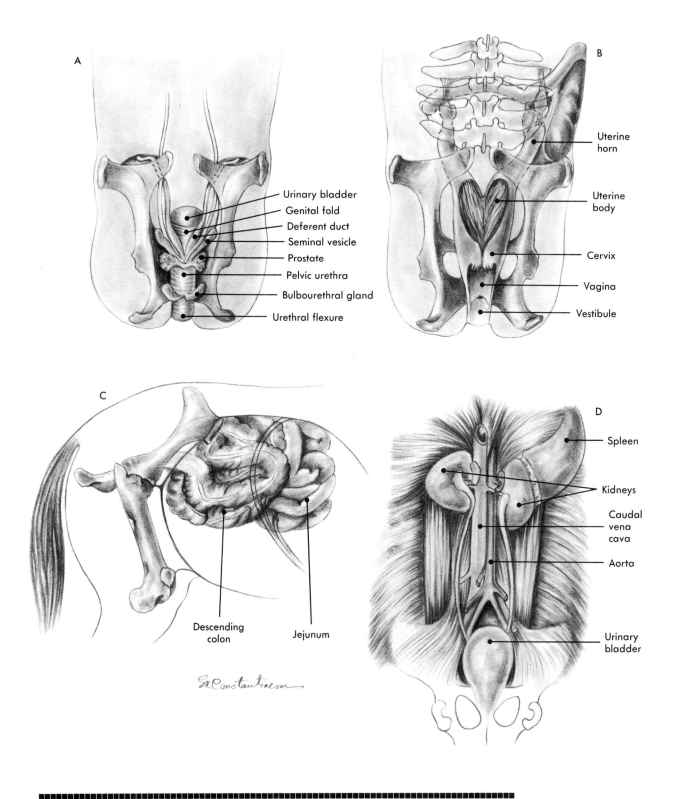

Fig. 9-7   **A,** Viscera of pelvic cavity in male. **B,** In female. **C,** Jejunum and descending colon (left view). **D,** Dorsal wall of abdominal cavity (ventrodorsal view).

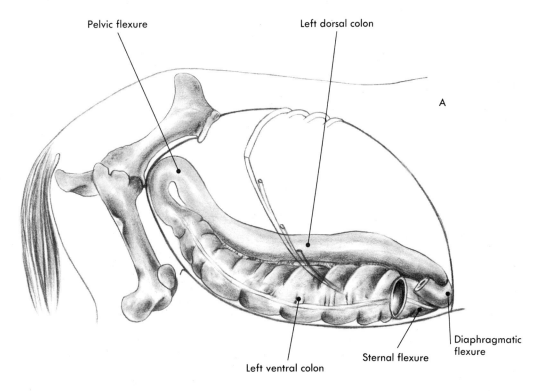

Pelvic flexure

Left dorsal colon

A

Diaphragmatic flexure

Sternal flexure

Left ventral colon

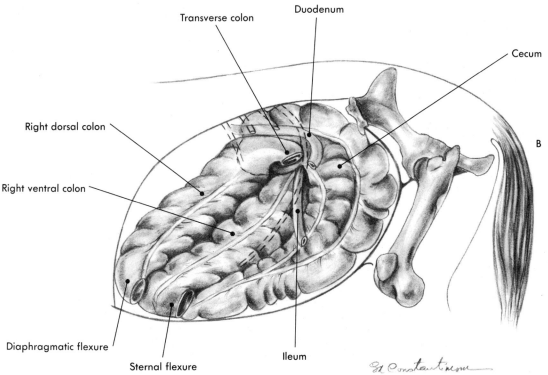

Transverse colon

Duodenum

Cecum

Right dorsal colon

B

Right ventral colon

Diaphragmatic flexure

Sternal flexure

Ileum

Gh Constantinescu

A, Left colon (right view). **B,** Cecum and right colon (left view).

Fig. 9-8

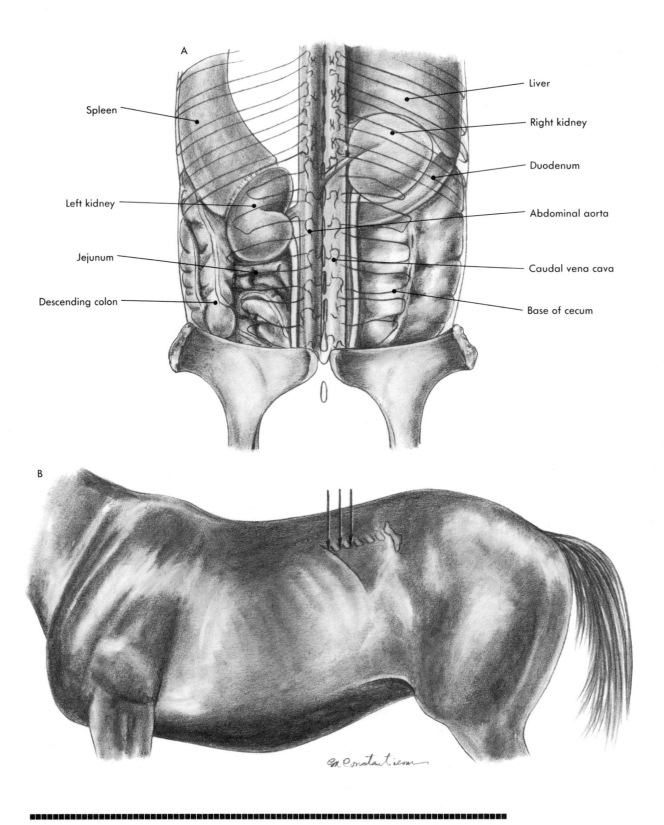

Spleen

Liver

Right kidney

Duodenum

Left kidney

Abdominal aorta

Jejunum

Caudal vena cava

Descending colon

Base of cecum

Fig. 9-9   **A,** Dorsal view of structures within abdomen. **B,** Arrows indicate paravertebral thoracolumbar approach.

I. Superficial perineal N. (Fig. 9-10, *A* and *B*).
   A. *Landmarks:* Tuber ischiadicum; arcus ischiadicus; pars spongiosa urethrae (in the male); and vulva (in female).
   B. *Approach:* Between the tuber ischiadicum and pars spongiosa urethrae (in the male) or the vulva (in female) on the arcus ischiadicus.

II. Middle caudal A.V. (Fig. 9-10, *C*).
   A. *Landmarks:* The midventral line of the tail.
   B. *Approach:* On the midventral line of the tail and between the two symmetrical sacrocaudalis ventralis medialis Mm.
      Caution! Dorsal to the anus and corresponding to the level of the third and fourth caudal vertebrae, the symmetrical rectococcygeus M. separates the sacrocaudalis ventralis medialis Mm. from the middle caudal A.V.

III. Dorsolateral caudal A.V. and dorsal branches of the caudal Nn. (Fig. 9-10, *D*)
   A. *Landmarks:* Transverse processes of caudal vertebrae.
   B. *Approach:* Dorsal to the transverse processes and deep, through the skin and caudal fascia, and between the sacrocaudalis dorsalis lateralis and intertransversarii dorsales caudae Mm.

IV. Ventrolateral caudal A.V. and ventral branches of the caudal Nn. (Fig. 9-10, *D*)
   A. *Landmarks:* Transverse processes of caudal vertebrae.
   B. *Approach:* Ventral to the transverse processes and deep, through the skin and caudal fascia, between the sacrocaudalis ventralis lateralis and intertransversarii ventrales caudae Mm.
      Caution! Dorsal to the anus and corresponding to the level of the third and fourth caudal vertebrae, the coccygeus M. lies superficially in a subcutaneous and subfascial position, between the two aforementioned muscles and protecting the vessels and nerves.

V. Subcutaneous bursae (Fig. 9-11, *A*).
   1. Bursa trochanterica.
      A,B. *Landmarks and approach:* Third trochanter.
   2. Bursa iliaca.
      A,B. *Landmarks and approach:* Tuber coxae.
   3. Bursa ischiadica.
      A,B. *Landmarks and approach:* Tuber ischiadicum.

VI. Muscular bursae (Fig. 9-11, *B*).
   1. Trochanteric bursa of superficial gluteal M.
      A. *Landmarks:* Third trochanter.
      B. *Approach:* Through the skin and fascia lata and deep between the insertion of superficial gluteal M. on third trochanter and the third trochanter.
   2. Trochanteric bursa of middle gluteal M.
      A. *Landmarks:* The caudal part of greater trochanter.
      B. *Approach:* Through skin and gluteal fascia, deep and in a cranial position between the attachment of the middle gluteal M. on the greater trochanter and the caudal part of the greater trochanter.
   3. Trochanteric bursa of accessory gluteal M.
      A. *Landmarks:* The cranial part of greater trochanter.
      B. *Approach:* Through the skin, gluteal fascia and middle gluteal M., deep between the tendon of the accessory gluteal M. and the lateral aspect of the cranial part of the greater trochanter.
   4. Trochanteric bursa of biceps femoris M.
      A. *Landmarks:* The caudal part of the greater trochanter.
      B. *Approach:* Through the skin and superficial lamina of fascia lata, deep

PELVIS, TAIL, AND EXTERNAL GENITALIA
(Figs. 9-10 to 9-12)

between the caudal border of the caudal part of the greater trochanter and the cranial border of the biceps femoris M.

    5. Ischiadic bursa of semitendinosus M.

      A. *Landmarks:* Tuber ischiadicum.

      B. *Approach:* Through the skin and superficial lamina of fascia lata, deep between the semitendinosus M. and the tuber ischiadicum.

VII. Lumbosacral puncture (Fig. 9-12, *A*).

    A. *Landmarks:* The spinous process of the last lumbar vertebra; the spinous process of the first sacral vertebra; and the two sacral tuberosities.

    B. *Approach:* Deep in the middle of these four landmarks; through the skin, supraspinous, interspinous and yellow ligg. to the vertebral canal; and in the same transverse plane as the caudal border of the tuber coxae and the cranial border of the tuber sacrale.

VIII. Epidural injection (Fig. 9-12, *A*).

    A. *Landmarks:* The spinous process of the last sacral vertebra and the spinous processes of the first and second caudal vertebrae.

    B. *Approach:* On the middorsal line between the sacrum and the first caudal vertebra (or between the first and second caudal vertebrae) and through the skin and caudal fascia to the vertebral canal.

      Caution! To easily palpate the intervertebral spaces, raise the tail and reflect it to the right and to the left, as well as dorsally and ventrally.

IX. Subsacral anesthesia (for pudendal N.) (Fig. 9-12, *B*).

    A. *Landmarks:* The base of tail; anus; ventral aspect of sacrum; promontory; and ventral sacral foramina.

    B. *Approach:* Keep the tail raised, introduce a hand with the palmar aspect oriented dorsally into rectum, and palpate the ventral aspect of the sacrum with the promontory and ventral sacral foramina (paired). Count them and choose the place for anesthesia (S_{3-4}). With the other hand, introduce a long gauge needle through the pelvic diaphragm inside of the cutaneous folds between the base of tail and the anus and up to the chosen ventral sacral foramina. Guide the tip of the needle with the hand still in the rectum.

X. Spermatic cord (Fig. 9-12, *C*).

    A. *Landmarks:* Superficial inguinal ring; testicle; and prepuce.

    B. *Approach:* Between the superficial inguinal ring and the corresponding testicle, lateral to prepuce.

XI. Lig. of the tail of epididymis (Fig. 9-12, *C*).

    A. *Landmarks:* Testicle; epididymis (tail); and scrotum.

    B. *Approach:* Through the scrotum on the dorsal aspect of the tail of epididymis.

XII. Inguinal approach for castration (Fig. 9-12, *C*).

    A. *Landmarks:* Superficial inguinal ring; testicle; and scrotum.

    B. *Approach:* Through the scrotum and external spermatic fascia at the level of the superficial inguinal ring.

XIII. Paramedian approach for castration (Fig. 9-12, *C*).

    A. *Landmarks:* Spermatic cord; testicle; and scrotal raphe.

    B. *Approach:* Parallel and 2 to 3 cm lateral to scrotal raphe through scrotum and external spermatic fascia.

XIV. Median approach for castration (Fig. 9-12, *C*).

    A. *Landmarks:* Scrotal raphe.

    B. *Approach:* The middle third of the scrotal raphe.

XV. Male urethra (pars spongiosa)

    1. Ischial approach (Fig. 9-10, *A*).

      A. *Landmarks:* Ischiadic arch and anus.

      B. *Approach:* Subanal at the level of the ischiadic arch and through the skin, the superficial perineal fascia, the retractor penis and bulbospongiosus Mm., and the corpus spongiosum penis.

2. Perineal approach (Fig. 9-10, *A*).
   A. *Landmarks:* Perineum.
   B. *Approach:* Through the skin, the superficial perineal fascia, the retractor penis and bulbospongiosus Mm., and the corpus spongiosum penis. ***Remember!*** The perineum in male is located between the anus and testicles.

3. Glans penis approach (Fig. 9-12, *C*).
   A. *Landmarks:* Prepuce and glans penis.
   B. *Approach:* Remove the glans penis from the prepuce, pass through the internal preputial lamina, the retractor penis and bulbospongiosus Mm., and the corpus spongiosum penis.

XVI. Penis (Fig. 9-12, *C*).
   A. *Landmarks:* Prepuce.
   B. *Approach:* With the penis in the prepuce, pass through the skin and the external and internal preputial laminae.

XVII. Perineal body of the mare (episiotomy) (Fig. 9-10, *B*).
   A. *Landmarks:* Anus and vulva.
   B. *Approach:* Between the anus and dorsal commissure of the vulva and through skin, the superficial perineal fascia, the continuation of the external anal sphincter with the constrictor vulvae, and the perineal septum.

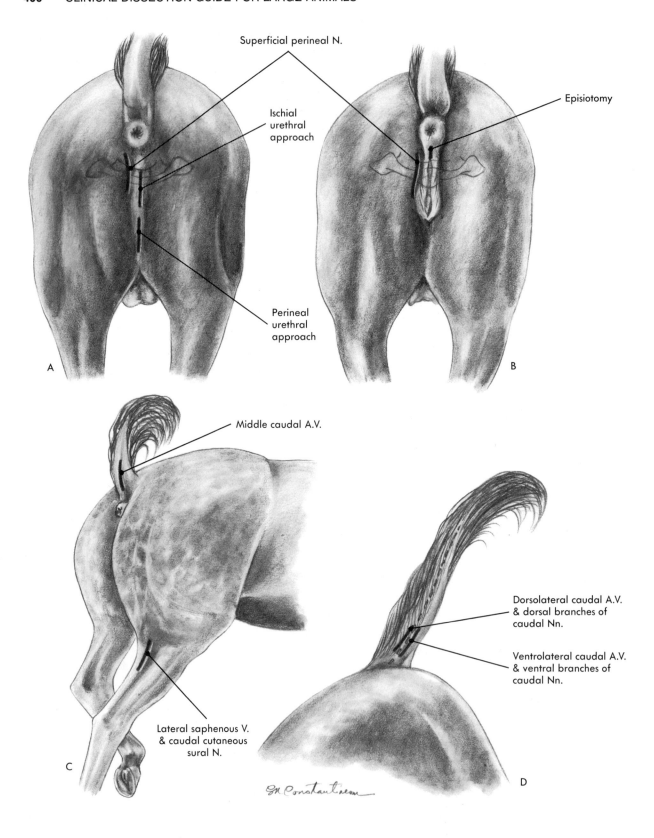

Superficial perineal N.

Ischial urethral approach

Episiotomy

Perineal urethral approach

Middle caudal A.V.

Dorsolateral caudal A.V. & dorsal branches of caudal Nn.

Ventrolateral caudal A.V. & ventral branches of caudal Nn.

Lateral saphenous V. & caudal cutaneous sural N.

Fig. 9-10   **A,** Male. **B,** Female. **C,** Structures in popliteal region and tail. **D,** Structures on tail.

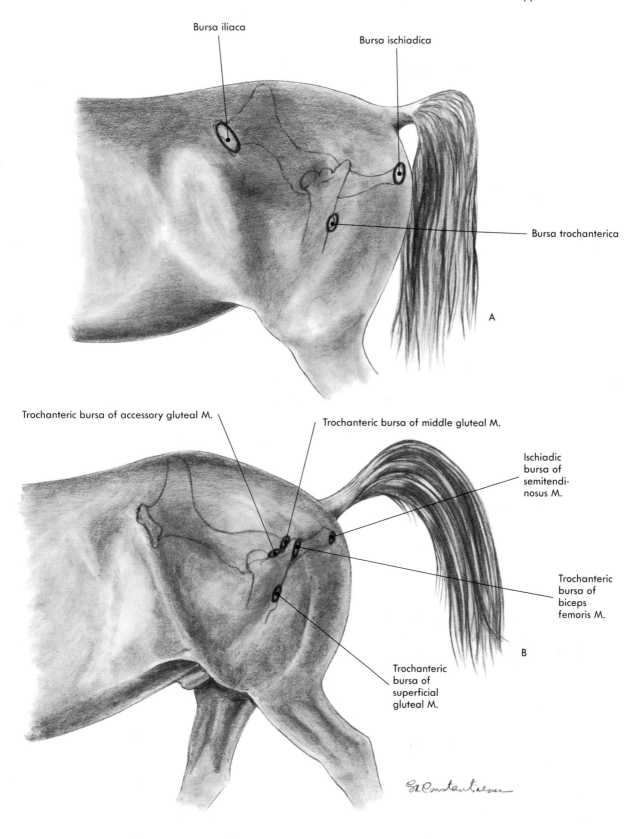

Bursa iliaca

Bursa ischiadica

Bursa trochanterica

A

Trochanteric bursa of accessory gluteal M.

Trochanteric bursa of middle gluteal M.

Ischiadic bursa of semitendinosus M.

Trochanteric bursa of biceps femoris M.

Trochanteric bursa of superficial gluteal M.

B

A, Subcutaneous bursae. **B,** Muscular bursae.

Fig. 9-11

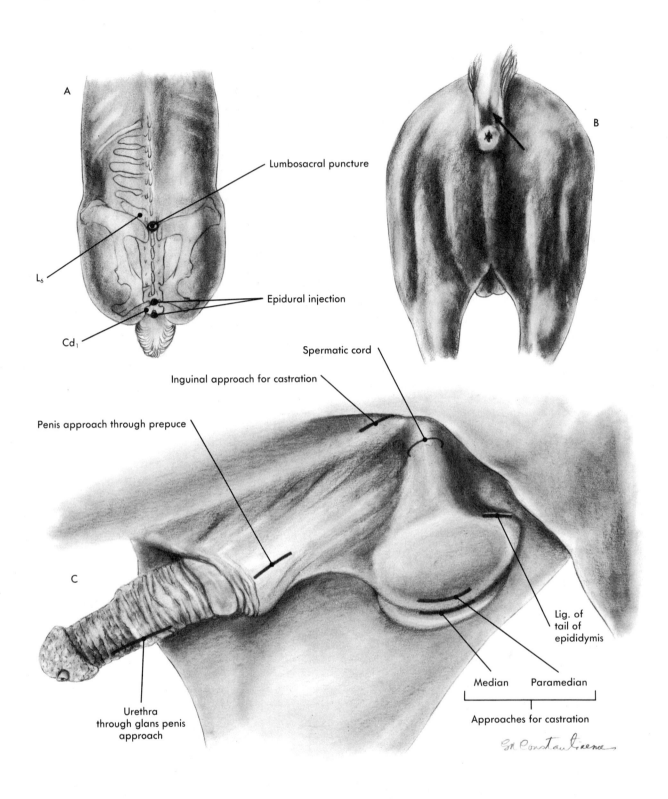

Lumbosacral puncture

L₆

Epidural injection

Cd₁

Spermatic cord

Inguinal approach for castration

Penis approach through prepuce

Lig. of
tail of
epididymis

Median   Paramedian

Approaches for castration

Urethra
through glans penis
approach

Fig. 9-12   **A,** Lumbosacrocaudal sites for punctures. **B,** Subsacral anesthesia. **C,** Approaches to male external genitalia.

I. Femoral A.V.N. and deep inguinal lnn. (Fig. 9-13, *A*).
   A. *Landmarks:* Cranial border of pubis; caudal border of sartorius M.; and cranial border of gracilis M.
   B. *Approach:* In the femoral (Scarpa) triangle, through the skin and lamina femoralis, between the sartorius, gracilis, and pectineus Mm., and toward the cranial border of pubis.

II. Saphenous A.N. and medial saphenous V. (in the thigh region) (Fig. 9-13, *A*).
   A. *Landmarks:* Cranial border of pubis; intermuscular space between sartorius and gracilis Mm.; medial femoral condyle.
   B. *Approach:* In the intermuscular space between the sartorius and gracilis Mm. in the distal half of a line between the cranial border of pubis and the medial femoral condyle.

III. Medial saphenous V. and saphenous N. (in the crus) (Fig. 9-13, *A*).
   A. *Landmarks:* The cranial and caudal borders of the tibia, on the medial aspect of the crus.
   B. *Approach:* On both borders of the tibia, through the skin, superficial crural fascia and crural fascia proper, and in the middle third of the crus.

IV. Caudal cutaneous sural N. and lateral saphenous V. (Fig. 9-10, *C*).
   A. *Landmarks:* The common calcanean tendon and the distal end of the intermuscular space between the biceps femoris and semitendinosus Mm.
   B. *Approach:* Between the distal end of the intermuscular space that separates the biceps femoris and semitendinosus Mm. and along the lateral aspect of the common calcanean tendon (under the skin and crural fascia proper).

V. Tibial N. and caudal femoral A. (Fig. 9-13, *A*).
   A. *Landmarks:* The medial groove of the crus separating the common calcanean tendon from the rest of the crus.
   B. *Approach:* With the pelvic limb in flexion, the groove separating the common calcanean tendon from the rest of the crus is widely exposed and allows for easy palpation of the nerve. Both nerve and artery are located under the skin and crural fascia proper one hand proximal to the tuber calcanei.
   Caution! The nerve may also be palpated in the lateral groove that separates the common calcanean tendon from the rest of crus.

VI. Cranial tibial A. (Fig. 9-13, *A*).
   A. *Landmarks:* The medial groove of the crus between the cranial border of tibia and long digital extensor M.
   B. *Approach:* Deep between the tibia and long digital extensor M. and through the skin, superficial crural fascia and crural fascia proper in the distal third of crus.

VII. Common fibular N. (Fig. 9-13, *B*).
   A. *Landmarks:* Lateral condyle of the tibia.
   B. *Approach:* The nerve is palpable on the lateral condyle of the tibia in an oblique direction (caudocranially and dorsoventrally) by rolling your finger firmly back and forth (pressing the skin, superficial crural fascia and crural fascia proper).

VIII. Superficial and deep fibular Nn. (sensory branches) (Fig. 9-13, *B*).
   A. *Landmarks:* The lateral groove of the crus between the long and lateral digital extensor Mm.
   B. *Approach:* At a point 10 cm proximal to the hock through the skin, superficial crural fascia and crural fascia proper (the superficial fibular N.) or deep (the deep fibular N.) in the lateral groove of the crus between the long and lateral digital extensor Mm.

PELVIC LIMB (Figs. 9-13 to 9-15)

IX. Dorsal metatarsal A. (Fig. 9-13, *B*).

    A. *Landmarks:* The dorsal groove between the metatarsal bones III and IV (lateral).

    B. *Approach:* Subcutaneous—in the dorsal groove between the metatarsal bones III and IV (one of the sites for taking an arterial pulse or obtaining arterial blood samples).

X. Dorsal common digital V. II (Fig. 9-13, *A*).

    A. *Landmarks:* Proximal end of the metatarsal bones III and II (medial).

    B. *Approach:* An oblique line crossing the proximal end of metatarsal bones III and II in a dorsoplantar and proximodistal direction under the skin.

XI. (Lateral) and medial plantar Nn. (Fig. 9-13, *C*).

    A. *Landmarks:* The tendons of the superficial and deep digital flexor Mm. in the metatarsal area.

    B. *Approach:* Through the skin and plantar fascia on both sides along these tendons.

XII. (Lateral) and medial plantar metatarsal Nn. (Fig. 9-13, *C*).

    A. *Landmarks:* The distal ends of the metatarsal bones IV and II (the splint bones).

    B. *Approach:* Subcutaneous—at the distal ends of the splint bones, which are palpable.

XIII. Dorsal metatarsal Nn.

    1. Lateral metatarsal N. (Fig. 9-13, *B*).

        A. *Landmarks:* Same as for the dorsal metatarsal A.

        B. *Approach:* Same as for the dorsal metatarsal A., but continues distally to the lateral aspect of the fetlock.

    2. Medial metatarsal N. (Fig. 9-13, *C*).

        A. *Landmarks:* Tendon of the long digital extensor M. and metatarsal bone II.

        B. *Approach:* In the tarsal area, the nerve parallels the medial border of the tendon of the long digital extensor M., then obliquely crosses the metatarsal bone III distoplantarly, running parallel to and between the metatarsal bones III and II under skin.

XIV. Subcutaneous bursae

    1. Prepatellar bursa (Fig. 9-14, *A*).

        A. *Landmarks and approach:* Cranial aspect of patella (palpable).

    2. Bursa of tibial tuberosity (Fig. 9-14, *A*).

        A. *Landmarks and approach:* Tibial tuberosity (cranioproximal extent of tibia—palpable).

    3. Bursae of lateral and medial malleoli (Figs. 9-14, *A* and *B*).

        A. *Landmarks and approach:* Lateral and medial tibial malleoli (palpable).

    4. Calcanean bursa (Fig. 9-14, *A*).

        A. *Landmarks and approach:* Tuber calcanei (palpable).

XV. Subfascial, subtendinous and articular bursae

    1. Subfascial prepatellar bursa (Fig. 9-14, *A*).

        A. *Landmarks:* Cranial aspect of patella.

        B. *Approach:* Through the skin and the superficial layer of the superficial lamina of the fascia lata.

    2. Subtendinous prepatellar bursa (Fig. 9-14, *A*).

        A. *Landmarks:* Patella.

        B. *Approach:* Deep between the base of the patella and the insertion of the quadriceps femoris M.

    3. Distal subtendinous bursa of biceps femoris M. (Fig. 9-14, *A*).

        A. *Landmarks:* Patella and stifle joint.

B. *Approach:* Deep on the lateral aspect of the stifle joint, 5 to 10 cm caudal to the patella, and between the tendon of the biceps femoris M. and the lateral femoral condyle.

4. Subtendinous bursa of semitendinosus M. (Fig. 9-14, *B*).
   A. *Landmarks:* Cranial border of tibia.
   B. *Approach:* Between the tendon of the semitendinosus M. and the cranial border of the tibia through the skin, superficial crural fascia and crural fascia proper.

5. Subtendinous bursa of cranial tibial M. (Fig. 9-14, *B*).
   A. *Landmarks:* Medial aspect of the hock; proximal end of the metatarsal bones III and II.
   B. *Approach:* Through the skin, between the cunean tendon of the cranial tibial M. and tarsal bones I and II, and 2 to 3 cm above the proximal ends of the metatarsal bones II and III on the medial aspect of hock.

6. Calcanean bursa of superficial digital flexor M. (Fig. 9-14, *B*).
   A. *Landmarks:* Tuber calcanei and the attachment of the common calcanean tendon on tuber calcanei.
   B. *Approach:* Between the attachments of the superficial flexor and triceps surae Mm. on the tuber calcanei.

7. Tendinous calcanean bursa (Fig. 9-14, *B*).
   A. *Landmarks:* Tuber calcanei and the attachment of the common calcanean tendon on tuber calcanei.
   B. *Approach:* Between the attachment of the triceps surae M. on the tuber calcanei and the tuber calcanei.

8. Tendinous bursa of long digital extensor M. (Fig. 9-14, *C*).
   A. *Landmarks:* Dorsal aspect of fetlock; tendon of long digital extensor M.
   B. *Approach:* Between the tendon of the long digital extensor M. and the dorsal aspect of the fetlock (distal end of the metatarsal bone III).

9. Bursa podotrochlearis pedis (Fig. 9-14, *D*).
   A. *Landmarks:* Navicular bone and the terminal tendon of the deep digital flexor M. within the hoof.
   B. *Approach:* Deep within the hoof, between the flexor surface of the navicular bone and the tendon of the deep digital flexor M. at the level of the coronary band and between the two bulbs of the heels.

10. Proximal infrapatellar bursa (Fig. 9-14, *B*).
    A. *Landmarks:* Patella and middle patellar lig.
    B. *Approach:* Between the apex of the patella and the proximal attachment of the middle patellar lig.

11. Distal infrapatellar bursa (Fig. 9-14, *B*).
    A. *Landmarks:* Tibial tuberosity and middle patellar lig.
    B. *Approach:* Between the tibial tuberosity and the distal attachment of the middle patellar lig.

XVI. Medial patellar lig. (Fig. 9-14, *B*).
   A. *Landmarks:* Patella; medial ridge of femoral trochlea; and tibial tuberosity.
   B. *Approach:* Deep between the patellar fibrocartilage and tibial tuberosity, along the medial aspect of the medial ridge of the femoral trochlea.

XVII. Intraarticular approach (injections)
   1. Coxofemoral (hip) joint (Fig. 9-15, *A*).
      A. *Landmarks:* The cranial and caudal parts of the greater trochanter.
      B. *Approach:* A needle may be introduced into the joint between the cranial and caudal parts of the greater trochanter through the trochanteric

incisure, along the femoral neck, and at a 45-degree angle with the sagittal plane.

   2. Femorotibiopatellar (stifle) joint.
      α. Femoropatellar synovial sac (Fig. 9-15, *A* and *B*).
         A. *Landmarks:* Patella; medial ridge of femoral trochlea; and middle patellar lig.
         B. *Approach:* In the angle between the apex of patella, the proximal attachment of the middle patellar lig., and the medial ridge of femoral trochlea.
      β. Lateral femorotibial synovial sac (Fig. 9-15, *A*).
         A. *Landmarks:* Tibial tuberosity; patella; lateral patellar lig.; lateral tibial condyle; extensor groove; lateral collateral lig. of stifle joint.
         B. *Approach:* Between the lateral collateral lig. of the stifle joint and the lateral patellar lig., midway between the patella and extensor groove (bordered by the tibial tuberosity and lateral tibial condyle).
      γ. Medial femorotibial sinovial sac (Fig. 9-15, *B*).
         A. *Landmarks:* Medial patellar lig. and medial condyle of tibia.
         B. *Approach:* Just dorsal to the medial condyle of the tibia and at the caudal border of the medial patellar lig. (palpable).
   3. Tarsal (hock) joint.
      α. Distal intertarsal joint space (Fig. 9-15, *B*).
         A. *Landmarks:* Cunean tendon of cranial tibial M. and medial aspect of hock.
         B. *Approach:* Dorsal to the cunean tendon, on the medial aspect of hock.
      β. Tarsocrural joint (dorsomedial pouch) (Fig. 9-15, *B*).
         A. *Landmarks:* Medial malleolus of tibia and medial ridge of trochlea of talus.
         B. *Approach:* In the dorsal angle between the two landmarks.
      γ. Tarsocrural joint (mediopalmar pouch) (Fig. 9-15, *B*).
         A. *Landmarks:* Distal extremity of tibia; sustentaculum tali; tendons of the lateral digital flexor and caudal tibial Mm.; and the tendon of the medial digital flexor M.
         B. *Approach:* Among the four landmarks.
      Δ. Tarsocrural joint (lateropalmar pouch) (Fig. 9-15, *D*).
         A. *Landmarks:* Lateral malleolus of tibia; calcaneus; tendon of lateral digital extensor M.; and tendon of lateral digital flexor M.
         B. *Approach:* In the angle between the lateral malleolus of the tibia and the calcaneus, in the space between the tendons of the lateral digital extensor and lateral digital flexor Mm.
XVIII. Vaginal (synovial) tendinous sheaths
   1. Sheath of the cranial tibial M. (Fig. 9-15, *C*).
      A. *Landmarks:* Dorsal aspect of hock and tendon of long digital extensor M. just above the tarsal area.
      B. *Approach:* Through the skin, around the tendon of cranial tibial M. (deep to fibularis tertius M.), on the distal third of the dorsal aspect of crus, medial to and parallel with the tendon of the long digital extensor M.
   2. Sheath of the long digital extensor M. (Fig. 9-15, *C*).
      A. *Landmarks:* Same as for 1.
      B. *Approach:* Through the skin and extensor retinaculum and around the tendon of the long digital extensor M. on the dorsal aspect of the hock.

3. Sheath of the lateral digital extensor M. (Fig. 9-15, *D*).
   A. *Landmarks:* Lateral aspect of hock; tendon of lateral digital extensor M.; lateral collateral lig. of hock joint; and lateral malleolus.
   B. *Approach:* Through the skin and tarsal fascia, around the tendon of the lateral digital extensor M. from the level of the lateral malleolus; along the lateral collateral lig. of the hock joint, and to the junction between the tendons of the long and lateral digital extensor Mm.
4. Sheath of the lateral digital flexor and caudal tibial Mm. (Fig. 9-15, *B* and *D*).
   A. *Landmarks:* Sustentaculum tali; tuber calcanei; proximal medioplantar and proximal lateroplantar aspects of hock; and the deep caudal muscles of crus.
   B. *Approaches:*
      i. Through the skin and flexor retinaculum on the medioplantar aspect of the hock, around the tendons of the lateral digital flexor and caudal tibial Mm., and close to the sustentaculum tali (dorsally) and tuber calcanei (laterally).
      ii. On the lateroplantar aspect of the hock, around the same tendons and between the cranial (dorsal) border of the tuber calcanei and the deep caudal muscles of the crus.
5. Sheath of the medial digital flexor M. (Fig. 9-15, *B*).
   A. *Landmarks:* Medial collateral lig. of hock joint.
   B. *Approach:* Through skin and tarsal fascia and around the tendon of the medial digital flexor M., caudal to and parallel with the medial collateral lig. of hock joint.

XIX. Accessory (check) lig. of deep digital flexor M. (See the similar structure of the thoracic limb, p. 416.)

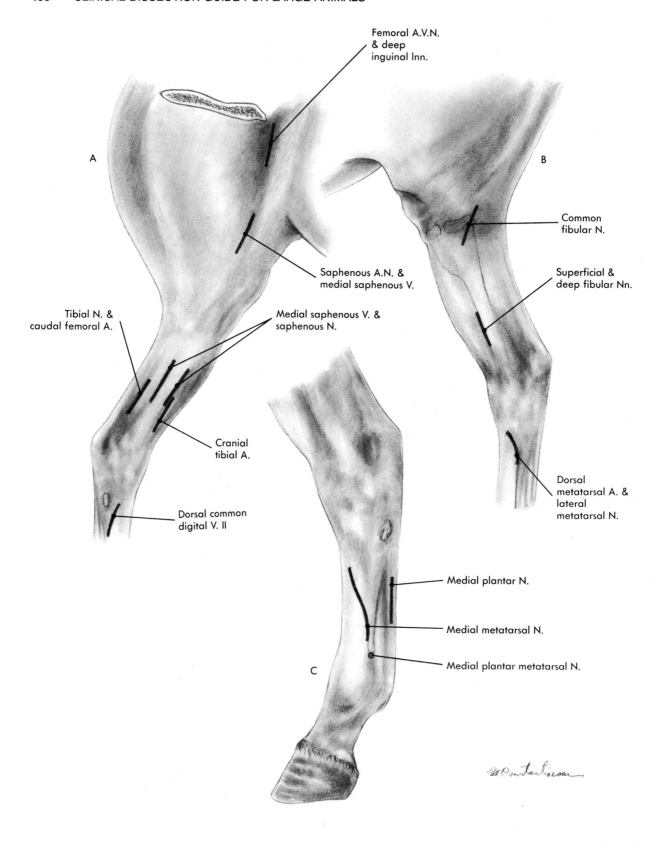

Femoral A.V.N.
& deep
inguinal lnn.

A

B

Common
fibular N.

Saphenous A.N. &
medial saphenous V.

Superficial &
deep fibular Nn.

Tibial N. &
caudal femoral A.

Medial saphenous V. &
saphenous N.

Cranial
tibial A.

Dorsal common
digital V. II

Dorsal
metatarsal A. &
lateral
metatarsal N.

C

Medial plantar N.

Medial metatarsal N.

Medial plantar metatarsal N.

Fig. 9-13   Pelvic limb. **A,** Medial aspect. **B,** Lateral aspect. **C,** Medial aspect of autopodium.

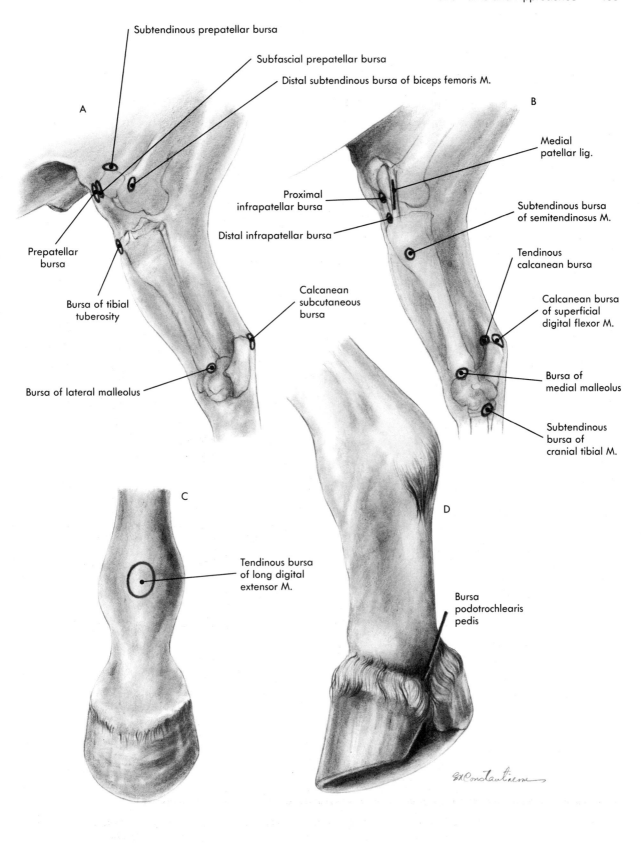

Subtendinous prepatellar bursa

Subfascial prepatellar bursa

Distal subtendinous bursa of biceps femoris M.

A

B

Medial patellar lig.

Proximal infrapatellar bursa

Subtendinous bursa of semitendinosus M.

Distal infrapatellar bursa

Prepatellar bursa

Tendinous calcanean bursa

Calcanean bursa of superficial digital flexor M.

Bursa of tibial tuberosity

Calcanean subcutaneous bursa

Bursa of lateral malleolus

Bursa of medial malleolus

Subtendinous bursa of cranial tibial M.

C

D

Tendinous bursa of long digital extensor M.

Bursa podotrochlearis pedis

Pelvic limb between stifle joint and hock. **A,** Lateral aspect. **B,** Medial aspect digit. **C,** Dorsal aspect. **D,** Plantar aspect.

Fig. 9-14

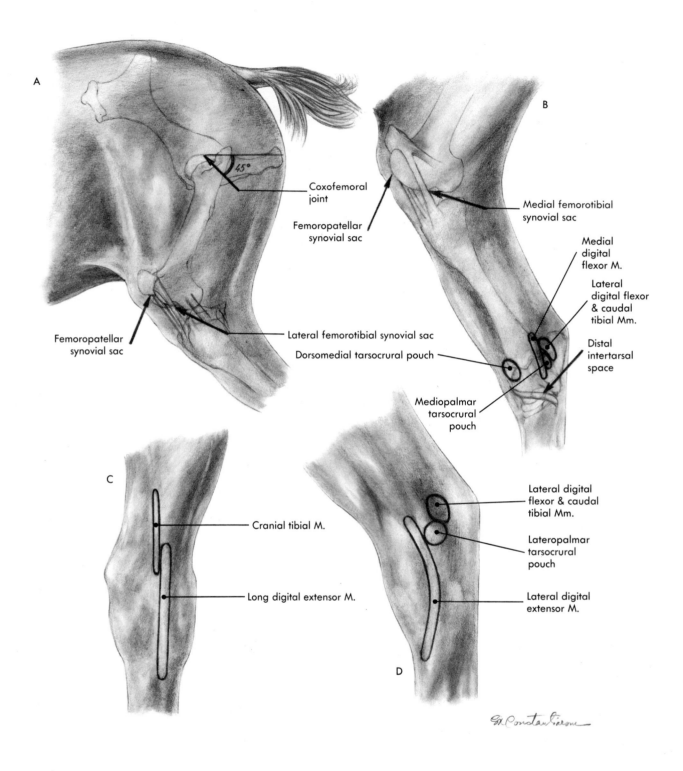

A

45°

Coxofemoral joint

Femoropatellar synovial sac

Femoropatellar synovial sac

Lateral femorotibial synovial sac

Dorsomedial tarsocrural pouch

Mediopalmar tarsocrural pouch

B

Medial femorotibial synovial sac

Medial digital flexor M.

Lateral digital flexor & caudal tibial Mm.

Distal intertarsal space

C

Cranial tibial M.

Long digital extensor M.

D

Lateral digital flexor & caudal tibial Mm.

Lateropalmar tarsocrural pouch

Lateral digital extensor M.

Fig. 9-15 Pelvic limb. **A,** Lateral aspect of thigh. **B,** Medial aspect of crus and hock. **C,** Dorsal aspect of hock. **D,** Lateral aspect of hock.

I. Median A.V.N. (Fig. 9-16, *A*).
  1. Proximal approach.
    A. *Landmarks:* Medial aspect of elbow.
    B. *Approach:* Through the skin, transverse pectoral M., and brachial fascia on the medial aspect of the elbow. The nerve and vessels produce a popping motion when rolled back and forth under a firmly pressing finger.
    Caution! The median N. lies caudal to the vessels.
  2. Distal approach.
    A. *Landmarks:* Chestnut on medial aspect of antebrachium; radius; and flexor carpi radialis M.
    B. *Approach:* On the medial aspect of the antebrachium deep through the skin, superficial and proper antebrachial fasciae, and between the radius and flexor carpi radialis M., one hand proximal to the chestnut. The median A.V.N. lie on the caudal aspect of the radius.
II. Ulnar N. and collateral ulnar A.V. (Fig. 9-16, *B*).
  A. *Landmarks:* Accessory carpal bone; olecranon; and flexor and extensor carpi ulnaris Mm.
  B. *Approach:* At a point one-hand width proximal to the accessory carpal bone, between the flexor and extensor carpi ulnaris Mm. on the line between the accessory carpal bone and the olecranon, and through the skin and proper antebrachial fascia.
  Caution! For an easy approach, flex the forelimb.
III. Cranial cutaneous antebrachial N. (Fig. 9-16, *C*).
  A. *Landmarks:* Deltoid tuberosity of humerus and deltoideus M.
  B. *Approach:* Emerging on the lateral aspect of the arm, caudodorsal to the deltoid tuberosity and caudal to the insertion of the deltoideus M. on the deltoid tuberosity in a dorsoventral direction through the skin and omobrachialis M. (and brachial fascia, if located more ventrally).
IV. Lateral cutaneous antebrachial N. (Fig. 9-16, *D*).
  A. *Landmarks:* The ventral border of lateral head of triceps brachii M.; deltoid tuberosity; olecranon; and common digital extensor M. (bordered cranially and caudally by two grooves on the lateral aspect of forearm).
  B. *Approach:* Midway between the deltoid tuberosity and the olecranon, emerging from under the ventral border of the lateral head of the triceps brachii M. in a dorsoventral direction, and along the long axis of the common digital extensor M. through the skin and proper antebrachial fascia.
  Caution! The nerve regularly has two branches close (and parallel) to one another.
V. Caudal cutaneous antebrachial N. (Fig. 9-16, *B*).
  A. *Landmarks:* Tuber olecrani; accessory carpal bone; and flexor and extensor carpi ulnaris Mm.
  B. *Approach:* Subcutaneous and 10 cm ventral to tuber olecrani on the line between the tuber olecrani and the accessory carpal bone, between the flexor and extensor carpi ulnaris Mm.
VI. Medial cutaneous antebrachial N. (Fig. 9-16, *A*).
  A. *Landmarks:* Lacertus fibrosus.
  B. *Approach:* On the medial aspect of lacertus fibrosus.
    Caution! The nerve is accompanied by the accessory cephalic V. and cephalic V. (the antebrachial segment) and is palpable.

THORACIC LIMB
(Figs. 9-16 to 9-19

VII. Deep br. of radial N. (sensory) (Fig. 9-16, *D*).

   A. *Landmarks:* Tendon of common digital extensor M. in the forearm; caudal groove of lateral aspect of forearm; and carpus.

   B. *Approach:* One hand proximal to the carpus on the tendon of the common digital extensor M. (first approach); and deep in the caudal groove of the lateral aspect of the forearm (second approach). In both approaches—through the skin and the proper antebrachial fascia.

   Caution! Both approaches must be considered in a nerve block.

VIII. Dorsal br. of ulnar N. (Fig. 9-16, *D*).

   A. *Landmarks:* Accessory carpal bone and the two branches of the distal tendon of the extensor carpi ulnaris M.

   B. *Approach:* Through the skin and on the lateral aspect of the carpus, just above the accessory carpal bone between the two branches of the distal tendon of the extensor carpi ulnaris M., where a small finger tip–like depression may be palpated.

IX. (Lateral) and medial palmar Nn. (Fig. 9-17, *A*).

   A. *Landmarks:* Interosseus medius M. (suspensory lig.) and tendons of deep and superficial digital flexor Mm. in the palmar metacarpal area.

   B. *Approach:* Through the skin (and palmar fascia) on both sides of the palmar metacarpal area between the interosseus medius M. and the tendons of the deep and superficial digital flexor Mm.

   Caution! The medial palmar N. is accompanied by the palmar common digital A.V. II, whereas the lateral palmar N. is accompanied by the palmar common digital A.V. III. The relationship between the nerves and vessels is the following: V.A.N., in a dorsopalmar direction.

X. Deep palmar br. of ulnar N. (Fig. 9-17, *A*).

   A. *Landmarks:* Accessory carpal bone; accessoriometacarpal lig. and the lateral border of the tendons of the deep and superficial digital flexor Mm. on the mediopalmar aspect of the carpus, just distal to flexor retinaculum (when palpable).

   B. *Approach:* A needle can be inserted in a proximodistal and palmodorsal direction deep through the skin between the medial aspect of the accessoriometacarpal lig. and the lateral border of the tendons of the deep and superficial digital flexor Mm., just distal to the flexor retinaculum.

   Caution! Repeated flexions and extensions of carpus are suggested.

XI. Lateral and medial palmar metacarpal Nn.

   A,B. *Landmarks and approach:* Similar to the lateral and medial plantar metatarsal Nn. (see p. 404).

XII. (Lateral) and medial palmar digital Nn. (Fig. 9-17, *A*).

   A. *Landmarks:* The abaxial (interosseus M.) aspect of proximal sesamoid bones.

   B. *Approach:* The nerves, accompanied by the corresponding arteries and veins in a subcutaneous position, produce a popping movement when rolled back and forth under a firmly pressing finger.

   Caution! The relationship between the nerves and vessels is as follows: V.A.N., in a dorsopalmar direction.

XIII. Palmar br. of palmar digital N. (Fig. 9-17, *A*).

   A. *Landmarks:* Pastern.

   B. *Approach:* In the long axis of the digit in the pastern region, subcutaneously between the middle and palmar thirds of the area, and deep to the lig. of ergot, which obliquely crosses the pastern in a distoproximal and dorsopalmar direction.

XIV. Intraarticular approaches
    1. Scapulohumeral (shoulder) joint (Fig. 9-17, *B*).
        A. *Landmarks:* The incisure between the cranial and caudal parts of the greater tubercle of the humerus (on the lateral aspect of shoulder).
        B. *Approach:* A needle is inserted in an approximately horizontal position and is directed craniocaudally in a 45-degree angle with the body. It penetrates the skin, the axillobrachial fascia, and the incisure between the cranial and caudal parts of the greater tubercle of humerus, which is bordered by the tendons of the supraspinatus and infraspinatus Mm.
    2. Humeroradial (elbow) joint (Fig. 9-17, *B*).
        A. *Landmarks:* The lateral prominence of the humeral trochlea; the lateral collateral lig. of elbow; and the caudal groove on the lateral aspect of forearm.
        B. *Approach:* Palpate the humeral trochlea and the lateral collateral lig. of the elbow while flexing the joint; reflect the caudal border of the common digital extensor M. from the caudal groove on the lateral aspect of the forearm cranially, and feel a depression between the first two landmarks. A needle is inserted in a horizontal position and is slightly obliquely directed craniocaudally and lateromedially through the skin and proper antebrachial fascia.
    3. Radiocarpometacarpal (carpus) joint (Fig. 9-17, *C*).
        α. Radiocarpal approach.
            A. *Landmarks:* Dorsal aspect of carpus and tendon of extensor carpi radialis M.
            B. *Approach:* Flexing the carpus, a deep depression can be felt between the radius and proximal row of carpal bones. Here is the site of the approach through the skin and extensor retinaculum, either at the medial or at the lateral border of the tendon of the extensor carpi radialis M.
        β. Intercarpal approach.
            A. *Landmarks:* Same as for the radiocarpal approach.
            B. *Approach:* Proceed in a manner similar to that for the radiocarpal approach and a depression can be felt between the two rows of carpal bones. This approach is through the skin and extensor retinaculum, either at the medial or at the lateral border of the tendon of the extensor carpi radialis M. (similar to that for the radiocarpal approach).
    4. Metacarposesamophalangeal or metatarsosesamophalangeal (fetlock) joint (Fig. 9-18, *A*, & 9-19, *C*).
        A. *Landmarks:* Metacarpal (metatarsal) bone III; proximal sesamoid bones; and interosseus medius M.
        B. *Approach:* Through the skin, proximal to the proximal sesamoid bones, between the metacarpal (metatarsal) bone III and the interosseus medius M.
    5. Proximal interphalangeal (pastern) joint (Fig. 9-18, *A*).
        A. *Landmarks:* Pastern and tendon of common (long) digital extensor M.
        B. *Approach:* A needle is inserted on either side of the tendon of the common (long in pelvic limb) digital extensor M. and directed toward the pastern joint through the skin in an oblique direction (proximodistally and dorsopalmarly/plantarly).
    6. Distal interphalangeal (coffin) joint (Fig. 9-18, *A*).
        A. *Landmarks:* Coffin; coronary border of hoof; and tendon of common (long) digital extensor M.

B. *Approach:* A needle is inserted on either side of the tendon of the common (long in pelvic limb) digital extensor M., through skin and toward the coffin joint at the coronary border of the hoof in a similar direction as performed for the pastern.

XV. Subcutaneous bursae

1. Prescapular bursa (Fig. 9-18, *B*).

A,B. *Landmarks and approach:* Tuberosity of scapular spine.

2. Olecranon bursa (Fig. 9-18, *B*).

A,B. *Landmarks and approach:* Tuber olecrani.

3. Dorsal carpal bursa (Fig. 9-18, *C*).

A,B. *Landmarks and approach:* Dorsal aspect of carpus.

XVI. Subtendinous bursae

1. Bursa of infraspinatus M. (Fig. 9-18, *B*).

A. *Landmarks:* The caudal part of the greater tubercle of humerus.

B. *Approach:* Through the skin and between the tendon of the infraspinatus M. and the lateral aspect of the caudal part of the greater tubercle of the humerus.

2. Intertubercular bursa (bursa of biceps brachii M.) (Fig. 9-18, *B*).

A. *Landmarks:* Deltoid tuberosity of humerus and cranial part of the greater tubercle of humerus.

B. *Approach:* A needle is inserted caudocranially and ventrodorsally at the level of the deltoid tuberosity, along the cranial aspect of the humerus toward the cranial part of the greater tubercle of the humerus, deep through the skin and brachial fascia, and between the humerus and the cleidobrachialis M. Finally, the needle is located between the humerus and the biceps brachii M.

3. Bursa of brachialis M. (Fig. 9-18, *D*).

A. *Landmarks:* Medial aspect of proximal extremity of radius and medial collateral lig. of elbow.

B. *Approach:* Through the skin, superficial and proper antebrachial fasciae at the cranial border of the medial collateral lig. of the elbow, deep between the brachialis M. and the medial aspect of the proximal extremity of the radius.

4. Bursa of extensor carpi radialis M. (Fig. 9-18, *C*).

A. *Landmarks:* Insertion of tendon of extensor carpi radialis M. and tuberosity of metacarpal bone III.

B. *Approach:* Through the skin and between the insertion of the extensor carpi radialis M. and the tuberosity of the metacarpal bone III, at the distal end of the extensor retinaculum.

5. Bursa of abductor pollicis longus M. (Fig. 9-18, *D*).

A. *Landmarks:* Insertion of tendon of abductor pollicis longus M. and base of metacarpal bone II.

B. *Approach:* Through skin and between the insertion of the tendon of the abductor pollicis longus M. and the base of the metacarpal bone II.

6. Bursa of common digital extensor M.

A,B. *Landmarks and approach:* Similar to those used for the tendinous bursa of long digital extensor M. in the pelvic limb (see p. 405).

7. Bursa of lateral digital extensor M. (Fig. 9-18, *A*).

A. *Landmarks:* Dorsal aspect of fetlock and tendon of lateral digital extensor M.

B. *Approach:* Through the skin and between the tendon of the lateral digital extensor M. and the dorsal aspect of the fetlock joint.

8. Bursa podotrochlearis manus.

   A,B. *Landmarks and approach:* Similar to those used for the bursa podotrochlearis pedis (see p. 405).

XVII. Vaginal (synovial) tendinous sheaths

1. Sheath of the extensor carpi radialis M. (Fig. 9-19, *A*).

   A. *Landmarks:* Dorsal aspect of carpus and tendon of extensor carpi radialis M.

   B. *Approach:* Through the skin around the tendon of the extensor carpi radialis M. proximal to the extensor retinaculum, and between the extensor retinaculum and dorsal carpal joint capsule.

2. Sheath of the abductor pollicis longus M. (Fig. 9-19, *A*).

   A. *Landmarks:* Distal tendons of the extensor carpi radialis and abductor pollicis longus Mm. proximal to the carpus.

   B. *Approach:* Through the skin between the tendons of the abductor pollicis longus and extensor carpi radialis Mm., and around the tendon of the abductor pollicis longus M. in an oblique position dorsoventrally and lateromedially.

3. Sheath of the common digital extensor M. (Fig. 9-19, *A*).

   A. *Landmarks:* Dorsal aspect of carpus and tendon of common digital extensor M.

   B. *Approach:* Through the skin around the tendon of the common digital extensor M. proximal to the extensor retinaculum and dorsal carpal joint capsule, and exceeding the extensor retinaculum by 2 to 3 cm proximally and distally.

4. Sheath of the lateral digital extensor M. (Fig. 9-19, *A*).

   A. *Landmarks:* Lateral aspect of carpus and tendon of lateral digital extensor M.

   B. *Approach:* Through the skin around the tendon of the lateral digital extensor M., between the superficial and deep fascicles of the lateral collateral lig. of the carpus, and exceeding the extensor retinaculum by 2 to 3 cm proximally and distally.

5. Sheath of the flexor carpi radialis M. (Fig. 9-19, *B*).

   A. *Landmarks:* Accessory carpal bone; chestnut; and distal extent of radius (on the medial aspect of the limb).

   B. *Approach:* Midway between the accessory carpal bone and the chestnut, parallel to the caudal border of the distal third of the radius, and through the skin and the superficial and proper antebrachial fasciae.

6. Common synovial sheath of superficial & deep digital flexor tendons (Fig. 9-19, *B*).

   A. *Landmarks:* Lateral and medial aspect of carpus; accessory carpal bone; tendons of lateral digital extensor, extensor carpi ulnaris, flexor carpi ulnaris and flexor carpi radialis Mm.; and metacarpal bones II and IV.

   B. *Approach:* Through the skin, proximal and distal to the limits of the extensor and flexor retinacula. On the lateral aspect, proximal to extensor retinaculum between the tendons of the lateral digital extensor M. (cranially) and extensor carpi ulnaris M. (caudally), distal to the extensor retinaculum, in a palmar position to the metacarpal bone IV. On the medial aspect, proximal to the flexor retinaculum between the tendons of the flexor carpi radialis M. (cranially) and flexor carpi ulnaris M. (caudally) and distal to the flexor retinaculum, in a palmar position to the metacarpal bone II.

7. Sheath of deep digital flexor tendon (Fig. 9-19, *C*).
    A. *Landmarks:* Fetlock; pastern; P. I; interosseus medius M.; tendons of superficial and deep digital flexor Mm.
    B. *Approach:* On the lateral and medial aspect of the fetlock, between the interosseus medius M. and the tendon of the superficial digital flexor M. (the pouches of this sheath are located proximal and distal to the proximal sesamoid bones). On the lateral and medial aspect of the pastern, between P. I and the tendon of the superficial digital flexor M. On the palmar (plantar in the pelvic limb) aspect of the pastern, between the bifurcation of the tendon of the superficial digital flexor M. and the distal digital annular lig. All pouches are subcutaneous.

XVIII. Accessory (check) lig. of superficial digital flexor M. (Fig. 9-19, *B*).
    A. *Landmarks:* Medial aspect of carpus; radius; and flexor carpi radialis and flexor carpi ulnaris Mm.
    B. *Approach:* Through the skin, the superficial and proper antebrachial fasciae, and 2 to 3 cm above the medial aspect of the carpus in a horizontal or slightly oblique position between the radius and the distal tendon of the flexor carpi ulnaris M. The tendon of the flexor carpi radialis M. crosses the direction of the ligament.

XIX. Accessory (check) lig. of deep digital flexor M. (Fig. 9-19, *B*).
    A. *Landmarks:* Lateral and medial palmar aspect of metacarpus (metatarsus in the pelvic limb); the splint bones; interosseus medius M.; and deep and superficial digital flexor tendons.
    B. *Approach:* Distal to the carpus (tarsus in the pelvic limb), on both sides of metacarpus (metatarsus) between the splint bones and corresponding border of the interosseus medius M. and the tendons of the deep and superficial digital flexor Mm. in the proximal third of the area. It is located deep and under the skin and palmar (plantar) fascia.

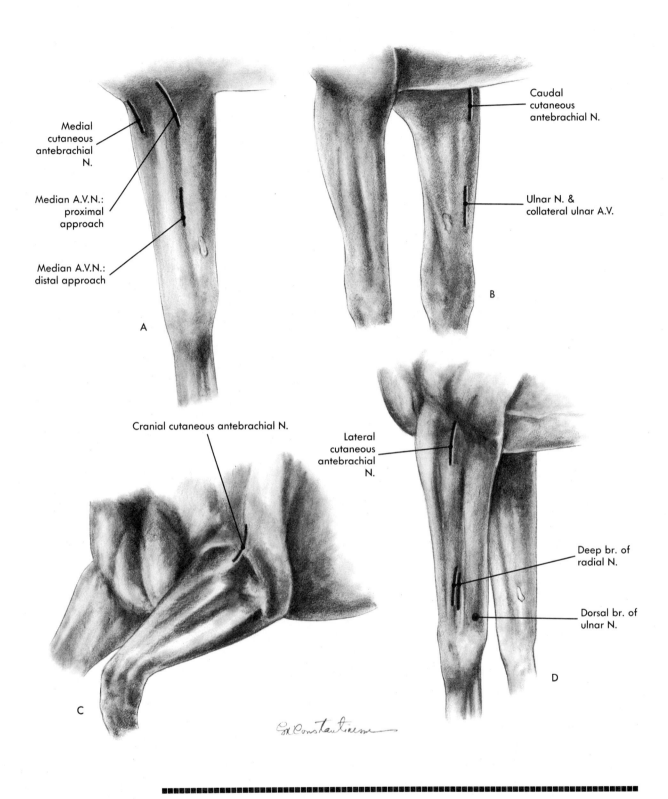

Medial
cutaneous
antebrachial
N.

Median A.V.N.:
proximal
approach

Median A.V.N.:
distal approach

A

Caudal
cutaneous
antebrachial N.

Ulnar N. &
collateral ulnar A.V.

B

Cranial cutaneous antebrachial N.

Lateral
cutaneous
antebrachial
N.

Deep br. of
radial N.

Dorsal br. of
ulnar N.

D

C

Thoracic limb. **A,** Medial aspect. **B,** Caudomedial aspect. **C,** Craniolateral aspect. **D,** Lateral aspect.   **Fig. 9-16**

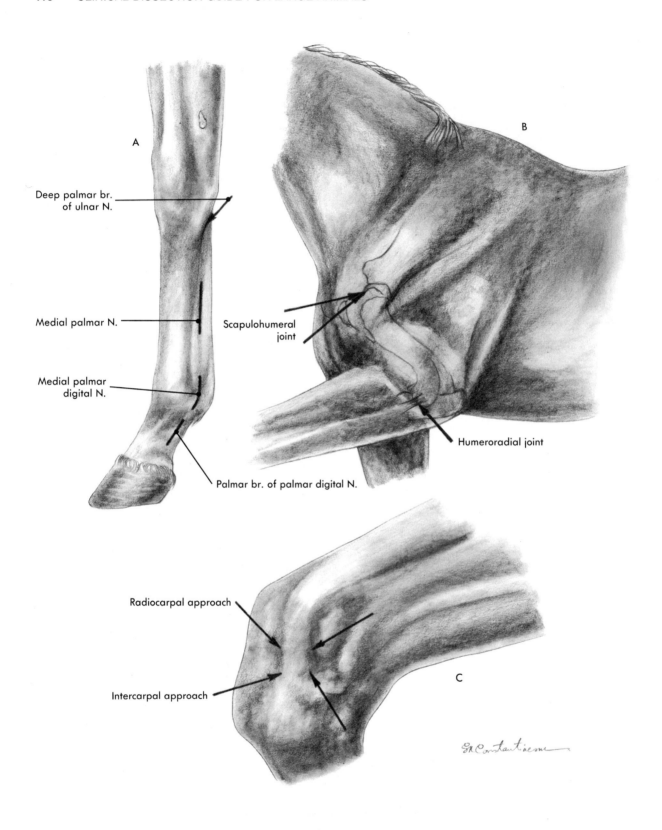

Deep palmar br.
of ulnar N.

Medial palmar N.

Medial palmar
digital N.

Scapulohumeral
joint

Humeroradial joint

Palmar br. of palmar digital N.

Radiocarpal approach

Intercarpal approach

Fig. 9-17   Thoracic limb. **A,** Medial aspect of autopodium. **B,** Lateral aspect. **C,** Dorsolateral aspect of carpus.

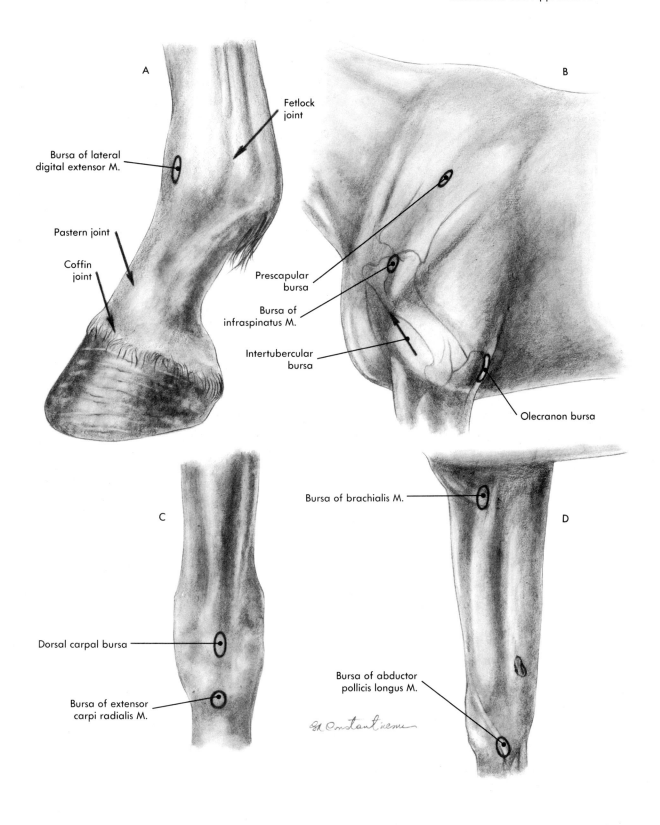

Thoracic limb. **A,** Lateral aspect of digit. **B,** Lateral aspect of shoulder. **C,** Dorsal aspect of carpus. **D,** Medial aspect of forearm.

Fig. 9-18

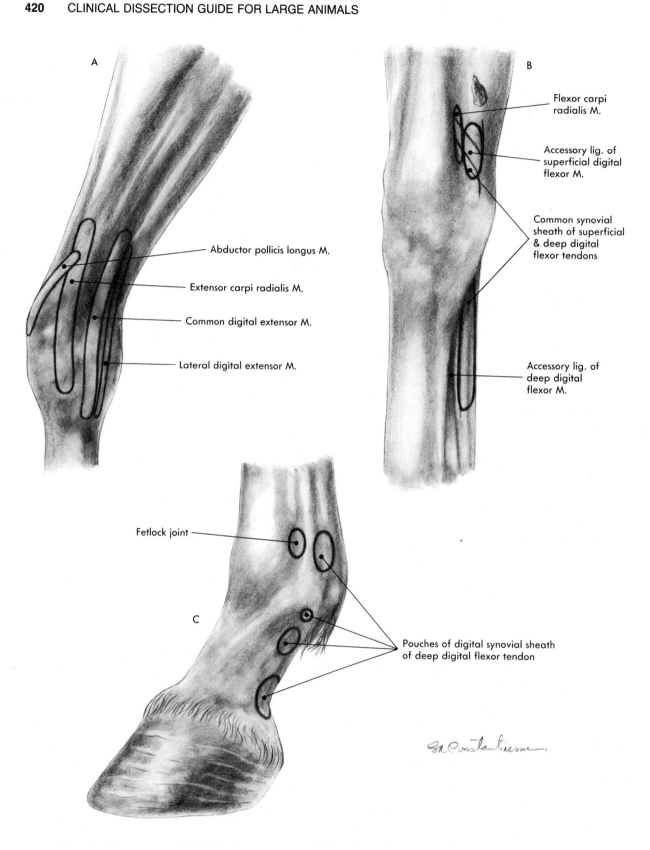

A

Abductor pollicis longus M.

Extensor carpi radialis M.

Common digital extensor M.

Lateral digital extensor M.

B

Flexor carpi radialis M.

Accessory lig. of superficial digital flexor M.

Common synovial sheath of superficial & deep digital flexor tendons

Accessory lig. of deep digital flexor M.

Fetlock joint

C

Pouches of digital synovial sheath of deep digital flexor tendon

Fig. 9-19   Thoracic limb. Synovial tendinous sheaths. **A,** Lateral aspect of carpus. **B,** Medial aspect of carpus. **C,** Digital area of carpus.

I. Facial A.V. and parotid duct (Fig. 9-20, *A*).
   A. *Landmarks:* Rostral border of masseter M.; vascular incisure of mandible; and facial tubercle.
   B. *Approach:* Through the skin, buccopharyngeal fascia, and cutaneus faciei M. along the rostral border of the masseter M., and from the vascular incisure of the mandible halfway to the facial tubercle.
   Caution! The relationship between the vessels and the duct is as follows: A.V. parotid duct, in a rostrocaudal direction.

II. Buccal V. (Fig. 9-20, *A*).
   A. *Landmarks:* Mandibular incisure; rostral border of masseter M.; and caudal extent of depressor labii inferioris M. (before entering under masseter M.).
   B. *Approach:* Through the skin, masseteric fascia, and masseter M. on the oblique line between the mandibular incisure and the intersection of the rostral border of the masseter M. with the depressor labii inferioris M.

III. Deep facial V. (Fig. 9-20, *A*).
   A. *Landmarks:* Facial crest and orbita.
   B. *Approach:* Through the skin, masseteric fascia, and masseter M., 2 cm ventral to and parallel with the facial crest from the orbita rostrally. The vein is located deep between the muscle and the maxilla.

IV. Lacrimal gland (Fig. 9-20, *A*).
   A. *Landmarks:* Zygomatic process of the frontal bone; orbital rim; and superior eyelid.
   B. *Approach:* Through the superior eyelid; along the orbital rim of the zygomatic process of the frontal bone; deep, on the medial aspect of the bone; and lying in the fossa for the lacrimal gland.

V. Mandibular lnn. (see Fig. 9-2, *B*).
   A. *Landmarks:* Intermandibular space and vascular incisure of mandible.
   B. *Approach:* Through the skin and buccopharyngeal fascia, in the intermandibular up to 10 to 15 cm caudal to the vascular incisure. Their deep aspect is related to the mylohyoideus, rostral belly of digastricus, medial pterygoid, omohyoideus, and sternohyoideus Mm., along the facial vessels.

VI. Maxillary N. (Fig. 9-20, *A*).
   A. *Landmarks:* Facial crest and lateral canthus of the eye.
   B. *Approach:* A needle can be inserted in a perpendicular direction through the skin, masseteric fascia, and masseter M., ventral to the facial crest at the level of the lateral canthus of the eye.
   Caution! The nerve travels between the foramen rotundum and pterygopalatine fossa alongside the maxillary A.

VII. Infraorbital N. (Fig. 9-20, *B*).
   A. *Landmarks:* Infraorbital foramen; facial crest and tubercle; and medial canthus of the eye.
   B. *Approach:* Reflect the levator labii superioris M. dorsally to insert a needle through the skin and levator nasolabialis M. in a rostrocaudal direction, parallel with the maxilla in the infraorbital groove and toward the infraorbital foramen.
   Notice that the infraorbital foramen is located at the rostral intersection of a line starting at the medial canthus of the eye and extending parallel to the facial crest in a rostral direction and a line extending dorsally from the facial tubercle. The infraorbital foramen can also be located approximately 2.5 cm dorsal to the midpoint of a line connecting the facial tubercle and the nasoincisive notch.

HEAD
(Figs. 9-20 to 9-22)

VIII. Inferior alveolar N. (Fig. 9-20, *B*).
   A. *Landmarks:* Viborg's triangle; lateral canthus of the eye; occlusal surface of the cheek teeth (premolars and molars); vascular notch of the mandible; and temporomandibular joint.
   B. *Approach:*
      1. A needle should be introduced through Viborg's triangle toward the mandibular foramen along the medial aspect of the mandible between pterygoid fossa and the medial pterygoid M. (α).
      2. A needle should be inserted in a ventrodorsal position on a line between the vascular notch of the mandible and the rostral limit of the temporomandibular joint between the pterygoid fossa and the medial pterygoid M. (β).
      3. A needle should be inserted medial to the ventral border of the mandible in the direction of the perpendicular line that is drawn from the lateral canthus and between the pterygoid fossa and the medial pterygoid M. (γ).
      Notice that the mandibular foramen is located at the intersection of the caudal extension of the occlusal surface of the cheek teeth with the perpendicular line drawn from the lateral canthus.
 IX. Mental N. (Fig. 9-20, *B*).
   A. *Landmarks:* Mental foramen; commissure of the lips; tendon of depressor labii inferioris M.; and orbicularis oris M.
   B. *Approach:* Reflect the tendon of the depressor labii inferioris M. ventrally at the level of the commissure of the lips (ventral to orbicularis oris M.) to insert a needle through the skin and cutaneus faciei M. in a rostrocaudal direction, parallel to the mandible, and toward the mental foramen.
  X. Facial and auriculotemporal Nn. and transverse facial A.V. (Fig. 9-21, *A*).
   A. *Landmarks:* Facial crest; temporomandibular joint; and caudal border of ramus mandibulae.
   B. *Approach:* Through the skin and masseteric fascia 2 to 4 cm ventral to the temporomandibular joint and emerging from under the parotid gland in a rostral direction, parallel to the facial crest.
      Notice that the auriculotemporal N. is situated dorsal to the facial N., whereas the vessels are located between the two nerves or along the auriculotemporal N. All of them lie on the masseter M.
 XI. Masseteric N. (Fig. 9-21, *A*).
   A. *Landmarks:* Temporomandibular joint; zygomatic arch; mandibular notch; and vascular notch of mandible.
   B. *Approach:* Through the skin, masseteric fascia, and masseter M., deep, and on a line between the rostral aspect of temporomandibular joint and a point situated 2 to 3 cm caudal to the vascular notch, starting from the zygomatic arch.
XII. Palpebral br. of auriculopalpebral N. (Fig. 9-21, *A*).
   A. *Landmarks:* Temporomandibular joint; zygomatic arch dorsocaudal to the temporomandibular joint; and the base of external ear.
   B. *Approach:* A needle should be inserted in a ventrodorsal and caudorostral direction between the skin and zygomatic arch dorsocaudal to the temporomandibular joint and midway between the joint and the base of external ear.
XIII. Internal auricular br. of facial N. (Fig. 9-21, *A*).
   A. *Landmarks:* External acoustic meatus and base of pinna (auricle).
   B. *Approach:* A needle should be inserted perpendicularly through the skin and caudal to the external acoustic meatus (palpable) at the base of the ear.

XIV. Supraorbital N. (from frontal N.) (Fig. 9-21, *B*).

    A. *Landmarks:* Zygomatic process of frontal bone and supraorbital foramen.

    B. *Approach:* A needle should be inserted in a vertical position through the skin, levator anguli oculi medialis M., and supraorbital foramen, located on the base of the zygomatic process of the frontal bone (palpable).

XV. Zygomaticotemporal N. (Fig. 9-21, *B*).

    A. *Landmarks:* Dorsal border of zygomatic arch and zygomatic processes of frontal and temporal bones.

    B. *Approach:* Through the skin and parallel to the dorsal border of the zygomatic arch, starting in a rostrocaudal direction from the angle (suture) between the zygomatic processes of the frontal and temporal bones.

XVI. Lacrimal N. (Fig. 9-21, *B*).

    A. *Landmarks:* Lateral canthus of the eye and superior eyelid.

    B. *Approach:* Through the skin of the superior eyelid at the level of the lateral canthus of the eye.

XVII. Zygomaticofacial N. (Fig. 9-21, *B*).

    A. *Landmarks:* Temporal process of zygomatic bone; ventral rim of the orbita; and inferior eyelid.

    B. *Approach:* Through the skin of the inferior eyelid and at the dorsal border of the temporal process of the zygomatic bone, which corresponds to the ventral rim of the orbita.

XVIII. Infratrochlear N. (Fig. 9-21, *B*).

    A. *Landmarks:* Medial canthus of the eye and caudal lacrimal process.

    B. *Approach:* Through the skin at the medial canthus of the eye, close to the caudal lacrimal process (palpable).

XIX. Ophthalmic, oculomotor and abducent Nn. (Fig. 9-21, *B*).

    A. *Landmarks:* Facial crest and lateral canthus of the eye.

    B. *Approach:*

        1. A needle should be inserted through the skin, masseteric fascia, and masseter M., ventral to the facial crest at the lateral canthus of the eye and toward the opposite temporomandibular joint.

        2. A needle should be introduced in a manner similar to that for the maxillary N. (see p. 421); however, the position of the needle is oblique ventrodorsally in a 10-degree angle (the nerves emerge from the orbital fissure, which is located 1 to 2 cm dorsal to the foramen rotundum).

XX. Mandibular cheek teeth (Fig. 9-22, *A*).

    A. *Landmarks:* Oral cleft; commissure of the lips; ventral commissure of the nostril; ventral border of mandible; rostral border of masseter M.; jugular process of occipital bone; and facial crest.

    B. *Approach:*

        1. The occlusal surface of the cheek teeth (both mandibular and maxillary) is projected *A*− on a slightly convex line that starts 3 to 5 cm caudal to the commissure of the lips and continues from between the ventral commissure of the nostril and the oral cleft in a caudal direction toward the temporomandibular joint (the convexity is oriented ventrally), or it is projected *B*− on an almost horizontal line that continues the oral cleft through the commissure of the lips and toward the ventral extent of the jugular process of the occipital bone.

        2. Depending on the age of the animal, the mandibular (inferior) premolars and molars are anchored more or less deeply into the alveoli, with the tops of their roots extending up to the ventral border of the mandible.

3. Premolars 1 and 2 are protected by the buccinator and depressor labii inferioris Mm. and are palpable; premolar 3 and the molars are also protected (by the masseter M.) but are not palpable.

   Notice that the rostral border of the masseter M. indicates the limit between the palpable and nonpalpable cheek teeth.

   Notice that the commissure of the lips is located approximately midway between the incisors and the premolars and occasionally should be ventrally oriented.

XXI. Maxillary cheek teeth (Fig. 9-22, *A*).

    A. *Landmarks:* Oral cleft; commissure of the lips; ventral commissure of the nostril; facial crest and tubercle; infraorbital foramen; rostral border of masseter M.; jugular process of occipital bone; and orbita.

    B. *Approach:*

      1. See paragraph on the occlusal surface of the cheek teeth (XX, entry #1).

      2. Depending on the age of the animal, the dorsal ends of the roots may extend up to a line parallel to the facial crest, starting from the orbita rostrally and passing through infraorbital foramen. The roots decrease in length so that, in older specimens, they do not even reach the facial crest.

      3. All the premolars are located rostral to the infraorbital foramen and the facial tubercle and are palpable (rostral to the rostral border of the masseter M.), whereas all the molars are located between the infraorbital foramen, facial tubercle, rostral border of the masseter M. and orbita and are not palpable because they are deep to the masseter M.

   Caution! The cheek teeth, which penetrate the rostral maxillary sinus, are the premolar 3 and molar 1. The cheek teeth that penetrate the caudal maxillary sinus are molars 2 and 3.

XXII. Ventral nasal meatus (Fig. 9-22, *A*).

    A. *Landmarks:* Ventral commissure of the nostril; nasal septum; and floor of nasal cavity.

    B. *Approach:* A flexible tube (esophageal or gastric) may be introduced into the ventral nasal meatus just above the ventral commissure of the nostril and firmly pressed against the nasal septum (medially) and the floor of the nasal cavity (ventrally).

   Caution! The tube must be introduced slowly and carefully to avoid entering the larynx and trachea. Passage of the tube into the esophagus is facilitated by flexing of the head and neck. Proper esophageal placement must be confirmed by palpating the tube within the esophagus.

XXIII. Paranasal sinuses (Fig. 9-22, *B*).

The only sinuses readily accessible to a clinical intervention are the conchofrontal and maxillary sinuses.

    1. Conchofrontal sinus

      A. *Landmarks:* Middorsal line of skull; base of zygomatic process of frontal bone; dorsal rim of orbita; medial canthus of the eye; and facial tubercle.

      B. *Approach:* Through the skin and in an area that extends from the middorsal line of skull to the base of the zygomatic process of the frontal bone along the dorsal rim of orbita and up to the medial canthus of the eye rostral to a point midway between the facial tubercle and orbita.

   Caution! In older specimens, the caudal extension of this sinus exceeds the orbita caudally.

2. Caudal maxillary sinus
   A. *Landmarks:* Rostral rim of orbita; medial canthus of the eye; and facial crest and tubercle.
   B. *Approach:* Through the skin, malaris, and levator labii superioris Mm. in an area outlined between the medial canthus of the eye, along the rostral rim of the orbita and the facial crest, a short distance caudal to the facial tubercle.
3. Rostral maxillary sinus
   A. *Landmarks:* Medial canthus of the eye, facial crest and tubercle and infraorbital foramen.
   B. *Approach:* Through the skin and levator labii superioris M. within the area located rostral to the previous sinus, between the medial canthus of the eye, infraorbital foramen, and facial tubercle.

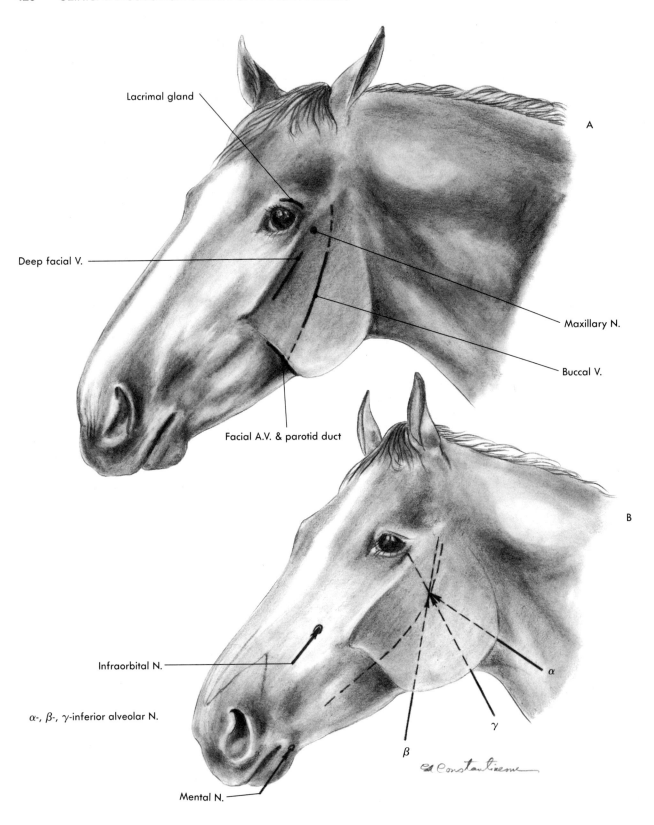

Lacrimal gland

Deep facial V.

Maxillary N.

Buccal V.

Facial A.V. & parotid duct

A

B

Infraorbital N.

α-, β-, γ-inferior alveolar N.

Mental N.

β

α

γ

Fig. 9-20　　　　**A,** Structures in ocular and masseteric areas. **B,** Infraorbital, inferior alveolar, and mental Nn.

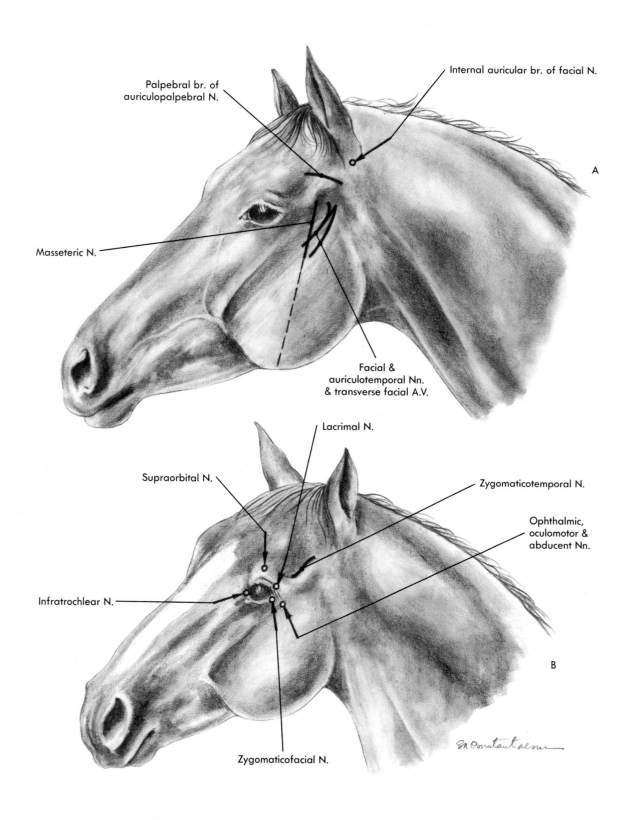

Palpebral br. of
auriculopalpebral N.

Internal auricular br. of facial N.

A

Masseteric N.

Facial &
auriculotemporal Nn.
& transverse facial A.V.

Lacrimal N.

Supraorbital N.

Zygomaticotemporal N.

Ophthalmic,
oculomotor &
abducent Nn.

Infratrochlear N.

B

Zygomaticofacial N.

A, Structures around temporomandibular joint. **B,** Structures around eye.

Fig. 9-21

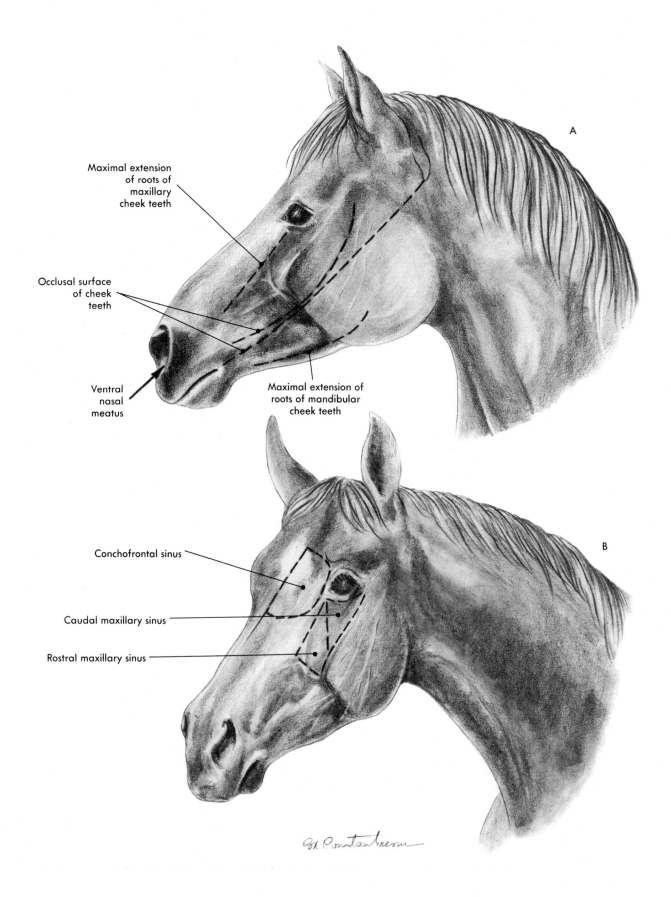

Maximal extension of roots of maxillary cheek teeth

Occlusal surface of cheek teeth

Ventral nasal meatus

Maximal extension of roots of mandibular cheek teeth

A

Conchofrontal sinus

Caudal maxillary sinus

Rostral maxillary sinus

B

Fig. 9-22

A, Approach to teeth and nasal cavity. **B,** Paranasal sinuses.

I. Linguofacial V. (Fig. 9-23)

    A. *Landmarks:* Jugular groove and angle of mandible.

    B. *Approach:* Through the skin and cutaneus colli M. between the cranial extent of the jugular groove and the angle of the mandible in a horizontal position.

II. Masseteric br. (A.V.) (Fig. 9-23)

    A. *Landmarks:* Caudal border of ramus mandibulae, the insertions of the occipitomandibular part of digastric and the sternomandibular Mm.

    B. *Approach:* Through the skin, masseteric fascia, and cutaneus faciei M.; at the intersection of the two insertions of the occipitomandibular part of the digastric and sternomandibular Mm. on the caudal border of the ramus mandibulae.

    Notice that the vessels emerge from under the parotid gland and lie on the masseter M.

III. Parotid lnn. (Fig. 9-23)

    A. *Landmarks:* Temporomandibular joint and caudal border of ramus mandibulae.

    B. *Approach:* Through the skin and the masseteric and parotid fasciae 2 to 3 cm ventral to the temporomandibular joint at the caudal border of the ramus mandibulae.

    Notice that the lymph nodes lie on the transverse facial A.

IV. Viborg's triangle (Fig. 9-23)

    A,B. *Landmarks and approach:* Linguofacial V.; caudal border of ramus mandibulae; and tendon of sternomandibular M.

    Caution! Holding the head of the horse in extension, this area is practically free of major vessels and nerves and corresponds to the most ventral extent of the guttural pouch in abnormal situations, such as empiema.

    Notice that Viborg's triangle constitutes the ventral approach (Whitehouse) in hyovertebrotomy.

V. Occipitohyoideus M. (Fig. 9-23)

    A. *Landmarks:* Jugular process of occipital bone and stylohyoid angle of stylohyoid bone.

    B. *Approach:* Through the skin and aponeurosis of cleidomastoideus M. between the jugular process of the occipital bone and the stylohyoid angle of the stylohyoid bone.

    Notice that the occipitohyoideus M. represents the dorsal approach (Chabert) in hyovertebrotomy. The head must be held in extension.

VI. Great auricular N. (Fig. 9-23)

    A. *Landmarks:* Wing of atlas.

    B. *Approach:* Through the skin and aponeurosis of the splenius M. along the border of the wing of atlas.

VII. Cranial cervical ggl. (Fig. 9-23)

    A. *Landmarks:* Jugular process of occipital bone; wing of atlas; and parotid salivary gland.

    B. *Approach:* A needle should be introduced in a dorsoventral and caudocranial direction to a depth of ~5 cm between the ventral extent of the jugular process of the occipital bone and the wing of atlas at the caudal border of the parotid salivary gland.

    Caution! Hold the head of the specimen in extension, and introduce the needle at a 60-degree angle with the longitudinal axis of the neck.

PAROTID REGION
(Fig. 9-23)

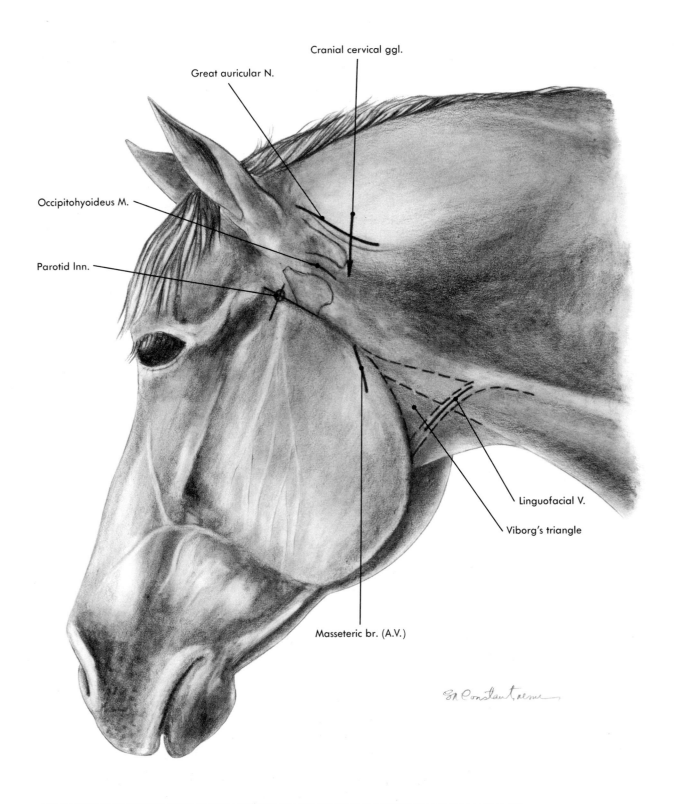

Great auricular N.

Cranial cervical ggl.

Occipitohyoideus M.

Parotid lnn.

Linguofacial V.

Viborg's triangle

Masseteric br. (A.V.)

Fig. 9-23

Structures in parotid region.

Index

![Index decorative bar]